CHANDIGARH UNIVERSITY
Discover. Learn. Empower.

NAAC GRADE A+
ACCREDITED UNIVERSITY

AA003979

# 2nd International Conference on Innovation and Regenerative Trends in Tourism and Hospitality Industry (IRTTHI 2025)

University Institute of Tourism &
Hospitality Management
Chandigarh University
Mohali, India
7-9 April 2025

*Editors:*

**Manish Sharma**
**Ajit Kumar Singh**
**Pankaj Kumar Tyagi**

ISBN: 979-8-3313-2574-9
DOI: https://doi.org/10.52202/081568

**Printed from e-media with permission by:**

Curran Associates, Inc.
57 Morehouse Lane
Red Hook, NY 12571

**Some format issues inherent in the e-media version may also appear in this print version.**



Printed with permission by Curran Associates, Inc. (2025)

For permission requests, please contact IRTTHI
at the address below.

University Institute of Tourism and Hospitality Management
Chandigarh University
Mohali, Punjab – 140413
India

srishti.uithm@cumail.in

**Additional copies of this publication are available from:**

Curran Associates, Inc.
57 Morehouse Lane
Red Hook, NY 12571 USA
Phone: 845-758-0400
Fax: 845-758-2633
Email: curran@proceedings.com
Web: www.proceedings.com

# TABLE OF CONTENTS

| | | |
|---|---|---|
| **Convener** | Prof. (Dr.) Manish Sharma | University Institute of Tourism and Hospitality Management, Chandigarh University , India |
| **Editors** | Prof. (Dr.) Manish Sharma | University Institute of Tourism and Hospitality Management, Chandigarh University , India |
| | Prof. (Dr.) Ajit Kumar Singh | University Institute of Tourism and Hospitality Management, Chandigarh University , India |
| | Prof. (Dr.) Pankaj Kumar Tyagi | University Institute of Tourism and Hospitality Management, Chandigarh University , India |
| **Organizing Secretary** | Prof. (Dr.) Pankaj Kumar Tyagi | University Institute of Tourism and Hospitality Management, Chandigarh University , India |
| | Prof. (Dr.) Ajit Kumar Singh | University Institute of Tourism and Hospitality Management, Chandigarh University , India |
| **Organizing Committee** | Mr. Bharpur Singh Abhi | University Institute of Tourism and Hospitality Management, Chandigarh University , India |
| | Mr. Chetan Chauhan | University Institute of Tourism and Hospitality Management, Chandigarh University , India |
| | Ms. Punita Thakur | University Institute of Tourism and Hospitality Management, Chandigarh University , India |
| | Ms. Arvinder Kaur | University Institute of Tourism and Hospitality Management, Chandigarh University , India |
| | Ms. Harjot Kaur | University Institute of Tourism and Hospitality Management, Chandigarh University , India |
| | Dr. Shaina Arora | University Institute of Tourism and Hospitality Management, Chandigarh University , India |
| | Mr. Sandeep Paatlan | University Institute of Tourism and Hospitality Management, Chandigarh University , India |
| | Mr. Sagar Bhatt | University Institute of Tourism and Hospitality Management, Chandigarh University , India |
| | Chef Saurabh Khurana | University Institute of Tourism and Hospitality Management, Chandigarh University , India |
| | Chef Gagandeep Passi | University Institute of Tourism and Hospitality Management, Chandigarh University , India |
| | Mr. Dhiraj Kalsi | University Institute of Tourism and Hospitality Management, Chandigarh University , India |
| | Mr. Manish Kumar Sharma | University Institute of Tourism and Hospitality Management, Chandigarh University , India |
| | Chef Preet Kamal Singh | University Institute of Tourism and Hospitality Management, Chandigarh University , India |
| | Mr. Amber M Mishra | University Institute of Tourism and Hospitality Management, Chandigarh University , India |

| | | |
|---|---|---|
| **Co Organizing Secretary** | Prof. (Dr.) Sandeep Guleria | University Institute of Tourism and Hospitality Management, Chandigarh University , India |
| | Dr. Aman Kumar | University Institute of Tourism and Hospitality Management, Chandigarh University , India |
| | Chef. (Dr.) Monika Rani | University Institute of Tourism and Hospitality Management, Chandigarh University , India |
| | Dr. Shikha Sharma | University Institute of Tourism and Hospitality Management, Chandigarh University , India |
| | Dr. Trilochan Kumar | University Institute of Tourism and Hospitality Management, Chandigarh University , India |
| **Technical Committee** | Prof. Nino Davidze | LLC University of Georgia |
| | Prof. Claudia | University of Coimbra, Portugal |
| | Assoc. Prof. Eslam Ahmed Fathy | Pharos University, Alexandria, Egypt |
| | Dr. Rubina Moniz Vieira | St. Marry University, United Kingdom |
| | Dr. Amr foaud | Pharos University, Alexandria, Egypt |
| | Dr. Arun Singh Thakur | Panjab University, India |
| | Dr. Mohd. Badruddoza Talukder | International University of Business, Agriculture and Technology, Bangladesh |
| | Chef (Dr.) Saurabh Sharma | Manipal University, Jaipur |
| | Dr. Ebru Kemer | Ömer Halisdemir University, Niğde, Turkey |
| | Dr. Ankit Shukla | RMUTT, Thanyaburi Bangkok, Thailand |
| | Dr. Juan Rodrigo Beldad Del Villar | St. Teresa International University, Thailand |
| | Dr Ahmad Albattat | Management and Science University Selangor, Malaysia |
| | Dr. Jeremy JeanClaude G Proces | Burapha University International College, Thailand |
| | Dr. Bhupinder Chahal | Yorkville University, BC, Canada & Granville College, BC, Canada |
| | Dr. Firdous Ahmad Malik | University of People, California |
| | Dr. Navdeep Kaur | Granville College, Vancouver BC, Canada |
| | Dr. Shad Ahmad Khan | University of Buraimi, Oman |
| | Dr. Amitabh Mishra | University of Technology and Applied Sciences, Oman |
| | Prof. (Dr.) Yashwant Singh Rawal | Amity University, India |
| | Dr. Jeetesh | Taylor's University, Malaysia |
| | Dr. Sahar Movahedi | Granville College, Canada |
| | Dr. Lisa Allen | Yorkville University NWM, BC, Canada |
| | Prof. (Dr.) Amrik Singh | Lovely Professional University, India |
| | Dr. Pawan Kumar | Graphic Era Deemed to be University, India |
| | Prof. (Dr.) Ashish Raina | CT University, India |
| | Prof. (Dr.) Sanjeev Sharma | Lovely Professional University, India |
| | Dr. Viana Hassan | University of Malta |
| | Dr. Suneel Kashyap | SBS College, New Delhi, India |

| | | |
|---|---|---|
| | Dr Kanwar Chanderdeep Singh | Central University of Himachal, India |
| | Prof. (Dr.) Deepak Tilgoriya | GNA University, India |
| | Dr. Nitin Vyas | Himchal Pradesh University, India |
| | Dr. Himanshu Malik | Shri Vishwakarma Skill University, India |
| | Dr. Preeti | Himchal Pradesh University, India |
| | Chef (Dr.) Umang Bhartwal | NIMS University, India |
| | Dr. Sandeep Raheja | Maharishi Markandeshwar (deemed to be) Unversity, India |
| | Dr. Amit Mathur | Central University Haryana, India |
| | Dr. Parikshit Sharma | Jagan Lakecity University, India |
| | Dr. Dixit Gupta | Lovely Professional University, India |
| | Prof. (Dr.) Gaurav Bathla | CT University, India |
| | Dr. Samriti Mahajan | Lingaya's Vidyapeeth, India |
| | Dr. Surjeet Kumar | Kurukshetra University, India |
| | Dr. Ankush Amberdar | Kurukshetra University, India |
| | Dr. Arun Bhatia | Central University of Himachal Pradesh, India |
| | Dr. Ranjeet Raman | Central University Jammu, India |
| | Dr. Subhadeep Chakraborty | Digboi College, Assam, India |
| | Dr. Vipin Singh Raghuvanshi | Vellore Institute of Technology, India |
| | Dr. Bharat Kapoor | Guru Nanak Dev University, India |
| | Dr. Amit Katoch | Govt. Degree College, Dharmasala, India |
| | Dr. Ashish Nag | Central University of Himachal Pradesh, India |
| | Dr. Vivek Deshwal | Dr. B R Ambedkar, National Institute of Technology, India |
| **Advisory Committee** | Prof. (Dr.) Sat Prakash Bansal | Central University of Himachal Pradesh, India |
| | Prof. (Dr.) Prashant Kumar Gautam | Panjab University, India |
| | Prof. (Dr.) Sunil Kumar Kabia | Bundelkhand University, India |
| | Prof. Christophe Martin | Sainte-Marie de Chavagnes-Institut Supérieur du Tourisme de Cannes, France |
| | Prof. Gavin R R Caldwell | Partner Schools Kastanienbaum, Lucerne, Switzerland |
| | Prof. Evangelos Moustakas | Canadian University, United Arab Emirates |
| | Ms Nidhi Chopra | Campus France, India |
| | Dr. María Jesús Jerez Jerez | University of Bedfordshire, United Kingdom |
| | Dr. Hazem Khairy | University of Sadat City, Egypt |
| | Dr. I Nengah Subadra | University of Triatma Mulya, Bali, Indonesia |
| | Dr. Shaik Rehana Banu | Windsor University, USA |
| | Dr. Vipin Nadda | University of Sunderland, United Kingdom |
| | Prof. Ekka Davis | International Black Sea University, Georgia |
| | Prof. Felix Bello | University of Pretoria, South Africa |
| | Prof. Kaitano Dube | Vaal University of Technology, South Africa |
| | Dr. Sunita Badhwar | Skill Council (MEPSC), India |

# Understanding the Vital Role of International Travel Insurance: A Case Study of Schengen Visa

Jagjit Singh[1] Shikha Sharma[2]

[1&2]Chandigarh University, Mohali, India

* *Corresponding author: Jagjitsingh.uithm@cumail.in*

**Abstract.** The Schengen Area consists of 27 European nations that have removed passport and other border controls. As part of the EU's freedom, security, and justice policies, it generally works as a single jurisdiction with a shared visa policy. The region is called after the 1985 and 1990 Schengen Agreements, and both signed in Luxembourg.

**Purpose** A Schengen visa is a short-stay visa that permits a person to travel to any Schengen Area member for tourist or business reasons for up to 90 days per stay.

Schengen visas are the most prevalent in Europe. It allows the bearer to enter, travel freely within, and exit the Schengen zone from any Schengen member state. No border restrictions exist inside the Schengen Area.

If someone wants to study, work, or reside in one of the Schengen nations for more than 90 days, you must apply for the national visa of that European country, not a Schengen Visa.

**Design/methodology/approach** This Article will assist prospective travelers outside the European Union in understanding the insurance requirements, valid insurance providers, charges, and services that are and are not covered. It will be accomplished by analyzing online official Visa sites, such as India's major VFS.

**Finding** Extract from "Regulation (EC) No. 810/2009 of the European Parliament and of the Council of July 13, 2009," which became effective on April 5, 2010:

"Applicants for a uniform visa for one or two entries must show they have appropriate and valid travel medical insurance to cover medical repatriation, urgent medical care, emergency hospital treatment, or death during their stay(s) on Member State territory.

The insurance must be valid across the Schengen Area and cover the entire stay or transit. Minimum coverage: €30,000."

**Originality/value** Many individuals desired to go to Europe after COVID-19 for various reasons. However, they may have yet to be aware that Travel Insurance or Medical Insurance is required before arranging an appointment in any foreign nation.

**Keywords:** Travel Insurance, Medical Insurance, Schengen Area, Visa, Hospitality & Tourism Industry

## 1. Introduction

International travel insurance is essential for providing financial protection and security to travelers. This study examines the importance of travel insurance within the framework of the Schengen Visa, which requires insurance coverage for admission. The research investigates the effects of this mandate on passengers, the insurance sector, and legislators, while evaluating the advantages and obstacles related to foreign travel insurance. The Schengen Area, consisting of 29 European nations, facilitates unrestricted travel by abolishing internal border barriers. Acquiring a Schengen Visa requires that applicants get travel insurance meeting defined coverage standards. This stipulation guarantees that tourists are financially safeguarded against medical problems, accidents, or trip interruptions. This article examines the significance of travel insurance in foreign travel, particularly its implications for Schengen Visa holders (Aygül & Mertek, 2024).

### 1.1 The Necessity of International Travel Insurance

International travel presents people with several hazards, such as medical problems, trip cancellations, and misplaced luggage. Travel insurance alleviates these dangers by offering cash reimbursement and assistance services. The Schengen Visa mandates travel insurance to provide medical coverage for international visitors, therefore alleviating the financial strain on the healthcare systems of host nations (Lee et al., 2018).

 https://doi.org/10.52202/081568-0001

### 1.2 Schengen Visa and Compulsory Travel Insurance

Applicants for a Schengen Visa must fulfill certain insurance criteria:

- Minimum coverage of €30,000 for medical expenditures.
- Provision for urgent medical care, inpatient services, and repatriation.
- Validity in all 27 Schengen nations for the whole of the journey (Bligh, 2010).

These stipulations guarantee that passengers have sufficient coverage, therefore mitigating possible liability for host countries and healthcare facilities.

### 1.3 Advantages of Travel Insurance for Schengen Visa Holders

- Financial Security: Addresses unexpected medical bills, safeguarding tourists from exorbitant healthcare costs (Szwedo, 2012).
- Trip Protection: Compensation for trip cancellations, delays, or disruptions.
- Peace of Mind: Guarantees access to emergency assistance services, including medical evacuation.
- Adherence to Visa Regulations: Ensures efficient visa processing and approval.

### 1.4 Challenges and Considerations

- Cost Consideration: Travelers may see insurance rates as an additional financial encumbrance.
- Policy Variability: Discrepancies in coverage across providers may induce misunderstanding.
- Claim Processing: Certain passengers encounter challenges in claim settlements and refunds.

### 1.5 The Effects of Schengen Travel Insurance Mandates

- An analysis of statistical data from Schengen visa applicants indicates:
- A substantial proportion of visa denials is attributable to insufficient insurance coverage.
- Heightened demand for foreign travel insurance companies serving Schengen passengers.
- Enhanced financial readiness among tourists, reducing dependence on foreign healthcare systems.

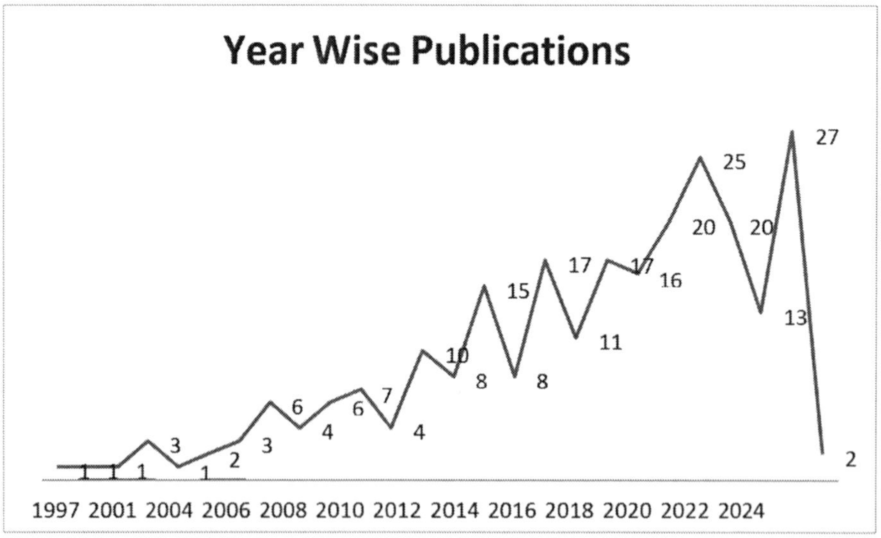


**Figure 1 Year Wise Publications (Source: Scopus Data Base)**

Fig 1 data represents yearly values from 1997 to 2025, showing an increasing trend in numbers over the years, with some fluctuations. The values remained low in the early years, with only one occurrence in 1997, 1998, 2001, and 2004. From 2006 onward, there was a noticeable increase, reaching a peak of 27 in 2024 before dropping to 2 in 2025. Significant growth was observed between 2012 and 2018, with numbers fluctuating between 8 and 17. A further spike occurred from 2020 onwards, peaking at 25 in 2021, then slightly dropping and rising again in 2024.

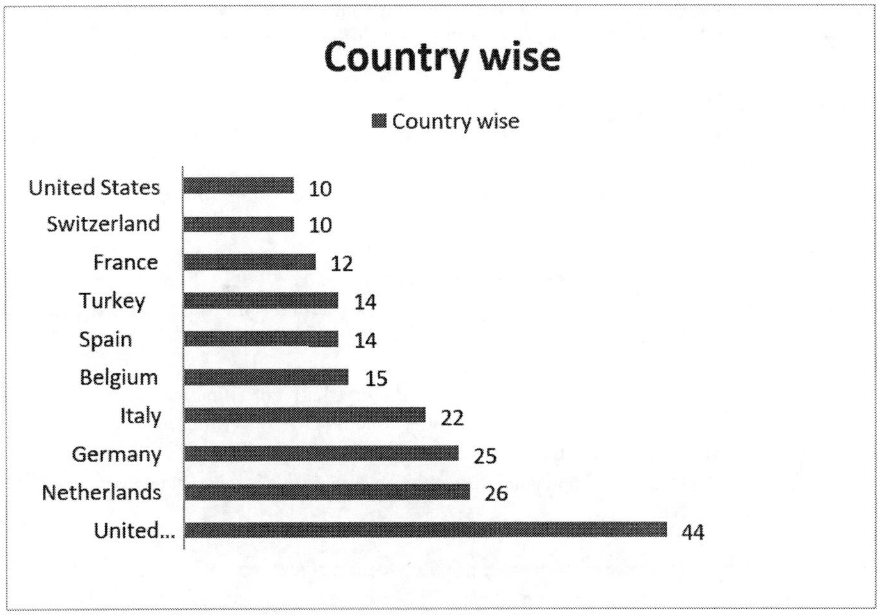


**Figure 2 Country Wise Data (Source: Scopus Data base)**

Fig 2 shows United Kingdom recorded the highest value at 44, followed by the Netherlands with 26 and Germany with 25. Italy registered 22, while Belgium had 15. Spain and Turkey both had 14, with France slightly lower at 12. Switzerland and the United States each accounted for 10.

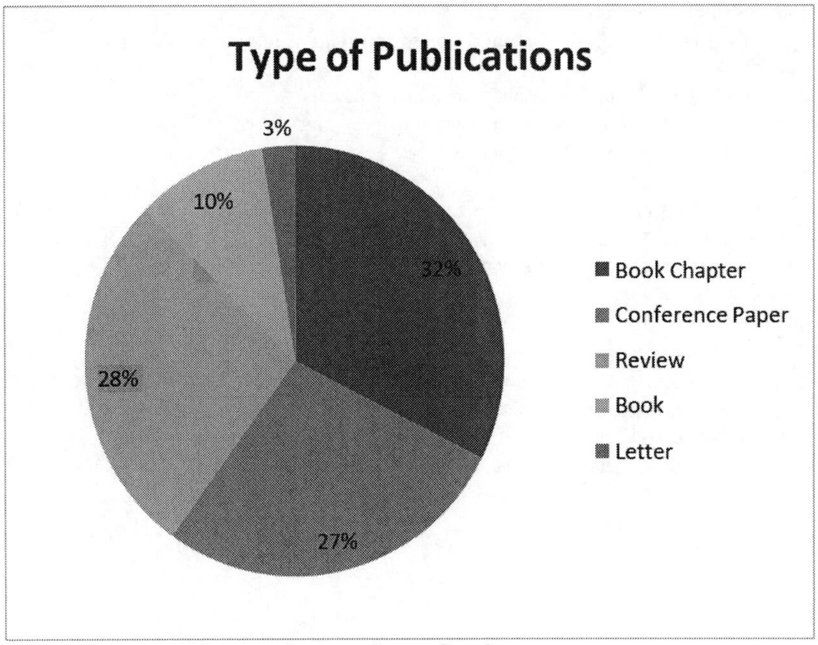


**Figure 3 Type of Publications Sources: Scopus Database**


https://doi.org/10.52202/081568-0001

Fig 3 shows categorize different types of publications based on their frequency. "Book Chapters" are the most common, with 13 recorded instances, indicating a significant contribution to academic literature in this format."Conference Papers" and "Reviews" both have 11 entries, suggesting that presenting research at conferences and publishing review articles are equally valued in this dataset. These types of publications help disseminate findings and provide critical analyses of existing research."Books" are less frequent, with only 4 entries, indicating that full-length publications are less common compared to shorter academic contributions. Lastly, "Letters" appear only once, suggesting that this form of publication is the least utilized in this dataset.

1. **Duration of Coverage**
   - Low Cost & Europe Travel Plans: Coverage is provided for the validity period of the Schengen Visa, up to 180 days .
   - Multi-Trip Plan: Provides coverage for up to 12 months, but each trip within that period must not exceed 90 consecutive days.
2. **Geographical Coverage**
   - Low Cost Plan: Covers only Schengen Area countries.
   - Europe Travel & Multi-Trip Plans: Covers all Schengen Area countries plus the UK, Ireland, and Cyprus.
3. **24/7 Medical Teleconsultation**
   - All three plans provide **round-the-clock medical consultation services**, ensuring travelers receive guidance on medical issues anytime, anywhere.
4. **Refund in Case of Visa Refusal**
   - If the Schengen Visa application is denied, all three plans guarantee a refund of the insurance fee.
5. **Payment of Medical Expenses**
   - Low Cost Plan: Maximum coverage of €30,000 for medical expenses.
   - Europe Travel & Multi-Trip Plans: Increased coverage of up to €100,000.
6. **Medical and Surgical Costs**

Covers medical fees, hospital stays, and prescribed medication:

   - Low Cost Plan: Up to €30,000.
   - Europe Travel & Multi-Trip Plans: Up to €100,000.
7. **Urgent Dental Care**
   - All plans include dental treatment coverage up to €150 per insured person.
8. **Assistance and Repatriation**
   - All plans cover medical repatriation or transportation in case of hospitalization or a serious medical event.
   - Europe Travel & Multi-Trip Plans extend this to cover:
   - Repatriation of other insured travelers accompanying the victim.
   - Assistance for travelers traveling alone.
   - Search and rescue costs.
9. **Prolongation of Stay & Return After the Expected Date**
   - Only the **Multi-Trip Plan** covers cases where a medical condition **forces an extended stay** or prevents return as initially planned.
10. **Assistance in Case of Death**

    All plans include:

    - Repatriation of remains.
    - Burial expenses.
    - Assistance with formalities related to the death of the insured traveler.
11. **Baggage & Personal Effects**
    - Repatriation of baggage is covered under all three plans.
    - Early repatriation of accompanying insured persons is covered only in
    - Europe Travel & Multi-Trip Plans.

**12. Intervention Threshold for Medical Expenses**

The **maximum threshold** for medical costs is:

- €30,000 for Low Cost Plan.
- €100,000 for Europe Travel & Multi-Trip Plans.

**13. Excess (Deductible) for Reimbursements Outside the EU**

- A €20 fee applies for any reimbursements processed via bank transfer outside the European Union.

**14. Additional Travel Assistance Services**

Europe Travel & Multi-Trip Plans offer additional services like:

- Travel assistance (not included in the Low Cost Plan).
- Sending urgent messages in case of emergencies.
- Covering communication expenses (up to €100).
- Providing essential medicines, glasses, contact lenses, and prostheses.
- Daily compensation in case of hospitalization (up to €500).
- Identity and travel document theft protection (coverage of€100).
- Urgent early return of an insured traveler due to a serious incident.

**How Long Can You Stay in Europe With a Schengen Visa (Bigo, 2014)?**

A Schengen visa allows you to stay in Europe's Schengen Area for a maximum of **90 days within a 180-day period**. The exact number of days you are permitted to stay is stated on your visa sticker. The consular officer handling your visa application determines the validity of your visa.

Types of Schengen Visas Based on Entries

- **Single-entry visa** – Allows you to enter the Schengen Area **once** and remain there for the permitted number of days. Once you leave, you cannot re-enter.
- **Double-entry visa** – Grants permission to enter the Schengen Area **twice**. The total number of days in both entries cannot exceed the validity stated on the visa sticker.
- **Multiple-entry visa** – Allows multiple entries into the Schengen Area during the visa's validity. However, the total days across all entries **cannot exceed the allowed duration** mentioned on the visa sticker.

Schengen Visa Purposes (Pedersen, 2015)
Schengen visas are issued for the following travel purposes:

- Tourism and holidays
- Business trips
- Visiting friends and family
- Attending cultural, religious, or sports events
- Airport transit (A-type visa)
- Official visits
- Medical treatment
- Short-term study courses or research purposes

**Schengen Visa Refusal Rates and Trends in 2023** (Trauttmansdorff, 2023)

Schengen visa refusal rates fluctuate annually based on the policies of Schengen member countries. In 2023, the top five countries rejecting the highest number of visa applications were:

- France: 26.7%
- Spain: 15.3%
- Germany: 12.7%


https://doi.org/10.52202/081568-0001

- Italy: 8.17%
- Netherlands: 6.9%

Other countries with high refusal rates included Finland, Austria, and Belgium.

The EU Council released Schengen visa statistics on 15 May 2024, providing a comprehensive analysis of visa issuance in 2023. A total of 10.3 million applications were received, with 8.5 million visas granted, including multiple-entry, uniform, on-border, and Limited Territorial Validity (LTV) visas. The overall refusal rate was 16% (1,632,984 applications), marking a slight decrease from 17.9% in 2022.

**Countries with the Highest Schengen Visa Refusal Rates** (Scheel, 2022)

The number of visa rejections largely depends on the volume of applications received. Although some countries had high refusal rates, they also processed a significant number of applications. For example, the Netherlands and Greece exhibited a 61.9% and 61.5% increase in their refusal rates, respectively, compared to previous years. This indicates fluctuations in the approval criteria rather than simply labeling them as the strictest countries. Among countries with lower application volumes, Malta had the highest refusal rate at 38%, followed by Estonia (33.1%) and Belgium (26.5%).

**Schengen Visa Refusal Statistics by Country** (Twigt, 2024)

Table.1 The following table outlines visa application and refusal statistics for 2023:

| **Country** | **Applications** | **Refusals** |
|---|---|---|
| Austria | 229,613 | 32,922 |
| Belgium | 225,951 | 60,148 |
| Croatia | 38,511 | 7,561 |
| Czech Republic | 138,840 | 24,218 |
| Denmark | 107,872 | 21,509 |
| Estonia | 13,190 | 4,347 |

| | | |
|---|---|---|
| Finland | 92,078 | 16,602 |
| France | 2,625,846 | 436,893 |
| Germany | 1,459,560 | 208,155 |
| Greece | 627,008 | 91,379 |
| Hungary | 215,799 | 32,641 |
| Iceland | 22,541 | 474 |
| Italy | 1,108,035 | 133,495 |
| Latvia | 18,897 | 2,208 |
| Lithuania | 30,668 | 3,911 |
| Luxembourg | 11,227 | 1,354 |
| Malta | 33,306 | 12,261 |
| Netherlands | 669,291 | 114,269 |
| Norway | 108,536 | 17,457 |
| Poland | 124,057 | 18,564 |
| Portugal | 207,654 | 33,220 |
| Slovakia | 13,468 | 1,752 |
| Slovenia | 19,990 | 3,097 |
| Spain | 1,413,222 | 251,470 |
| Sweden | 169,151 | 38,680 |
| Switzerland | 603,261 | 64,397 |
| **Total** | **10,327,572** | **1,632,984** |

**Source:** SchengenVisa.news & Europa.eu

**Schengen Visa Rejection Rates for Third Countries**

Visa rejection rates for applicants from **third countries** depend on various factors, including:

- Incomplete or missing required documents
- Lack of previous travel history
- Bilateral relations and cooperation on asylum seeker repatriation
- Political policies and EU-wide strategies (e.g., recent visa sanctions on Ethiopia under **quid pro quo** rules)

In 2023, the visa refusal rates increased for:

- Belarus: 3.4%
- Russia: 10.6%
- Turkey: 16.1%
- Iran: 30.3%

Meanwhile, historically high-rejection countries, including Pakistan, Bangladesh, Nigeria, and Senegal, continued to face overall refusal rates of approximately 75%.

**1.5 Immigration Expert Opinion**

Immigration expert Umer Rasib, based in Germany, highlighted key factors affecting Schengen visa refusals in 2023:


https://doi.org/10.52202/081568-0001

"The visa policies of Schengen member states are influenced by quotas (unofficial), trust in third-country nationals, and diplomatic relations. Some countries, such as India, Turkey, Malaysia, UAE, Qatar, and Kuwait, are generally trusted to follow visa rules. In contrast, applicants from countries with higher immigration risks, such as Pakistan, Bangladesh, Nigeria, and Senegal, often face higher refusal rates due to concerns about overstaying."

He further emphasized that modernized and effective evaluation techniques should be implemented to differentiate between genuine travelers and those intending to migrate illegally. This would prevent undue refusals while maintaining border security and facilitating tourism.

### 1.6 Visa Application Guidelines

When to Apply

- Submit your application at least 15 days before your intended journey and no earlier than 6 months beforehand (Studzińska, 2019).
- Some consulates may require an appointment before lodging your application.

Required Documents

- Valid passport (expiry date at least 3 months after departure from the Schengen area; for multiple-entry visas, 3 months after leaving the last Schengen country visited).
- Visa application form.
- Passport photo (ICAO-compliant) (Lin et al., 2001).
- Medical insurance (covering emergency care, hospitalization, and repatriation, including death).
- Supporting documents (proof of purpose of stay, financial means, accommodation, and intent to return to home country).
- Fingerprint collection (exemptions apply to specific applicants).
- Additional documents may be requested by consulates.

Visa Fees (Hobolth, 2014)

- €90 for adults
- €45 for children aged 6-12
- €35 for applicants from Armenia, Azerbaijan, and Belarus
- €67.50 for applicants from Cabo Verde
- Extra fees may apply if using visa service centers.
- Some applicants qualify for a fee waiver.

Processing Time (Everling, 2009)

- Standard processing: 15 days
- Extended processing (if additional review required): up to 45 days
- Family members of EU/EEA citizens may qualify for a free and expedited visa process under Free Movement Directive.

## Conclusion

Schengen visa refusal rates in 2023 showed a minor decrease overall (16% vs. 17.9% in 2022), with specific countries experiencing significant variations. Policy changes, diplomatic relations, and migration risks remain key factors in shaping visa approval trends. Ongoing improvements in visa assessment processes could ensure fairness for genuine travelers while maintaining border integrity across the Schengen zone. Schengen visa refusal rates in 2023 showed a minor decrease overall (16% vs. 17.9% in 2022), with specific countries experiencing significant variations. Policy changes, diplomatic relations, and migration risks remain key factors in shaping visa approval trends. Ongoing improvements in visa assessment processes could ensure fairness for genuine travelers while maintaining border integrity across the Schengen zone.The Schengen Visa's obligatory travel insurance stipulation highlights the essential function of international travel insurance in protecting both visitors and host nations. Despite existing hurdles, the advantages of travel insurance far surpass the disadvantages, making it an essential component of

https://doi.org/10.52202/081568-0001

international trip preparation. Subsequent study may investigate the economic implications of these regulations on the travel insurance sector and proposed modifications to improve accessibility and cost for international passengers.

## References:


1. Aygül, C., & Mertek, S. (2024). Visas and mobility: The EU's 'great shutdown' that won't stop population flows. *Mediterranean Politics*.
2. Bigo, D. (2014). The (in)securitization practices of the three universes of EU border control: Military/Navy - border guards/police - database analysts. *Security Dialogue*, *45*(3), 209 – 225.
3. Bligh, A. (2010). Electronic digital passport as a means of partial response to the lack of intelligence in the field of border control. *9th European Conference on Information Warfare and Security 2010, ECIW 2010*, 19 – 27.
4. Everling, U. (2009). The European Union as a Federal Association of States and Citizens. In *Principles of European Constitutional Law*. Bloomsbury Publishing Plc.
5. Hobolth, M. (2014). Researching Mobility Barriers: The European Visa Database.
6. *Journal of Ethnic and Migration Studies*, *40*(3), 424 – 435.
7. Lee, Y.-H., Lu, C.-W., Wu, P.-Z., Huang, H.-L., Wu, Y.-C., & Huang, K.-C. (2018).
8. Attitudes and awareness of medical assistance while traveling abroad. *Globalization and Health*, *14*(1).
9. Lin, P. H., Sun, W. S., & Chen, J. P. (2001). Inventory of aircraft emissions over the Taiwan area. *TERRESTRIAL ATMOSPHERIC AND OCEANIC SCIENCES*, *12*(1), 41–62.
10. Pedersen, M. J. (2015). The intimate relationship between security, effectiveness, and legitimacy: a new look at the Schengen compensatory measures. *European Security*, *24*(4), 541 – 559.
11. Scheel, S. (2022). Reconfiguring Desecuritization: Contesting Expert Knowledge in the Securitization of Migration. *Geopolitics*, *27*(4), 1042 – 1068.
12. Studzińska, D. (2019). Selected aspects of the transformation in function, and permeability, of the Polish-Russian border; [Wybrane aspekty transformacji funkcji i stopnia przenikalności granicy Polsko-Rosyjskiej]. *Przeglad Geograficzny*, *91*(4), 553 –571.
13. Szwedo, P. (2012). Programme of the polish presidency of the EU; [Programme de la présidence polonaise de l'UE]. *Revue Du Marche Commun et de l'Union Europeenne*, *555*, 101 – 107.
14. Trauttmansdorff, P. (2023). The fabrication of a necessary policy fiction: the interoperability 'solution' for biometric borders. *Critical Policy Studies*, *17*(3), 428 –446.
15. Twigt, M. (2024). Doing Refugee Right(s) with Technologies? Humanitarian Crises and the Multiplication of "Exceptional" Legal States. *Refugee Survey Quarterly*, *43*(1),



https://doi.org/10.52202/081568-0001

# Potential and opportunity of mice tourism in India: Case study of Delhi

Praveen Thakur[1*], Dr. Shikha Sharma[2]

[1&2]Chandigarh University, Mohali, India

*Corresponding author: 23BTT10005@cuchd.in[1]

**Abstract.** MICE tourism is one of the fastest-growing sectors in the global travel and tourism industry, with a huge potential for economic growth and development. In the Indian context, Delhi is one of the prime destinations for MICE activities, as it is strategically located, boasts world-class infrastructure, rich cultural heritage, and a vibrant economic landscape. It goes on to further explore the MICE tourism that has untapped potential and opportunity in India by particularly focusing on a case study basis of Delhi. Delhi, the capital city, boasts an extensive network of state-of-the-art convention centers, luxury hotels, and efficient connectivity via air, road, and rail. Its unique blend of modernity and tradition attracts business travelers seeking not only professional engagements but also cultural enrichment. Additionally, Delhi's positioning as a political and economic hub adds to its appeal for hosting international conferences, government summits, and corporate events. This paper evaluates existing MICE infrastructure in Delhi, sheds light on trends in the recent meetings industry, and appraises the impact of MICE tourism on local and national economy. It identifies important challenges, which are sustainability issues, competition with other global destinations, and enhancement requirements for marketing strategy.

**Key words**- MICE Tourism, Economic Growth, Business Travel, Convention Centers, Cultural Heritage, Sustainability Challenges, Global Competition, Marketing Strategies, Tourism Development.

## 1. Introduction

The Meetings, Incentives, Conferences, and Exhibitions (MICE) industry is among the world's fastest-developing sectors in the tourism business, making valuable contributions to the economic development of a nation, business growth, and global cooperation. MICE tourism is one of the essential drivers of elevating a nation's global status by drawing in corporate events, trade exhibitions, and global conferences. In India, the country's fast-paced business infrastructure growth, increasing corporate networking, and the government's efforts to boost the industry underpin the huge scope for MICE tourism. But even with its economic and strategic import, India has not yet solidified itself as an international leader in this market.

Among Indian cities, Delhi is a prominent MICE destination based on its strategic location, state-of-the-art infrastructure, world-class convention facilities, and well-connected air, road, and rail links. Delhi being the political and economic capital of India often hosts high-level international conferences, government summits, corporate events, and large exhibitions. The city's heritage, historic sites, and contemporary business center also add a special attraction to business travelers looking for professional interactions combined with cultural enrichment. Though it is strong, Delhi has a number of challenges that it needs to overcome in its quest to be a globally competitive MICE destination. Sustainability issues, rising competition from global MICE destinations like Dubai and Singapore, and more effective marketing and branding efforts hold it back from realizing its full potential. This research paper seeks to investigate the available MICE infrastructure in Delhi, analyze trending trends, assess its economic contribution, and determine challenges and opportunities for increased expansion. By tackling these elements, Delhi has the potential to emerge as a premium MICE destination, making significant contributions to India's economic and tourism growth.

https://doi.org/10.52202/081568-0002

## 2. Objective

2. To analyze the current status and growth potential of MICE tourism in India, with a special focus on Delhi.
3. To examine the existing MICE infrastructure in Delhi, including convention centers, hotels, and transportation facilities.
4. Toassess the economic impact of MICE tourism on Delhi's local and national economy.

## 3. Literature Review

MICE (Meetings, Incentives, Conferences, and Exhibitions) tourism has emerged as a key segment within the global travel industry, contributing significantly to economic development, international business networking, and infrastructure growth. Several studies have examined the impact, challenges, and opportunities of MICE tourism in different countries, highlighting its potential for revenue generation and urban development. This literature review explores existing research on MICE tourism globally, its role in India's tourism sector, and Delhi's position as a leading MICE destination.

**3.1 MICE Tourism in India**

India's MICE tourism industry has been increasing continuously because of its developing economy, improving infrastructure, and government policies. The Federation of Indian Chambers of Commerce and Industry (FICCI, 2019) stated that India can become a world-class MICE destination, and Delhi, Mumbai, Bengaluru, and Hyderabad are at the forefront of the industry. But it has faced setbacks in terms of lack of exclusive MICE promotion, issues related to sustainability, and rivalry from global locations. Research conducted by Chaturvedi (2020) indicates that development of infrastructure, marketing, and policy reforms will help improve the competitiveness of India in MICE tourism considerably.

**Delhi as a MICE Destination**

Delhi, being the capital and financial hub of India, has emerged as a major MICE destination. Studies by Sharma (2021) reveal that the location advantage of Delhi, state-of-the-art convention halls, five-star hotels, and good connectivity make it a first choice for holding global conferences, corporate meets, and government conferences. Popular places like Pragati Maidan, India Habitat Centre, and Vigyan Bhawan have organized large-scale conferences like the India International Trade Fair and Auto Expo.

**Economic Significance of MICE Tourism in India**

The MICE industry is projected to grow at a compound annual growth rate (CAGR) of over 7% globally, and India is no exception. According to Srivastava and Chauhan (2023), the MICE sector contributed approximately ₹37,567 crore in 2017, with expectations to reach ₹1,99,990 crore by 2025. India has also improved its ranking in the International Congress and Convention Association (ICCA) rankings, moving up 11 positions from 2014 to 2017. Delhi, as the capital city, has been at the forefront of this growth, hosting several international summits, trade fairs, and government conferences (Tembhare et al., 2022).


https://doi.org/10.52202/081568-0002

**MICE Tourism in Delhi: An Emerging Hub**

Delhi is a leading MICE destination due to its well-developed infrastructure, world-class venues (Pragati Maidan, India Habitat Centre, Vigyan Bhavan, and Aerocity convention centers), and seamless international connectivity (Jaiswal, 2019). Studies indicate that Delhi's strategic location, strong transport network, and extensive hospitality services make it a preferred city for global MICE events (Singh & Parkash, 2016). However, competition from global MICE hubs like Dubai, Singapore, and Bangkok necessitates continuous improvement in infrastructure and service delivery (Jamgade, 2018).

### 3.2 Government Policies and Strategic Initiatives

The Indian government has recognized MICE tourism as a high-growth sector, with initiatives such as:

- National Tourism Policy 2015, which focuses on infrastructural development and global marketing strategies (Srivastava & Chauhan, 2023).
- E-visa and visa-on-arrival facilities, which have enhanced accessibility for international business travelers (Tembhare et al., 2022).
- Smart City Projects and PPP (Public-Private Partnerships) aimed at boosting MICE-related infrastructure in Delhi and other metro cities (Singh & Parkash, 2016).
- Despite these efforts, policy gaps and lack of coordination between central and state governments remain significant challenges in fully realizing the potential of MICE tourism in Delhi (Srivastava & Chauhan, 2023).

### 3.3 Challenges in MICE Tourism Development

Several studies highlight major challenges hindering the growth of MICE tourism in India, particularly in Delhi:

- Lack of a dedicated MICE policy: Unlike Singapore or Thailand, India lacks a structured and centralized approach to developing MICE tourism (Singh & Parkash, 2016).
- Infrastructure gaps: While Delhi has top-tier venues, inadequate digital integration, traffic congestion, and high operational costs affect its global competitiveness (Jaiswal, 2019).
- Branding and marketing deficiencies: India's MICE sector lacks aggressive international marketing campaigns, which limits its visibility among global event organizers (Jamgade, 2018).

### 3.4 Future Prospects and Opportunities

Despite the challenges, Delhi has the potential to become a global MICE leaderthrough:

- Technological advancements in event management, such as AI-driven planning, hybrid conferences, and VR-based interactions (Tembhare et al., 2022).
- Expanding beyond Delhito nearby cities like Gurgaon and Noida, which offeraffordable and modern MICE facilities (Singh & Parkash, 2016).
- Encouraging Public-Private Partnerships (PPP) to enhanceevent planning, sponsorships, and hospitality standards (Srivastava & Chauhan, 2023).

## 4. Methodology

This study employs a secondary data analysis approach to assess the potential and opportunities of MICE tourism in India, with a case study of Delhi. The research draws on existing literature, industry reports, and government publications to examine the current status, challenges, and future opportunities of Delhi as a MICE tourism hub.

## 5. Data collection

The study is based on secondary data sources from peer-reviewed journals, government reports, and industry publications. The key sources reviewed include:

- Karnavati Journal of Multidisciplinary Studies (2023)
- MICE tourism is a major contributor to India's GDP and is projected to grow at a CAGR of 7.09% until 2028.
- India ranked 24th in the ICCA (International Congress and Convention Association) ranking in 2017, moving up 11 positions from 2014.
- The sector is expected to grow from ₹37,567 crore (2017) to ₹1,99,990 crore (2025), indicating its high growth potential.
- Atna – Journal of Tourism Studies (2022)
- Research on MICE tourism in India has been increasing, with major studies focusing on conference management, venue perception, tourism marketing, and social media promotion.
- Delhi is identified as a key MICE destination, with key events hosted in Pragati Maidan, India Expo Centre & Mart, and Vigyan Bhawan.
- International Journal of Agricultural Invention (2019)
- The total contribution of tourism and hospitality to India's GDP is projected to rise from $234.03 billion (2017) to $492.21 billion (2028).
- MICE tourism is a high-growth sector within India's broader hospitality industry.
- IJRE (2016) - MICE Tourism in India: Challenges and Opportunities
- Discusses how Delhi, Mumbai, Bangalore, and Hyderabad are leading MICE destinations in India, with Delhi being the preferred choice for government and international events.
- The report highlights challenges such as high taxation, visa issues, infrastructure constraints, and the need for better promotional strategies.
- Scope of MICE Tourism as an Emerging Hospitality Industry (2018)
- MICE tourism in Delhi accounts for a significant share of India's business travel due to its connectivity, historical significance, and world-class conference venues.
- The Delhi government's focus on developing more convention centers, improving hospitality infrastructure, and promoting the city globally is expected to strengthen its position as a MICE hub.
- Strategic Initiatives for the Empowerment of Women in Bangalore MICE Sector (2022)
- This study emphasizes the need for a gender-inclusive workforce in the MICE sector and highlights the importance of training and skill development.
- Delhi's MICE industry could benefit from similar strategies to boost employment and service quality.
- Development Opportunities of MICE Tourism in Developing Countries (2017) - Case of Georgia
- While focusing on Georgia, this study provides comparative insights into how developing nations can enhance their MICE industry through strategic marketing, infrastructure investments, and international collaborations.
- Lessons from Georgia's MICE growth strategies can be applied to Delhi to attract more global events.

## 6. Findings & Discussion

Current status and growth potential of mice tourism in India ( with a focus on Delhi)-The MICE tourism industry in India has been experiencing steady growth, driven by increasing global business engagements, government initiatives, and improved infrastructure. India has been improving its ranking in the International Congress and Convention Association (ICCA) listings, indicating its growing significance in the MICE sector.


https://doi.org/10.52202/081568-0002

**6.1 Growth Trends**:

- The Indian MICE industry has been growing at a CAGR of 8–10%, supported by corporate events, trade fairs, and international summits.
- Delhi, being the political and economic capital, hosts a significant share of India's business tourism, making it a key player in the industry.
- The post-pandemic rise in hybrid and virtual events has further expanded India's MICE opportunities, integrating digital platforms with physical event infrastructure.
- Challenges Hindering Growth:
- Strong competition from global hubs like Singapore, Dubai, and Bangkok, which offer incentives and seamless event planning services.
- High operational costs in Delhi, including venue rentals and accommodation.
- Visa and bureaucratic challenges, despite improvements like e-visa and visa-on-arrival services.
- Lack of a dedicated MICE policy, unlike Thailand's MICE Capabilities Development Plan or Singapore's Business Events in Asia (BEiA) Strategy.

**6.2 Existing mice infrastructure in Delhi: strengths &limitation**

- Delhi is a leading MICE destination in India, equipped with modern venues, high-end hotels, and strong connectivity. However, there are infrastructural gaps that need to be addressed.
- Strengths of Delhi's MICE Infrastructure:
- Top-tier convention centers: Pragati Maidan, India Habitat Centre, Vigyan Bhavan, and Aerocity's Integrated Exhibition Centre.
- Growing hotel industry: Luxury hotels (e.g., Taj Palace, ITC Maurya, Leela Ambience) cater to corporate travelers.
- Excellent connectivity: Delhi's Indira Gandhi International Airport (IGIA) is among the top airports globally, offering direct flights to major MICE destinations.
- Metro and road network expansion improves accessibility to MICE venues.
- Infrastructure Gaps and Challenges:
- Limited high-capacity venues: Compared to international hubs like
- Shanghai and Singapore, Delhi lacks ultra-large convention centers.
- Traffic congestion: Despite metro expansion, road congestion affects event accessibility.
- Technology limitations: Some venues still lack integrated AI, smart event planning tools, and virtual event capabilities.
- Shortage of mid-range business hotels: While luxury hotels are abundant, affordable accommodations for business travelers are still developing.

**6.3 Economic impact of mice tourism on Delhi local& national economy**

MICE tourism plays a pivotal role in Delhi's economy, contributing significantly to business revenues, employment generation, and infrastructure development.

**Direct & Indirect Contributions:**

- The MICE sector generates revenue through event hosting, hotel bookings, airline and transport services, catering, and tourism packages.
- MICE-related businesses such as event management companies, technology providers, and hospitality services see significant growth.

**Employment Generation:**

- MICE tourism supports thousands of direct and indirect jobs, from event planners and hospitality staff to transportation and security personnel.
- Growth in MICE events encourages skill development in conference management, business services, and digital event solutions.

**Foreign Exchange Earnings & Investments:**

- International business events bring foreign exchange inflows, boosting national GDP.

https://doi.org/10.52202/081568-0002

- Increased foreign direct investment (FDI) in hospitality and real estate due to MICE growth, particularly in Delhi NCR regions like Gurgaon and Noida.
- Post-Pandemic Economic Recovery:
- Delhi's MICE sector played a crucial role in reviving the hospitality and travel industries after COVID-19.
- Hybrid events, government-backed business summits, and international trade fairs have helped accelerate economic recovery.

### 6.4 Emerging trends & future potential

**Digital Transformation in MICE Tourism:**

- Rise in AI-driven event management tools, virtual conference platforms, and smart registration systems.
- Blockchain-based ticketing and contactless event experiences are becoming increasingly popular.

**Sustainability Initiatives**:

- Shift towards eco-friendly event planning, carbon-neutral venues, and green energy use in conference facilities.
- Delhi's sustainability challenges (pollution, waste management at events) need urgent policy intervention.

**Growth of MICE Hubs beyond Delhi**:

- Expansion into Gurgaon, Noida, and Greater Noida for more affordable and modern MICE facilities.
- Plans for new exhibition centers and integrated convention districts in the Delhi NCR region.

## 7. Conclusion

MICE tourism is a crucial driver of economic growth, global business networking, and infrastructure development in India, with Delhi emerging as a prime destination. The city's strategic location, world-class convention centers, luxury accommodations, and robust connectivity make it an attractive hub for corporate events, government summits, and international exhibitions. Additionally, Delhi's rich cultural heritage and historical significance enhance its appeal to business travelers looking for both professional engagements and leisure experiences.

Despite its strengths, Delhi faces several challenges in establishing itself as a global leader in MICE tourism. Sustainability concerns, competition from international MICE hubs like Singapore and Dubai, and infrastructural limitations hinder its full potential. High operational costs, bureaucratic procedures, and limited mid-range accommodations further restrict growth. Addressing these issues requires a comprehensive strategy focusing on infrastructure enhancement, policy reforms, and sustainable event management.

The Indian government has taken significant steps to boost MICE tourism, including the National Tourism Policy (2015), the Incredible India 2.0 campaign, and initiatives like Smart City projects and Public-Private Partnerships (PPP). The successful hosting of international events, including the G20 Summit (2023), has showcased Delhi's ability to manage large-scale conferences and highlighted its potential as a leading MICE destination. However, for long- term success, dedicated MICE policies, better marketing strategies, and integration of emerging technologies such as AI, blockchain, and smart venues are essential.

The economic impact of MICE tourism on Delhi's local and national economy is significant. The sector generates employment, stimulates foreign investments, and contributes to the hospitality, transportation, and technology industries. The post-pandemic era has seen a rise in hybrid and virtual events, further expanding MICE opportunities. With planned expansions into Gurgaon, Noida, and Greater Noida, Delhi NCR is poised to become a more competitive and versatile MICE hub.


https://doi.org/10.52202/081568-0002

In conclusion, while Delhi has immense potential to become a top-tier global MICE destination, sustained growth will depend on strategic investments, international collaborations, policy improvements, and sustainable practices. By leveraging its existing strengths and addressing key challenges, Delhi can strengthen its position in the global MICE industry, significantly contributing to India's tourism and economic landscape.

## References


1. Chaturvedi, R. (2020). Strategic Analysis of MICE Tourism Potential in India. IJRE – International Journal for Research in Education, 4(5).
2. Grover, P. S. (n.d.). MICE Tourism in Delhi [PowerPoint slides]. SlideShare.
3. Jamgade, S. (2018). Scope of MICE Tourism as an Emerging Hospitality Industry. UAS Journal of Management and Commerce, 4(2), 26–29.
4. Jaiswal, R. (2019). MICE Tourism in India – A Case Study of Delhi. International Journal of Agricultural Invention, 4(3).
5. Joshi, P. (2018). MICE Tourism: Trends and Challenges. Research Horizons, 8(2).
6. Sharma, S. (2021). Delhi as an Emerging MICE Hub: An Analytical Study. Atna Journal of Tourism Studies, 16(2).
7. Singh, J., & Parkash, R. (2016). MICE Tourism in India: Challenges and Opportunities. International Journal for Research in Education, 3(9), 36–43.
8. Srivastava, A. K., & Chauhan, A. (2023). MICE Tourism in India: Unleashing the Potential of Tourism Sector. Karnavati Journal of Multidisciplinary Studies, 1(1).
9. Tembhare, V., Khare, A., & Naidu, L. (2022). Digital Trends in MICE Events: Post-Pandemic Resurgence. Atna Journal of Tourism Studies, 17(1).
10. Tembhare, V., Khare, A., & Naidu, L. (2022). Progress in MICE Industry Research: Case of India. Atna Journal of Tourism Studies, 17(1), 33–60.
11. Anonymous. (2022). Strategic Initiatives for the Empowerment of Women in Bangalore MICE Sector. Conference Proceedings – Indian Tourism Development Forum.


https://doi.org/10.52202/081568-0002

# Adventure Tourism in India: A Comprehensive Case Study of Himachal Pradesh

Narender Singh[1], Dr. Shikha Sharma[2]

[1&2]Chandigarh University, Mohali, India

*Correspondence author: shikhasharma.uithm@cumail.in

**Abstract.** Adventure tourism is a type of niche tourism where exploration or travel involves some degree of risk (real or perceived) and may require special skills, and physical effort. The activities include in adventure tourism are hiking, mountaineering, river rafting, scuba diving, wildlife safaris. Himachal Pradesh is one of the most famous state of India to experience all these thrilling activities. Himachal has various famous trek such as :Kheerganga trek, Triund trek, Hampta pass trek, Indrahar trek. And the other activites like river rafting in Kullu, paragliding in Bir Billing, skiing in Manali and Kufri and camping in Dalhousie. Himachal Pradesh offers a wide range of adventure activities for all interests and skill levels. From the thrill of white-water rafting to the serenity of high-altitude trekking, this region offers a unique blend of adrenaline-pumping challenges and soul-stirring experiences.

**Keywords:** niche tourism, risk, special skills, adrenaline pumping, and soul stirring.

## 1 Introduction

### 1.1 Adventure Tourism in Himachal Pradesh

Adventure tourism has increasingly become a significant segment of the global tourism industry, attracting travellers.

**Geographical and Cultural Context**

**Geographical Highlights**
Himachal Pradesh boasts a varied topography that includes high-altitude plateaus, snow-capped peaks, dense forests, and meandering rivers. The state's climate ranges from temperate in the lower altitudes to alpine in the higher regions, allowing for a diverse array of adventure activities throughout the year. Key regions such as Shimla, Manali, Bir Billing, Solang Valley, and Spiti Valley each offer unique landscapes and terrain suited for different types of adventure sports.

**Cultural Heritage**
The state's indigenous cultures and traditions add a distinctive flavor to the adventure tourism experience. Visitors not only engage in thrilling activities but also get the opportunity to immerse themselves in local customs, festivals, and cuisines, providing a holistic and enriching travel experience.

## 2 Review of Literature

(ArchnaTyagi ,1989) in study entitled, ―A Study of Tourism Marketing in Himachal Pradeshǁ has pointed out that there are so many possibilities to develop tourism industry in the state of Himachal Pradesh. This has also highlighted many tourist attractions such as cultural traditions, natural beauty, natural environment, etc. which may be managed and marketed to boost the tourism industry in the state of Himachal Pradesh.
According to Heeson (2004), the World Tourism Organization has pinpointed several obstacles hindering national tourism industries. A key focus for the tourism sector is the development of eco-tourism and rural tourism. However, progress is hampered by


https://doi.org/10.52202/081568-0003

issues such as ineffective tourism policies, inadequate planning, limited financial viability, deficient infrastructure, poor transportation links, and restrictive government regulations. Specifically in India, a significant challenge identified by experts is the lack of meaningful community engagement in the tourism industry."

Kumar& Vinod (2014) was observed that the most notable effect of tourism on traditional values and culture is a certain social and human interactions change into the commercialization of cultural values and traditions at the destination for the purpose of earning of a living. It was indicated that with the introduction of tourism at large-scale in developing countries, goods and/or services that used to be part of people's individual and social life have been commercialized and presented as commodities. He suggested that if there is a proper share of community involvement at a different level of tourism development and people retain control, original social tradition and norms will be less affected and the negative impacts of cultural tourism will be reduced.

In the 2018 article, "Potential of Tourism in Himachal Pradesh," Dr. Subhash Chandra explores the diverse opportunities for adventure tourism and other tourism types in the state. This highlights that the tourism sector in Himachal Pradesh holds immense potential to thrive, offering a fresh impetus to the state's economy. According to author, tourism ranks as the second most significant economic resource after hydroelectric power. The region provides ample prospects for wildlife enthusiasts and sports lovers alike. With its unique topography and terrain, it serves as an ideal destination for trekkers, climbers, and adventure enthusiasts. Activities such as trekking, mountaineering, rock climbing, heli-skiing, hang gliding, skiing, and water sports like water skiing, river rafting, and fishing are all viable pursuits here.

In the 2020 research paper titled "Role of Hospitality Sector in Promoting Adventure Tourism – Study of Himachal Pradesh," Ashutosh Sharma and Dr. Praveen Chauhan present insights into the current state of the tourism industry. It discuss the rising popularity of adventure tourism and its growing significance within the hospitality and tourism sectors. The authors attribute this trend to an increasing desire for adventure and a surging interest in pristine natural environments, which are seen as key drivers benefiting such regions. The study emphasizes how the hospitality sector plays a vital role in supporting and promoting adventure tourism, particularly by catering to the influx of adventure-seeking tourists.

("S. R. Chugh ,2012) identified tourism as a growing industry in Himachal Pradesh, emphasizing the importance of effective promotional strategies to cultivate a positive tourist image.

(Jeena Gupta ,2015), focusing on Kullu-Manali, examined the socio-economic and political impacts of tourism, including infrastructure development such as roads, transportation, and communication networks. Gupta's study aimed to assess tourist facilities and conduct a SWOT analysis of Himachal Pradesh's tourism potential, while also addressing challenges encountered during peak tourist seasons.

## 3 Objectives

1. To assess the current status and potential of adventure tourism in Himachal Pradesh.
2. To examine the role of infrastructure, hospitality, and marketing in promoting adventure tourism.
3. To identify challenges and opportunities for sustainable development of adventure tourism.

## 4 Methodology

This study adopts a mixed-method research design, combining both qualitative approaches to comprehensively explore adventure tourism in Himachal Pradesh. The qualitative aspect focuses on understanding the perceptions, experiences, and challenges faced by stakeholders.

### 4.1 Key Adventure Activities in Himachal Pradesh

**Paragliding**
Bir Billing in the Kangra district is globally recognized as the "Paragliding Capital of India." The region's favourable thermal currents and scenic vistas make it an ideal spot for both beginners and experienced paragliders. Other notable paragliding sites include BundlaDhar and Solang Valley, each offering unique aerial perspectives of the Himalayan landscapes.

**Trekking**
Himachal Pradesh offers an extensive network of trekking routes catering to different skill levels. Popular treks such as Triund, Kheerganga, Beas Kund, Indrahar Pass, and Hampta Pass attract adventurers with their breathtaking views and challenging terrains. These treks not only provide physical challenges but opportunities to experience remote Himalayan villages and untouched natural beauty.

**River Rafting and Water Sports**
The state's numerous rivers, including the Beas, Sutlej, Spiti, and Ravi, are prime locations for white-water rafting and kayaking. Kullu, in particular, has gained fame for its water-based sports, offering rapids of varying difficulty levels that attract thrill-seekers from around the world. These activities are complemented by well-regulated safety protocols and trained guides to ensure a safe and enjoyable experience.

**Skiing and Heli-Skiing**
Solang Valley and Kufri are the primary destinations for skiing in Himachal Pradesh. The region offers pistes for both beginners and advanced skiers, with the added appeal of heli-skiing for those seeking untouched snow terrains. The winter months transform these areas into snowy playgrounds, drawing enthusiasts eager to embrace the cold climate and snowy slopes.

**Rock Climbing and Mountaineering**
Manali serves as a hub for rock climbing and mountaineering activities. The rugged terrains and towering peaks provide an ideal setting for climbers to test their skills and endurance. Mountaineering expeditions often include guided climbs of notable peaks, offering both challenge and the reward of stunning panoramic views.

**Mountain Biking and Camping**
Adventure seekers also enjoy mountain biking through the rugged trails and scenic routes of Himachal Pradesh. Camping under the starry Himalayan skies allows tourists to connect with nature intimately, providing a sense of tranquillity and adventure.

### 4.2 Infrastructure and Support Systems

**Government Initiatives**
The Himachal Pradesh government plays a pivotal role in promoting and sustaining adventure tourism. Through the Himachal Pradesh Tourism Department, various training programs and events are organized to enhance the skills of adventure sports enthusiasts and professionals. Facilities such as the Adventure Sports Hostel in Dharamsala and the Aero Sports Complex in Bir Billing provide essential support and training for adventure activities.


https://doi.org/10.52202/081568-0003

**Private Sector Involvement**
Private enterprises collaborate with the government to develop infrastructure, offer guided tours, and ensure safety standards are met. Investments in accommodations, transport, and adventure gear facilities are crucial in making the region accessible and appealing to international and domestic tourists alike.

**Training and Certification**
To maintain high safety standards, the state emphasizes training and certification for adventure guides and operators. Accredited training programs ensure that adventurers are well-prepared and that guiding services meet international safety benchmarks.

### 4.3 Economic and Social Impacts

**Economic Contributions**
Adventure tourism significantly bolsters the local economy of Himachal Pradesh. It contributes approximately 2.5% to the state's GDP and creates numerous employment opportunities in sectors such as hospitality, guiding, transportation, and retail. The influx of tourists stimulates local businesses and fosters entrepreneurial ventures, enhancing the overall economic landscape.

**Community Engagement**
Local communities are actively involved in the adventure tourism value chain. Community- based tourism initiatives empower residents by providing jobs and promoting cultural exchange. This engagement fosters a sense of ownership and ensures that the benefits of tourism are distributed equitably among the population.

**Sustainable Practices**
Emphasis on sustainable tourism practices helps mitigate the environmental impact of adventure activities. Initiatives include waste management programs, conservation efforts, and the promotion of eco-friendly adventure packages. Sustainable practices ensure that Himachal Pradesh's natural beauty and ecological integrity are preserved for future generations.

### 4.4 Challenges and Constraints

**Environmental Concerns**
The surge in adventure tourism can lead to environmental degradation if not managed properly. Issues such as over-tourism, inadequate waste management, and disturbances to local wildlife pose significant threats. Implementing robust regulatory frameworks and promoting eco- friendly practices are essential to address these challenges.

**Safety and Risk Management**
Adventure sports inherently carry risks. Ensuring the safety of tourists requires stringent certification for guides, quality equipment,and effective emergency response mechanisms. Continuous monitoring and enforcement of safety standards are crucial in minimizing accidents and ensuring a secure environment for adventurers.

**Infrastructure Gaps**
While major tourist hubs are well-equipped, remote areas often lack essential infrastructure such as reliable access roads, healthcare facilities, and robust communication networks. Bridging these gaps is vital for enhancing accessibility and ensuring the safety and comfort of tourists in less developed regions.

### 4.5 Cultural Sensitivity
The influx of tourists can lead to the commercialization of local traditions and potential disruption of traditional lifestyles. Maintaining cultural sensitivity and ensuring that tourism development respects and preserves local customs and heritage is imperative for sustainable growth.

https://doi.org/10.52202/081568-0003

**Sustainable Tourism Practices**
Adopting eco-friendly infrastructure and enforcing regulations that minimize the ecological footprint of adventure activities can preserve Himachal Pradesh's natural environment. Initiatives such as promoting zero-waste policies, encouraging the use of sustainable materials, and implementing conservation projects are pivotal in achieving sustainable tourism.

**Diversification of Adventure Packages**
Creating innovative adventure packages that blend physical activities with cultural immersion, local cuisine experiences, and wellness retreats can attract a broader audience. Diversification not only enhances the tourist experience but also distributes the economic benefits across different sectors.

**Skill Development and Local Engagement**
Investing in training programs for locals ensures that the workforce is skilled and capable of delivering high-quality adventure tourism services. Programs focused on adventure tourism management, first aid, and hospitality enhance the overall visitor experience and contribute to socio-economic upliftment.

**Improved Marketing and International Collaborations**
Leveraging digital marketing strategies and forming partnerships with international adventure tourism agencies can expand Himachal Pradesh's global reach. Enhanced visibility through online platforms, social media campaigns, and international tourism fairs can attract a diverse range of tourists.

**4.6 Case Studies and Success Stories Paragliding in Bir Billing**

Bir Billing has successfully established itself as a premier paragliding destination globally. The region's optimal weather conditions and stunning landscapes attract thousands of paragliding enthusiasts each year. The presence of experienced instructors and well-maintained facilities has solidified Bir Billing's reputation as a top adventure tourism spot.

**Trekking Routes in Spiti and Kinnaur**
The trekking circuits in Spiti and Kinnaur have significantly contributed to local economies by promoting community-based tourism. These treks offer authentic cultural interactions and allow tourists to experience the serene beauty of remote Himalayan villages, fostering a deeper appreciation for the region's natural and cultural heritage.

**River Rafting and Water Sports in Kullu**
Kullu's river rafting and kayaking have gained national and international acclaim. The establishment of safety protocols and well-regulated activities have made Kullu a sought-after destination for water sports enthusiasts. The economic benefits from these activities have bolstered local businesses and infrastructure development.

**4.7 Policy Implications and Recommendations Developing Regulatory Frameworks**

Enforcing stringent safety and environmental standards is crucial for sustainable adventure tourism. Policies should mandate regular audits, certifications for adventure operators, and adherence to eco-friendly practices. This ensures the safety of tourists and the preservation of natural resources.


https://doi.org/10.52202/081568-0003

**Infrastructure Revitalization**
Improving transportation networks, enhancing communication systems, and upgrading healthcare facilities, especially in remote adventure sites, are essential steps. These improvements reduce risk factors associated with high-adrenaline tourism and enhance the overall tourist experience.

**Training and Capacity Building**
Structured training programs for local youth and stakeholders in adventure tourism management, first aid, and hospitality services are vital. Building capacity ensures that the industry grows sustainably and that service quality meets international standards.

**Promoting Community Participation**
Engaging local communities as key partners in the adventure tourism value chain ensures equitable distribution of economic benefits and fosters a sense of ownership and responsibility towards sustainable tourism development.

## Conclusion

Himachal Pradesh embodies the essence of adventure tourism in India, offering a unique blend of natural beauty, diverse adventure activities, and rich cultural experiences. The state's commitment to developing robust infrastructure, promoting sustainable practices, and engaging local communities has positioned it as a leading adventure tourism destination. However, addressing challenges related to environmental conservation, safety, and infrastructural gaps is imperative for sustaining growth and ensuring that adventure tourism continues to benefit both the economy and the local population. Through strategic planning, continuous investment, and a focus on sustainable development, Himachal Pradesh can maintain and enhance its reputation as a premier hub for adventure tourism, balancing economic prosperity with environmental stewardship and cultural preservation.

## References


1. S. R. Chugh (2012). ―Tourism Promotion in Himachal Pradesh‖, International Journal of Hospitality & Tourism Systems.
2. Gupta, Jeena (2015), Tourism in Himachal Pradesh: A Study of KulluManali. International Journal of Engineering Research & Management Technology.
3. Dr. Chandra, Subhash (2018). Potential of Tourism in Himachal Pradesh: An Analysis of Tourist's Opinions. International Journal of Creative Research Thoughts Sharma, Ashutosh&Dr. Chauhan, Praveen (2020). Role of Hospitality Sector in Promoting Adventure Tourism – Study of Himachal Pradesh. International Journal of Scientific & Technology Research,
4. Tyagi Archana, "A study on development potential of Tourism in H.P," M. Phil Thesis submitted to H.P.U. Shimla-1989.
5. HeesonWarher, "Editorial by FICCI on Tourism," Travel Talk, Dec, 2004.
6. Kumar Sandeep and Kumar Vinod, " Perception of Socio-Culture Impacts of Tourism: A Sociological Review", International Research Journal of Social Sciences,

# Wellness beyond the Ganges: Special Interest Tourism In Uttar Pradesh

Tanya Srivastava[1], Shikha Sharma[2*]

[1&2]Chandigarh University, Mohali, India

*Correspondence author: shikhasharma.uithm@cumail.in*

**Abstract.** This research paper signifies the continuous growth of "special interest tourism" in Uttar Pradesh, focusing on wellness tourism. As Uttar Pradesh is rich in cultural, spiritual and natural heritage, it also offers traditional healing practices, spiritual experiences and beautiful environments, making it popular for wellness tourism. The paper examines how the religious and spiritual sites of Uttar Pradesh, such as Varanasi, Prayagraj and Ayodhya also attract wellness-focused travellers (seeking not only physical wellness but also mental and spiritual wellness).
The study depicts the role of yoga, meditation, Ayurveda Therapies and Naturopathy in promoting wellness tourism in Uttar Pradesh. It focuses on the combination of wellness practices and religious pilgrims, offering a complete experience combining health, spirituality and culture. The research also discusses on the wellness retreats, spiritual centres that focuses on the growing demand of personalized wellness experiences.
Moreover, the research also depicts the growing demand of natural resources such as Ganges River and other scenic landscapes providing beautiful experience for wellness and meditation. Through a comprehensive analysis, the aim of the study is the expansion of wellness tourism in Uttar Pradesh, it's potential growth in sustainable growth and its role in broader context of special interest tourism.
The paper concludes by offering insights into how Uttar Pradesh can influence its spiritual and cultural aspects to further develop wellness tourism contributing both to local economy growth and global tourism trends.

**Keywords:**
Special Interest Tourism, Uttar Pradesh, Wellness Tourism, Meditation, Spiritual Healing

## 1 Introduction

Tourism in India is almost every time being included with the holy river Ganges, especially in places like Varanasi and Prayagraj, famous for spirituality and traditional aspects. However, with the global trend, Special Interest Tourism (SIT) is soaring high, considering interest of tourists in different types of tourism in different regions. While the Ganges continues to offer spiritual and cultural experiences, Uttar Pradesh have a diverse range of experiences to delve into. It includes wellness tourism as the major offering with wellness retreats, yoga centers, wellness-resorts, Ayurveda-based therapies existed in mostly underrated regions focusing more on value.
This research paper explores the growth of wellness tourism as in Special Interest Tourism (SIT) in Uttar Pradesh. After the pandemic is over, the focus on wellness tourism has increased much higher. Wellness tourism, as a niche market helping in mental, physical and spiritual healing has gained high popularity in destinations like Uttar Pradesh. The location with high cultural and spiritual interest, with natural landscapes is attracting the need of wellness tourism in the state.
By going through various wellness tourism initiatives and exploring how Special Interest Tourism (SIT) is impacting the local communities and its economies, this research paper will


https://doi.org/10.52202/081568-0004

focus on how Uttar Pradesh have more to offer as in experience of a tourist than only a cultural and spiritual focused state throughout. This research aims to explore how Uttar Pradesh could become a leading wellness tourism destination, going beyond its usual connection to the Ganges by looking at wellness spots, tourism strategies, and consumer trends.

## 2 Literature Review

Table 1. Literature review

| Author'sYear | Topic | Methodused | Keyresults |
|---|---|---|---|
| Shukla & Sharma (2017)[1] | Revival of the "Golden Bird" : A gap analysis of governments initiatives and tourist expectations in Uttar Pradesh using system's approach. | Surveys and evaluation | Improving special interest tourism. |
| Moin (2024) [2] | Role of Tourism in economic development of India : A study of Uttar Pradesh. | Web-scrapping | Improving niche tourism but cleanliness and sanitation become a concern to be noted. |
| Pandey & Agarwal (2021) [3] | How tourism industry has taken revival strategies Post-Pandemic : A case study of domestic study in Uttar Pradesh. | Web-scrapping | Observed that there is an increase of niche and special interest tourism. |
| Singh et al. (2022) [4] | Tourism and sustainability in times of Covid -19 : the case study of Uttar Pradesh. | Web-scrapping and secondary data sources | Focusing on health safety and economy building by tourism after pandemic through government initiatives. |
| Agarwal & Pandey (2022) [5] | Domestic Tourism : Post-Pandemic revival of travel and tourism industry in Uttar Pradesh. | Observation and secondary data sources. | Focusing on building economy by tourism after pandemic to improve sustainability. |
| Sharma & Pandey (2022) [6] | Religious tourism in Uttar Pradesh : A case study of Varanasi. | Case studies and surveys | It shows how religious tourism can also help in the development of wellness and leads to enhance Wellness Tourism |
| Chawla & Jha (2019) [7] | Avenues and problems of tourism in Uttar Pradesh. | Web-scrapping | It shows how tourism initiatives contribute in increasing the GDP of India. |
| Jacob & Bhalla (2024) [8] | Exploring inclusiveness in tourism sector of Uttar Pradesh, India. | Survey and examination of secondary data | The policies of tourism have to be improved in terms of cleanliness and others. |
| Chauhan (2024) [9] | Exploring the potential and challenges of sports tourism in Uttar Pradesh : emerging trends, strategic initiatives and sustainable | Survey and web-scrapping | The enhancement of sports integrates wellness tourism in Uttar Pradesh. |

| | | | |
|---|---|---|---|
| | development. | | |
| Gill (2024) [10] | A Review : Health and wellness promotion in context with Indian Tourism. | Web-scrapping and evaluation of existing data and initiatives by government. | Government has taken initiatives to promote wellness and to enhance online business. |

Shukla & Sharma (2017) [1] researched together through surveys & evaluations to find out the government initiatives in tourism. It resulted in finding that it increases the demand of special interest tourism.
Moin (2024) [2] researched through web-scrapping to find out role of tourism in economy development in Uttar Pradesh. It resulted in that as tourism promotes different niche tourism, cleanliness & sanitation is still a major concern.
Pandey & Agarwal (2021) [3]researched together through web-scrapping to find out the revival strategies of tourism in Uttar Pradesh Post-Pandemic. It resulted in with different government initiatives, there is a high-level of increase in niche & special interest tourism especially focusing on wellness tourism.
Singh et al. (2022) [4] researched with his teammates through Web-scrapping and evaluating secondary data to find out the tourism and sustainability during Covid -19 in Uttar Pradesh. It resulted in increase of health & safety and economy building Post-Pandemic in tourism.
Agarwal & Pandey (2022) [5] researched together through observation on subjects and evaluating secondary data to find out about domestic Tourism Post-Pandemic. It resulted in finding out tourist focused on sustainability and economy building.
Sharma & Pandey (2022) [6]researched together through case-studies and survey to find out about religious tourism in Uttar Pradesh, especially in Varanasi. It resulted in finding out its connection with wellness tourism that it improves the importance of wellness mentally and physically promoting wellness tourism.
Chawla & Jha (2019) [7] researched together through web-scrapping to find out the avenues and problems in tourism in Uttar Pradesh. It resulted in with the need of cleanliness and sanitation in the state as a problem, tourism is helping highly in raising the GDP of the state.
Jacob & Bhalla (2024) [8]researched together through surveys and evaluation of secondary data to find out the inclusiveness of tourism sector in Uttar Pradesh it resulted in finding out that government initiatives are increasing the sanitation level in the state.
Chauhan (2024) [9] researched through surveys and web-scrapping to find out potential and challenges of sports tourism in Uttar Pradesh. It resulted in founding that the interest of sports in Uttar Pradesh is increasing the level of wellness in people.
Gill (2024) [10] researched through web-scrapping and evaluation of secondary data and government initiatives to find out health & wellness in Indian Tourism. It resulted in finding out that through government initiatives & online business, wellness tourism is growing globally.

## 3 Methodology:

In this research, I have used the qualitative approach as a methodology to find the result of my findings, used only secondary data sources to comprehensively explore the growth and potential of wellness tourism in Uttar Pradesh. This methodology is designed to examine various aspects of wellness tourism i.e., from its current offerings to the spiritual, cultural and economic factors that contributes in the development.

**3.1 Data Collection:-**This study is based completely on secondary data. The data is collected from credible and authentic source i.e. from Google scholar and other online databases by completely analyzing it. The key sources of the secondary data include :

**3.2 Literature Review** – It includes an extensive review of academic literature,


https://doi.org/10.52202/081568-0004

government reports, tourism-based publications and previous researched studies (as mentioned in literature review) that is directly/indirectly connected to my research, i.e. conducted to get a deeper understanding of wellness tourism in Uttar Pradesh, as well as the government initiatives that helped in promotions, trends and strategies that influenced the growth of wellness tourism within the state.

**3.3 Tourism Industries and Government Reports** – It includes data from tourism boards, Uttar Pradesh Government Reports and Minister of Tourism publications are used to analyze the trends of tourism, wellness growth in the state, policies and initiatives made in the region. Reports of wellness tourism worked upon by the private sectors are also analyzed.

**3.4 Statistical data analysis** –The other secondary data sources include reports from National Sample Survey Office (NSSO), India Tourism Statistics and World Travel & Tourism Council (WTTC), were examined thoroughly to evaluate the trends in tourism growth, economic contributions and the demographics of wellness tourists in Uttar Pradesh.

**3.5 Tourism data analysis:-**

It includes the secondary data from government tourism departments, tourism boards and industry reports were analyzed to know and store the data that focuses on the growth and trends of wellness tourism within the state, the impact of Covid-19 pandemic on tourism sector and also the role of spiritual, cultural and religious destinations and landmarks that helps in attracting the tourists.

The secondary data was used to analyze by qualitative approach and statistical analysis. The key themes related to wellness tourism trends and updates include cultural, spiritual influence, economic interpretation policies were mentioned accordingly. Also, the analysis of trends were conducted to determine the after and before effect/ value and demand of wellness tourism in the state.

**3.6 Limitations:-**

As secondary data provides deeper insights, this research have some limitations too. Such as :-

**Lack of Primary data:** - This study doesn't include data collected directly from the stakeholders of tourism industry and other wellness tourism initiatives using samples so it lacks in-depth data.

**Data Availability and Accuracy:** - This study only includes the data from government reports and literature reviews which might not provide comprehensive data and information.

**General data:** - The data collected and initiatives made in the state may not be used to analyze any other region/state.

**Ethical Considerations:-**

As this study is completely based on secondary data and there are no direct interaction with the stakeholders and participants. All the data retrieved are ethically cited and gave the credit to the author in literature review. Additionally, the effort is made to get the credible and authentic data from credible sources.

# 4 Results and Findings

The data and findings of my research are based on only secondary data sources, which includes government reports and initiatives, academic reports and literature reviews, tourism industry analysis, and tourism-based publications. The results we have got gives us the in-depth information on growth, challenges and economic impact of wellness tourism in Uttar Pradesh.

Growth Trends in Wellness Tourism in Uttar Pradesh :- With the increase of spirituality and holistic well-being globally leads to growth of wellness tourism in Uttar Pradesh over the past decade. Such trends came in notice are as follows :-

Wellness travel post-pandemic – The emergence of COVID-19 globally, the need and demand for wellness tourism has increased, with the travelers seeking the services like mental well-being, stress-relief and immunity-building activities which includes yoga, meditation and Ayurveda therapies.

Wellness Tourism through Government Initiatives –Different initiatives and programs initiated by government such as "*Dekho ApnaDesh*" campaign and "Uttar Pradesh Tourism Policy" puts focus on wellness tourism in private sector.

Connection with pilgrimage tourism –Popular pilgrimage sites in Uttar Pradesh such as *Varanasi, Ayodhya*, and *Prayagraj* have seen a increase in the travelers seeking wellness activities which builds a connection between religion and wellness programs such as yoga, meditation and Ayurveda therapies.

Rise of ecofriendly wellness and luxury retreats – The rise of special interest tourism including personalized and authentic experiences such as services provided by boutique wellness *resorts* providing organic food, and meditation programs which contributed in the growth in wellness tourism sector.

**Promotion of wellness tourism through digital tools** –Wellness tourism retreats rapidly increased through social media, travel vlogs, blogs and wellness travel influencers took Uttar Pradesh to the position where the state is considered as one of the major destination for wellness tourism.

***4.1 Role of Spiritual and Cultural Landmarks :-*** As Uttar Pradesh is a major location for rich spiritual and cultural heritage destinations which leads it to as an ideal location for wellness tourism. Key findings are as follows :-

**Varanasi –** It is one of wellness hub and also known as "spiritual capital of India", therefore it is not only a pilgrimage site but also a center famous for yoga, meditation and Ayurveda therapies retreats. It includes the presence of Ganga Aarti, ghats for meditation and traditional healing practices which attracts wellness tourists.

**Ayodhya and Prayagraj** –In Uttar Pradesh, these cities is majorly known for their religious significance, are also coming into focus as destination for wellness-based tourism. There are many retreats near spiritual and religious destinations including ashrams focusing on spiritual rejuvenation and stress-relief activities with a structured programs.

**Wellness through natural resources –**The natural resources include the Ganges River, forests, and rural landscapes provide perfect landscapes providing eco-friendly wellness tourism. The experiences getting from these sources such as *river meditation, herbal healing* and *organic- farming* stays attracting wellness tourists.

**4.2 Economic Impact of Wellness Tourism in Uttar Pradesh :-** As through religious and spiritual, the GDP of the state has gradually increased significantly,


https://doi.org/10.52202/081568-0004

through wellness tourism also there is direct and indirect impact on the economy of Uttar Pradesh. We can get to know about it through government reports on wellness tourism. It follows :-

**Generating Revenue** – Wellness tourism has significantly contributed to the GDP of Uttar Pradesh through difference wellness retreats such as *luxury wellness retreats, Ayurveda centers* and *yoga tourism*.

**Job creation and employment opportunities** – The wellness tourism sector leads to different employment opportunities for the locals such as *Ayurveda practitioners, tour guides* and *hospitality workers*. They are being trained through different government initiative programs, and wellness tourism promoting private companies.

**Growth of allied industries** –As wellness tourism provides well-being soulfully, there are many other demands that boosts in originating different industries as stakeholders for wellness tourism such as *organic food, herbal medicine, handicrafts* and *eco-friendly resorts*, which leads to supporting local businesses.

***Challenges in the development of Wellness Tourism :-*** Despite the high potential and demand of wellness tourism in the state, there are several challenges that are being faced :-

**Lack of infrastructure** – In Uttar Pradesh there are many wellness retreats and tourism sites that lack proper roads, sanitation and public facilities, making it difficult for tourists to accessible.

**Regulatory concerns** –For the wellness centers, they doesn't have any standardized certifications for wellness centers yet, which leads to the concerns about the authenticity and the quality of the services provided.

**Marketing limitations** –The wellness tourism of Uttar Pradesh is not as well-known and well- promoted globally compared to the states like Kerala or Uttarakhand, which are well-known for Ayurveda and yoga tourism even internationally.

## 5 Discussion and Analysis:

How does Wellness Tourism fit into Special Interest Tourism (SIT) in Uttar Pradesh?

As we know that the need and demand of personalized and authentic experience are increasing, it leads to the growth of Special Interest Tourism globally, and wellness tourism is considered as the key component of this trend in tourism. As we know the major offerings in tourism in Uttar Pradesh includes traditional religious tourism, with that, over the decade, wellness tourism in Uttar Pradesh increased the demand of spiritual, cultural heritage and modern wellness practices and retreats, making it an unique offering.

Wellness extending religious tourism – Tourists visiting cities in Uttar Pradesh like Varanasi, Ayodhya, and Prayagraj are now combining pilgrimage with wellness retreats leading to spiritual healing and health-based tourism.

Diverse experiences in tourism – With developing trends and growth of wellness tourism, there are diverse offerings we can provide tourists such as detox retreats, digital detox camps, stress relief programs, and holistic healing as the part of their travel experiences.

5.1 The Role of Government and Private Sector in Promoting Wellness Tourism :-

https://doi.org/10.52202/081568-0004

Government initiatives and policies – To promote wellness tourism in Uttar Pradesh, the government launch many tourism initiatives, including *yoga festivals, Ayurveda promotion,* and *wellness tourism circuits.*

Private sector contributions – With the help of private sector, the industry launch luxury wellness resorts and independent yoga centers that helps in contributing to the development of local community and it's economy.

Potential for Sustainable Wellness Tourism :-

Eco-friendly retreats – To promote sustainable tourism practices, there are many wellness resorts that works upon it, including eco-lodges, organic farming and nature-based healing experiences.

**Community Involvement** – The contribution of locals in promoting wellness retreats such as local healers, yoga teachers, and Ayurveda practitioners ensures providing tourists cultural authenticity and economic sustainability.

## Conclusion:-

In the conclusion of this research study, our study encountered the growth, potential, and challenges of wellness tourism in Uttar Pradesh. The key points are :-

The growth of wellness tourism in Uttar Pradesh is due to spiritual heritage, natural landscapes, and post-pandemic demand in well-being.

For holistic-wellness experiences, the sites ideal for this are The Ganges River, cultural sites, and traditional healing practices.

Despite the demand and potential of wellness tourism in the state, there are various challenges such as infrastructure challenges, marketing gaps, and regulatory concerns.

To promote sustainable and high-quality wellness tourism in the state, the government and private sector must collaborate.

For the future reference, the researchers could explore digital wellness tourism, Ayurveda education, and international collaborations to enhance and analyze Uttar Pradesh's position as a global wellness tourism destination.

## 6 Recommendations:

To enhance and improve wellness tourism in Uttar Pradesh, the following recommendations are proposed :-

**Infrastructure Development :**

To improve sanitation, road and transportation facilities around wellness sites.


https://doi.org/10.52202/081568-0004

To develop dedicated wellness tourism hubs with certified wellness practitioners, resorts and retreats.

**Quality Control and Certification:**

To ensure the authenticity and credibility of the offerings, provide official certifications for wellness centers.

For Ayurveda therapies and Yoga centers, establish standardized guidelines to build credibility.

**Digital Marketing and International Promotion:**

To attract international tourists by using actively social media, travel influencers, and global wellness platforms.

To organize conferences with stakeholders and global wellness summits and retreats in Uttar Pradesh.

**Sustainable and Community-based Wellness Tourism :**

To encourage sustainable tourism initiatives and eco-friendly retreats.

To create economic benefits for local community and locals in rural areas and to involve locals in wellness tourism projects.

## References:


1. Agarwal, S., &Pandey, R. (2022). Domestic Tourism : Post-Pandemic revival of travel and tourism industry in Uttar Pradesh. Journal of Tourism Research, 18(4), 78- 94.
2. Chawla, M., Jha, P. (2019). Avenues and problems of tourism in Uttar Pradesh. Indian Tourism Review, 12(3), 112-130.
3. Chauhan, R. (2024). Exploring the potential and challenges of sports tourism in Uttar Pradesh : Emerging trends, strategic initiatives, and sustainable development. Journal of Sports and Tourism, 21(1), 55-70.
4. Gill, A. (2024). A review : Health and wellness promotion in context with Indian Tourism. International Journal of Wellness Studies, 9(2), 34- 50.
5. Jacob V., Bhalla, T. (2024). Exploring inclusiveness in the tourism sector of Uttar Pradesh, India. Tourism Policy Journal, 17(1), 89-105.
6. Minister of Tourism, Government of India. (2023). India Tourism Statistics 2023.https://tourism.gov.in/statistics
7. Moin, A. (2024). Role of Tourism in the economic development of India : A study of Uttar Pradesh. Economic perspectives of in Tourism, 16(1), 23-41.
8. Pandey, S. & Agarwal, R. (2021). How the tourism industry has taken revival strategies Post-Pandemic : A case study of domestic study in Uttar Pradesh. Journal of Indian Tourism, 14(2), 65-80.
9. Sharma, P. & Pandey, V. (2022). Religious Tourism in Uttar Pradesh : A case study of Varanasi. Spiritual TourismJournal, 10(4), 55-72.
10. Shukla, P. & Sharma, R. (2017). Revival of the "Golden Bird" : A gap analysis of government initiatives and tourist expectations in Uttar Pradesh using a systems approach. Journal of Tourism Development, 9(3), 88-105.
11. Singh K., Verma D., &Shah L. (2022). Tourism and sustainability in times of COVID-19 : The Case Study of Uttar Pradesh. Sustainable Tourism Review, 15(1), 98-112.
12. World Travel and Tourism Council. (2024, January 15). Tourism growth trends in India. WTTC Reports. https://wttc.org/india-tourism

# MASS TOURISM IN INDIA: CASESTUDY ON KUMBH MELA

Tanoy Biswas[1], Shikha Sharma[2*]

[1&2]Chandigarh University, Mohali, India

*Correspondence author: shikhasharma.uithm@cumail.in

**Abstract.** The term "mass tourism" has been a fixture in tourism research for decades, yet a universally accepted definition remains elusive.Fundamental mass tourism involves significant numbers of travelers converging on a single destination for leisure or recreation.While acknowledged for its potential to stimulate local economies, create employment, and foster cultural exchange, mass tourism also carries inherent risks.Environmental degradation and the disruption of local communities are among the well-documented concerns.The KumbhMela, a profoundly significant Hindu pilgrimage occurring at intervals of 6, 12, and 144 years, provides a compelling case study.Correlated with the astrological positions of Jupiter, it stands as the largest human gathering on Earth.Analyzing the KumbhMela offers valuable insights into the multifaceted nature of mass tourism, its socio-economic impacts, and the challenges of managing such immense human congregations. By examining this unique event, researchers can gain a deeper understanding of the complexities and consequences of mass tourism, ultimately informing more sustainable and responsible tourism practices.

**Key words:** Leisure, Local Economy, Human Gathering,Community Disruption

## 1 INTRODUCTION

Mass tourism refers to the movement of a large number of organized tourists to popular holiday destinations for recreational purposes. It is a phenomenon which is characterized by the use of standardized package products and mass consumption. Conceptually, this type of tourism features standardized leisure products and experiences packaged for mass tourist consumption.Religious tourism is an integral part of the tourism industry.Several developing economies offer religious tourism activities on a larger scale and religious destinations attract more than 300 million visitors each year worldwide (Gayathri, Gulhare, & Verma, 2020). The 2025 Prayag Maha Kumbh Mela, also referred to as the 2025 Prayag Kumbh Mela, is the ongoing iteration of the Kumbh Mela, a Hindu pilgrimage festival celebrating a full orbital revolution of Jupiter around the Sun. It is scheduled from 13 January to 26 February 2025, at the Triveni Sangam in Prayagraj, Uttar Pradesh, India. It is the world's largest gathering, with official data released on 12 February counting 48.83 million people taking a dip in the river so far. This event marks the completion of a 12-Kumbh Mela cycle and is officially termed a Maha Kumbh Mela, spanning 45 days.The Kumbh Mela has been organized since many years with its commencement date unknown (Onnela & Khanna, 2015). As per astronomical calculations, the 2025 edition is unique since the constellation alignment seen is witnessed once in 144 years. The site of each Kumbh Mela is decided by the astrological alignment of Jupiter, the Sun and the Moon. The Kumbh Mela has been recognised as part of Intangible Cultural Heritage by UNESCO.

## 2 LITERATURE REVIEW

Mass tourism refers to large-scale travel where tourists experience staged attractions rather than authentic local culture, often staying within a tourist bubble.The beginning of mass tourism is largely related to various changes, which occurred in the industrial societies in the second half

 https://doi.org/10.52202/081568-0005

of the 20th century. In fact, history of mass tourism begins from 1851 when Thomas Cook led a mass of tourists to the Great Exhibition to London. Mass gatherings, including concerts, religious pilgrimages, sports events, and political rallies, have been widely studied for their various social, economic, environmental, and public health impacts (Barnett, Khanna, & Onnela, 2016). The existing literature provides insights into the challenges and benefits associated with these large-scale events.

**2.1 STUDY ON KUMBH MELA**

The Kumbh Mela is recognized for its ability to bring together people from diverse backgrounds, fostering a sense of social unity and cohesion. Research highlights its role in reinforcing cultural and religious identity.The Kumbh Mela's profound socio-cultural impact. They examine the event's role in strengthening social bonds by bringing together devotees from diverse backgrounds. The Mela fosters community spirit through shared rituals and cultural exchange, creating a temporary yet vibrant social space (Sridhar, Gautret, & Brouqui, 2015). There are huge pedestrian's movements in form of crowd during Mela period. Managing the movement of people in a hassle-free manner, managing crowds on peak days, ensuring the safety and security of pilgrims, and effective crowd management planning, should be the key focus areas to ensure an incident-free mass gathering events. Large-scale human activities in the form of mass gathering event have become more frequent especially massive entertainment events, religious gatherings, sports events, political rallies, product promotions, etc. Crowd safety, security and surveillance have become a critical and challenging (Dwivedi & Cariappa, 2015). During Kumbh Mela (the festivals of the sacred Pitcher), pilgrims take bath or take a dip in a sacred river. Devotees believe that by bathing in the river, one is freed from sins liberating her/ him from the cycle of birth and death. Millions of people reach the place without any invitation.The festival is held at Prayagraj (Allahabad), Haridwar, Ujjain and Nasik every four years by rotation and is attended by millions of people irrespective of caste, creed or gender. There are four type of kumbh melas: purna Kumbh , ardh kumbh , Maha kumbh and Kumbh Mela. Currently in the maha kumbh was held in prayagraj. Which was coming after 144 years.

**2.2 TOURISM IN INDIA**

India is a large and growing market for tourism, offering a diverse range of experiences, including cultural, religious, adventure, and wellness tourism. India's tourism sector, while experiencing robust growth, faces challenges like infrastructure deficits, seasonal fluctuations, and the need for sustainable practices (International Society for Infectious Diseases, 2016), while also presenting opportunities for niche tourism products and economic growth. There's increasing interest in niche tourism products like cruises, adventure, medical, wellness, sports, MICE, eco-tourism, film, rural, and religious tourism. Domestic tourism is a major driver of the sector, with increasing numbers of Indians traveling within the country. India attracts international tourists, particularly for its cultural heritage and spiritual tourism. The tourism sector is a significant contributor to India's economy, generating foreign exchange and creating employment

Religious tourism is a significant aspect of India's tourism industry, drawing millions to sacred sites and locations of historical spiritual significance, with destinations like Ayodhya, Varanasi, Tirupati, and Shirdi being popular pilgrimage sites. Many tourists seek a deeper connection to their faith and a chance to experience spiritual growth. Visiting religious sites allows travelers to explore India's rich cultural heritage and historical past. Witnessing religious rituals and festivals can be a meaningful and fulfilling experience for tourists.

https://doi.org/10.52202/081568-0005

## 2.3 ENVIROMENTAL IMPACT

### WASTE MANAGEMENT:

The Kumbh Mela presents significant waste disposal challenges due to the massive influx of pilgrims, generating large quantities of solid and liquid waste, including non-biodegradable materials, which strains local infrastructure and poses environmental risks (Kanaujiya & Tiwari, 2025). The Kumbh Mela attracts millions of pilgrims, leading to a surge in waste generation, including food packaging, waste is non-biodegradable, such as plastic, which poses a long-term environmental threat. Pooja offerings, often left in the river or on the banks, are difficult to clean manually and contribute to pollution and ecosystem disruption. The influx of pilgrims also leads to increased sewage generation, which can contaminate the river water and pose health risks.

Waste dumped in rivers and the surrounding environment causes severe pollution, harming aquatic life, contaminating water sources, and impacting human health and ecosystems through the spread of diseases and the degradation of habitats. Waste, including industrial and domestic sewage, chemicals, and plastics, directly pollutes rivers and waterways, making them unsafe for drinking, swimming, and supporting aquatic.

### WATER RESOURCES

Mass gatherings can significantly impact water quality and availability, especially in areas with inadequate infrastructure, leading to pollution from sewage, waste, and increased demand, potentially causing waterborne diseases and straining local water resources (Tiwari, 2025). Mass gatherings generate large amounts of human waste, including sewage and solid waste, which can contaminate nearby water sources if not managed properly. Religious activities, like mass bathing or offering flowers, can introduce pollutants like soap, detergents, and organic matter into water bodies, affecting water quality. A sudden influx of people can strain existing water infrastructure, leading to shortages and potentially impacting the quality of the water supply.Contaminated water can lead to the spread of waterborne diseases, posing a significant health risk to attendees and local populations.

The Maha Kumbh Mela, a massive religious gathering, presents significant challenges to providing clean drinking water and sanitation due to the sheer scale of the event and the temporary nature of the infrastructure, potentially leading to water pollution and sanitation issues. The Kumbh Mela attracts millions of pilgrims, far exceeding the capacity of the local infrastructure, leading to strain on water resources and sanitation systems. Lack of adequate sanitation facilities and hygiene practices during the event can lead to the spread of waterborne diseases and other health problems.

### AIR POLLUTION:

The Kumbh Mela, a massive religious gathering, exacerbates air and water pollution due to increased vehicular traffic, waste generation, and religious practices, leading to challenges in sanitation and public health. The influx of millions of pilgrims results in a surge in vehicular traffic, leading to congestion and increased emissions of pollutants like carbon monoxide, particulate matter, and nitrogen oxides. The use of firewood for cooking and other purposes, along with religious rituals like "hawans" and "yajnas," contributes to smoke and particulate matter pollution. Illegal dumping and burning of waste, including plastic and garbage, further worsen air quality. The combination of vehicular emissions and smoke from burning fuel can lead to the formation of smog, a type of air pollution that can negatively impact respiratory health.


https://doi.org/10.52202/081568-0005

The Kumbh Mela, a massive religious gathering, presents significant public health challenges, including increased risk of respiratory and other infections, due to high population density, poor sanitation, and environmental pollution, particularly air pollution (Som, Gupta, Mishra, & Shivani, 2025). The large crowds and dust from the ground, along with potential burning of waste, contribute to high levels of particulate matter in the air. Exposure to these pollutants can lead to respiratory problems like cough, cold, fever, and exacerbation of conditions.

## 2.4 SOCIO-ECONOMIC IMPACTS

### ECONOMIC BENEFITS:

The Kumbh Mela, a massive Hindu pilgrimage, generates significant economic benefits for local businesses and communities through increased tourism, job creation, and infrastructure development, with projections for 2025 estimating a ₹2 lakh crore economic impact. The Kumbh Mela attracts millions of pilgrims, leading to a surge in demand for accommodations, food, transportation, and other services, benefiting hotels, restaurants, and local businesses. The event creates numerous temporary and permanent jobs in various sectors, including hospitality, transportation, retail, event management, and sanitation. The Kumbh Mela often triggers infrastructure upgrades, such as improved roads, bridges, and sanitation facilities, which can have long-term benefits for the host city and region. The event generates substantial revenue for the local economy through tourism, retail sales, and government taxes. Local artisans, vendors, and small businesses benefit from increased demand for religious items, handicrafts, and local products. The Kumbh Mela can position the host city as a tourism hub, attracting future investments and promoting cultural exchange. Enhanced infrastructure and renewed cultural pride can sustain economic momentum post-festival.

These festivals enrich the cultural capital of Uttar Pradesh and at the same time provide an impetus to the economic activity in the region. The Kumbh Mela has been a huge success, attracting lakhs of visitors and creating considerable income for the state exchequer. Tourism generates jobs in transportation, including taxi drivers, tour bus operators, and airport staff.Hotels, resorts, and guesthouses employ a large workforce in roles like receptionists, housekeeping staff, chefs, and restaurant staff.

### INFRASTRUCTURE DEVELOPMENT

For the Maha Kumbh Mela, temporary infrastructure includes roads, bridges, health facilities, and security measures, with a focus on managing large crowds and ensuring the safety of devotees. Temporary roads have been extended, and new Road Over Bridges (ROBs) and flyovers have been constructed. Pontoon bridges are used to facilitate access across the river and key areas (Kumar et al., 2025). A total of 92 roads have been renovated, with 17 major roads undergoing beautification. A comprehensive health arrangement is in place, including a 100-bed central hospital, two 25-bed sub-central hospitals, eight 20-bed sector hospitals, and two 20-bed infectious diseases hospitals. Bamboo and fabric structures are erected, and metal plates are laid down for roads. Temporary ghats and riverfront roads are constructed. The infrastructure is designed to facilitate smooth and fast movement of devotees, especially during the peak periods of the festival.

The Maha Kumbh Mela, a large religious gathering, necessitates significant infrastructure development, which can have long-term positive impacts on the host region, including improved connectivity, urban amenities, and economic growth, but also potential negative impacts like environmental degradation and increased health risks. The Maha Kumbh drives substantial economic activity, attracting pilgrims, boosting tourism, and creating jobs in various sectors like hospitality, transportation, and retail. The event often triggers investments in infrastructure, including roads, bridges, sanitation, and water supply, which benefit the local population even after the event concludes. Improved infrastructure, such as new roads and public transportation, can enhance connectivity and make the region more accessible for tourism and business.

**SOCIAL AND CULTURAL IMPACTS:**

The Kumbh Mela, a massive religious gathering, significantly impacts local culture and traditions by serving as a melting pot of Indian culture, reinforcing religious practices, and fostering unity and spiritual exchange, while also presenting challenges like environmental impact and overcrowding. The Kumbh Mela acts as a platform for cultural exchange and learning. Scholars, researchers, and tourists from around the world visit the festival, not just as observers but as participants (Buzinde, Manuel-Navarrete, Kalavar, & Kohli, 2022), eager to learn about and experience the cultural and spiritual traditions of India. This exchange fosters mutual respect and understanding, bridging cultural gaps and promoting global harmony.

The Kumbh Mela, a massive Hindu pilgrimage, significantly impacts the daily lives of local residents in Prayagraj, India, with both positive and negative consequences, including economic opportunities, increased traffic and sanitation challenges, and a temporary disruption of normal routines. The influx of pilgrims boosts local businesses, creating jobs in tourism, hospitality, transportation, and retail. The Mela fosters a sense of community and cultural exchange, bringing people from diverse backgrounds together (Udachan, 2025). The event often leads to improvements in infrastructure, such as roads, sanitation facilities, and public transport, benefiting the local population in the long run. The sheer volume of people at the Kumbh Mela creates fertile ground for opportunistic crimes, particularly theft and snatching. Data from the 2025 event, as reported by The Wire, indicates that over 315 First Information Reports (FIRs) were filed at two police stations—Daraganj and Kumbh Mela Kotwali—since the festival began on January 13. The overwhelming majority of these cases involved theft and snatching, with victims losing mobile phones, cash, jewelry, and bags. This aligns with historical patterns; for example, during the 2013 Maha Kumbh in Allahabad, similar petty crimes spiked due to crowded conditions and distracted pilgrims focused on rituals like bathing at the Sangam.

# 3METHODOLOGHY

The research on mass tourism in India, focusing on the Kumbh Mela, took a deep dive into the subject using a bunch of different approaches. They didn't just stick to one method—they mixed it up with focus groups, on-the-ground observations, official government reports, and even checked out the environmental impact. The team got chatting with pilgrims from all over India and beyond, asking them what drew them to the Mela, what they got up to while there, and how it all felt for them.

To really get the full picture, some researchers went all-in and lived in the temporary camps at the Kumbh Mela site. They soaked up the atmosphere, watching the religious ceremonies, seeing how people interacted, and getting a front-row seat to how everything was run. On top of that, they dug into reports from the Prayagraj Mela Authority and other government sources to get the lowdown on stuff like crowd management, healthcare setups, and security plans.

A big part of the study was figuring out what this huge gathering does to the environment. The researchers rolled up their sleeves and tested the water quality, sifted through the waste, and looked at how things like mass bathing, offerings, and just millions of people being there affected the surroundings. They also zoomed in on the economic side of things—how the flood


https://doi.org/10.52202/081568-0005

of visitors boosted local businesses, shifted tourism trends, and put pressure on the area's infrastructure. It was a real hands-on effort to see the Mela from every angle!

## CONCLUSION

The Kumbh Mela is an incredible blend of faith, culture, and the sheer spectacle of millions of people coming together, making it a fascinating case study of mass tourism in India. It's more than just a religious gathering—it's a vibrant celebration of heritage that brings economic boosts and a deep sense of spiritual connection for those who attend. But with all that energy and devotion comes some real challenges. Picture this: millions of pilgrims flooding into one place, putting a massive strain on things like sanitation and clean water. Efforts to manage waste and upgrade infrastructure are underway, but the scale of it all is so huge that it's tough to keep everything under control. Then there's the issue of safety—crowd management and security are no small feat when you're trying to prevent stampedes or other risks. Thankfully, technology like surveillance and crowd monitoring has stepped in to help organizers stay on top of things. The event draws people from across India and the world, which is amazing, but it also sparks some debate about whether this sacred tradition is becoming too commercialized. The way the media frames the Kumbh Mela shapes how it's seen globally, adding another layer to its story. At its core, it's a beautiful mix of devotion and logistics, but keeping it sustainable means finding a balance—holding onto its spiritual roots while tackling modern issues like environmental impact, safety, and responsible tourism that honors both the faith and the land it's built on.

## REFERENCES


1. Gayathri, H., Gulhare, S., & Verma, A. (2020). Understanding crowd dynamics in processions during mass religious gatherings: A case study of Shahi Snan in Kumbh Mela. *Collective Dynamics, 5*, 403–412.
2. Onnela, J.-P., & Khanna, T. (2015). Investigating population dynamics of the Kumbh Mela through the lens of cell phone data. *arXiv preprint* arXiv:1505.06360.
3. Barnett, I., Khanna, T., & Onnela, J.-P. (2016). Social and spatial clustering of people at humanity's largest gathering. *arXiv preprint* arXiv:1605.06898
4. Sridhar, S., Gautret, P., & Brouqui, P. (2015). A comprehensive review of the Kumbh Mela: Identifying risks for spread of infectious diseases. *Clinical Microbiology and Infection, 21*(2), 128–133.
5. Dwivedi, S., & Cariappa, M. P. (2015). Mass-gathering events: The public health challenge of the Kumbh Mela 2013. *Prehospital and Disaster Medicine, 30*(6), 621–624.
6. International Society for Infectious Diseases. (2016). Public health perspectives from the biggest human mass gathering on earth: Kumbh Mela, India. *International Journal of Infectious Diseases, 47*, 42–45.
7. Kanaujiya, A. K., & Tiwari, V. (2025). A comprehensive analysis of the multifaceted significance of the Kumbh Mela in India: An emerging epicenter of global Hindu religious tourism. *World Journal of Advanced Research and Reviews, 25*(2), 561–569
8. Tiwari, S. (2025). Beyond pilgrimage: The socio-economic and cultural impact of the Prayagraj Kumbh Mela. *International Journal of Research in Finance and Management, 8*(1), 333–338.
9. Som, G., Gupta, P., Mishra, M., & Shivani, S. (2025). Business opportunities in Maha Kumbh: A study on the impact of Maha Kumbh Mela on business entrepreneurs. *Journal of Neonatal Surgery, 14*(30S), 335–341.
10. Kumar, A., Kumar, S., Maurya, A., Bansal, A., Verma, A., & Binwal, S. (2025). Kumbh Mela at Prayagraj: A narrative review. *International Journal of Healthcare Education & Medical Informatics, 12*(1–2), 1–4.
11. Buzinde, C., Manuel-Navarrete, D., Kalavar, J. M., & Kohli, N. (2022). Social production and consumption of space: A Lefebvrian analysis of the Kumbh Mela. *International Journal of Religious Tourism and Pilgrimage, 10*(1), Article 15.
12. Udachan, V. (2025). Urban form and cultural symbolism in religious gatherings – A case of Kumbh Mela. *Cuestiones de Fisioterapia, 54*(1), 179–192.

# UNLOCKING THE POTENTIAL OF ADVENTURE TOURISM IN LESOTHO

Bongani Hoohlo[1], Shikha Sharma[2*]

[1&2]Chandigarh University, Mohali, India

*Correspondence author: shikhasharma.uithm@cumail.in

**Abstract.** Adventure tourism in Lesotho offers a unique blend of outdoor experiences, including hiking, mountain biking, and cultural immersion. This study investigates the potential of adventure tourism, emphasizing its growing popularity, economic benefits, and challenges related to infrastructure and sustainable development. The research also explores how tourism benefits local communities and contributes to the preservation of Lesotho's natural environment. With targeted marketing and investment, Lesotho can establish itself as a key destination for adventure seekers.
This chapter analyzes the tourism assets in Lesotho, exploring ways to improve the performance and sustainability of the tourism industry. The study highlights the need for better utilization of existing infrastructure and the development of new tourism products. In 2011, Lesotho welcomed 398,000 tourists, generating USD 35 million in revenue, representing 1.4% of the country's GDP. Tourism is recognized by the government and international organizations as a sector that can reduce poverty and absorb low-skilled labor.
Adventure tourism has the potential to drive economic growth in Lesotho, utilizing its unique landscapes and cultural heritage. However, challenges such as poor service quality, lack of digital marketing, inadequate infrastructure, and environmental mismanagement hinder its full development. This study emphasizes the need for improved tourism management, including investments in workforce development, digital marketing, sustainable tourism, and marketing strategies. Addressing these issues will help position Lesotho as a leading adventure destination, fostering sustainable development and enhancing its competitiveness in the global tourism market.

**Key words**: Adventure tourism, Lesotho tourism, economic benefits, sustainable development, infrastructure, local communities, cultural heritage, tourism development, environmental management, digital marketing, ecotourism.

## 1 Introduction

Lesotho is situated between 28° and 30° South latitude and between 27° and 30° East Longitude. It is surrounded by the Republic of South Africa. It has a land area of 30 355 Square kilometers. The country is commonly divided into four zones, known as the mountain zone which comprises 65 percent of the land area, the lowlands (19 percent), the foothills (8 percent), and the Senqu Valley (8 percent) (Figure 1.1). Lesotho is distinct in that it is the only country in the world where the lowest elevation is 1500m above sea level (Gill,1994). With its highest elevation at 3500m, the country also boasts the highest Elevation in Southern Africa and the second highest in Africa as a whole – a feature which has earned it a number of epithets, for example, "Kingdom in the Sky", "Roof of Africa" and "Switzerland of Africa".
Tourism is regarded as a fundamental element that fosters economic growth, serves as a transformative agent, and provides a means for reducing poverty. Its importance is anticipated to increase in the coming years World Travel and Tourism Council 2019; (Țîțu, R, 2016). The tourism sector encompasses a diverse array of services, and the caliber of these services plays a crucial role in influencing tourists' choices to return to a destination. This situation presents a challenge for decision-makers, who must

 https://doi.org/10.52202/081568-0006

enhance service quality to maintain competitiveness in the market (Haghkhah, Nosratpour, Ebrahimpour & Hamid, 2011; Jovanovic & Ivana, 2016). Adventure tourism, one of the fastest- growing sectors in global travel, has seen significant development in recent years, especially in outdoor activities and exploration.

Lesotho, often referred to as the "Kingdom in the Sky," is a small, landlocked country in the southernmost part of Africa, entirely surrounded by South Africa. Renowned for its distinctive geography, Lesotho features high elevations and divers landscapes, with its altitude surpassing 1,000 meters, making it the only country in the world to lie entirely above this height. The potential for adventure tourism in Lesotho is considerable due to its distinctive geographical characteristics, which feature high-altitude terrains, rivers, and a variety of ecosystems. Adventure tourism includes activities such as trekking, mountain biking, rock climbing, and pony trekking, all of which are especially favored in this area. Lesotho's economy is primarily rural, and its ability to generate income is closely tied to how effectively it can utilize its natural resources. Key resources for the country include land, water, and labor.Lesotho faces significant challenges with high rates of underemployment and unemployment, particularly in its rural regions. Over 85 percent of households in Lesotho are located in rural areas, where poverty levels are notably higher compared to urban settings. The country's climate, characterized by mountainous terrain and inadequate agricultural practices, results in minimal economic returns from the primary sector, which is agriculture. Consequently, households that primarily depend on farming or livestock for their income experience the highest levels of poverty (World Bank, 1995).

The government has been using various methods to alleviate poverty in rural areas, and one of the approaches is to use tourism projects to enhance employment and income opportunities for rural communities. This approach has been seen as an effective way to promote economic development in peripheral regions that have limited resources. Recent efforts in tourism, such as the establishment of pony trekking associations, serve as examples of tourist activities located inrural regions. These initiatives aim to provide rural communities with opportunities to enhance their income and elevate their social standing (Government of Lesotho, 1996).
However, there has been a growing apprehension regarding the negative impacts associated with mass tourism, which has prompted advocates to push for more sustainable approaches to tourism development (Cater and Goodall, 1994; Craik, 1991; 2 Murphy, 1994; Oppermann, 1997; Sharpley, 1997; Theobald, 1994). This method is pertinent to the ongoing tourism developments in Lesotho, considering the environmental sensitivity of many rural areas. Additionally, it aligns well with the values held by local communities.

Lesotho boasts breathtaking mountainous landscapes, rich cultural heritage, and diverse outdoor activities, making it a prime destination for adventure tourism. From rugged hiking trails and pony trekking to off-road adventures and winter sports, the country holds immense potential to position itself as a leading adventure tourism hub in Southern Africa. However, despite these natural advantages, Lesotho's adventure tourism sector remains largely underdeveloped due to challenges such as inadequate infrastructure, limited digital marketing, low service quality, and poor environmental management. Unlocking the full potential of this sector requires strategic interventions to enhance tourism services, improve accessibility, and promote the country's unique adventure experiences to global markets.

This study explores the key factors influencing adventure tourism in Lesotho, identifying both opportunities and barriers to its growth. By employing a qualitative research approach, including interviews with stakeholders and analysis of existing tourism practices, this research aims to provide insights into improving the adventure tourism landscape. The findings will offer practical recommendations for policymakers, tourism operators, and investors to drive sustainable growth, enhance visitor experiences, and position Lesotho as a premier adventure tourism destination.

With over 80% of its landmass lying above 1,800 meters (5,900 feet) above sea level, Lesotho is home to unique adventure landscapes, including the Maloti-Drakensberg Trans frontier Conservation Area and Sehlabathebe National Park, a UNESCO World Heritage Site. Despite these natural assets, the country's adventure tourism sector remains underdeveloped, limiting its contribution to economic growth and job creation.

The United Nation's World Tourism Organization (UNWTO) research reports that tourist arrivals Surpassed 1 billion for the first time in 2012. Despite occasional shocks, international tourist Arrivals have enjoyed virtually uninterrupted growth – from 277 million in 1980 to 528 million In 1995, and 1.035 billion in 2012. Developing countries are playing an increasingly prominent Role in this growing sector. Tourism is one of the top three exports for the majority of developing counties and for at least 22 LDCs (including Lesotho) therefore,Tourism development has been identified by the Lesotho government, International Monetary Fund as well as the United Nations World Tourism Organization as a sector that will reduce Poverty as well as absorb many low or semi skilled labor. Boosting the tourism sector in the Kingdom will inevitably work towards alleviating poverty in the Kingdom overall. If developed Well, tourism could effectively lead a LDC country like Lesotho out of poverty like it did for other developing countries.

Tourism in Lesotho currently contributes about 5.1% to the country's GDP and employs thousands of people, but the sector has not reached its full potential due to inadequate infrastructure, limited global marketing, poor service quality, and environmental mismanagement. Comparatively, neighboring South Africa attracts millions of adventure tourists annually, largely due to better promotional strategies and investment in tourism infrastructure. Unlocking the potential of adventure tourism in Lesotho could significantly boost local economies, increase employment opportunities, and enhance revenue generation through eco-tourism, hospitality, and cultural tourism.

## 2 Literature Review

According to (Theobald, 1994; Murphy, 1983)The tourism sector is among the most rapidly expanding industries globally and has the potential to enhance both directly and indirectly the living standards of host communities. The adventure travel sector is experiencing rapid growth within the tourism industry, particularly in developing countries. However, this swift expansion has prompted significant concerns regarding the potential harm to the social, cultural, and environmental aspects of the host destinations However, some initial methods are applied in order to reduce the significant impact of these discussed matters such as issuing and presenting knowledge about eco-tourism and sustainable practices to the travelers as well as the local communities. This act will not only improve the betterment life standard of the local communities but it will also improve the eco-system. The term "sustainable" can be defined as approaches or practices that can be upheld over an extended period without exhausting resources or inflicting considerable damage on the environment. It includes multiple aspects, such as ecological well-being, social fairness, and economic stability. The Brundtland Report, published by the World Commission on Environment and Development in 1987, highlighted a key concern regarding sustainable development. It defined sustainable development as a process that fulfills the current needs of society while ensuring that future generations have the capacity to satisfy their own needs.


https://doi.org/10.52202/081568-0006

In various studies related to tourism, there has been significant attention given to the sustainability of the resources that tourism relies on (Saarinen, 2006). According to Tosun (2001), for tourism development to be considered sustainable, it must fulfill certain criteria:Enhance not only the economic development of the nation but also that of regions and localities, ensuring that the benefits are equitably distributed among all segments of society.Involving local residents in the planning and decision-making processes related to tourism development ensures that their voices are heard. This engagement fosters a sense of ownership and pride in their community, which is vital for building self-esteem.Contribute to the fulfillment of essential and perceived needs of individuals previously marginalized in local tourist areas.Sustainable development that meets the needs of the present without compromising the ability of future generations to meet their own needs, by balancing social equity, economic growth, and environmental protection.Assist communities in achieving liberation from oppressive material circumstances and from subservience to nature, ignorance, other individuals, suffering, institutions, and rigid beliefs.

Due to explorative and descriptive nature of this study, review of relevant literature was Ideal for the study to obtain new insights into a phenomenon and determine priorities for Further research. The research was guided by a theoretical framework based on tourismModels drawn from sustainable development paradigms. In this regard, considerationsOf sustainable tourism, sustainable development and community participation are central To the study. The relevant literature review was done before commencement of the field Research. This facilitated better understanding of critical issues and served to guide the Interviews and data gathering process more effectively. In Kumar's (1989, p. 7) words Well formulated flexible framework helps to sharpen the focus of an inquiry and keep the interviews focus on critical issues." This helps in the formulation of concepts and hypotheses before field research, thus providing a guideline for the selection of data Collection instruments (Bailey, 1987; Garbers, 1996).Relevant background information to the study was collected from primary and secondary Sources. Primary sources were personal communication with concerned parties and Minutes of meetings. Secondary sources included books, government records and Documents such as government development plans and reports, project reports, monthly And yearly reports, minutes of meetings and statistical data from government Departments. Moreover, published and unpublished studies were also used. Publicity Brochures. Leaflets produced by the pony trekking centers, particularly Basotho Pony and Malealea and magazines were sourced.

This report aims to illustrate the economic benefits of adventure tourism in Lesotho, and showcasing the different adventure tourism activities of the nation also the popularity of the tourism sector is throughout recent years and how it has changed. The report will also look at the obstacles faced by Lesotho's infrastructure and sustainable development of how these factors impact and benefit the nations tourism industry.

### 2.1 Challenges Facing Adventure Tourism Development in Lesotho

Adventure tourism in Lesotho, while holding significant potential for economic growth and community development, faces several challenges that hinder its full realization. These challenges can be categorized into environmental, socio-economic, infrastructural, and governance-related issues.

### 2.2 Environmental Challenges

Lesotho's unique mountainous terrain and biodiversity are both an asset and a challenge for adventure tourism. The country is prone to environmental degradation due to overgrazing, deforestation, and climate change. These factors threaten the natural landscapes that attract tourists. For instance, increased soil erosion can lead to loss of trails used for trekking and other outdoor activities. Additionally, the impact of tourism itself can exacerbate these environmental issues if not managed sustainably. The construction of tourism facilities often encroaches on natural habitats, leading to biodiversity loss.

**2.3 Socio-Economic Challenges**

The socio-economic context in Lesotho presents significant hurdles for adventure tourism development. Many local communities depend on subsistence agriculture and have limited access to education and resources necessary for engaging with the tourism industry effectively. This dependency creates a scenario where local populations may not benefit from the economic opportunities presented by adventure tourism. Furthermore, there is often a lack of awareness or understanding of how to market local attractions or engage with tourists effectively.

Additionally, the dependency theory suggests that tourism can perpetuate inequalities rather than alleviate them (Oppermann, 1993). This means that while foreign investors may profit from adventure tourism ventures, local communities might remain marginalized without adequate support or involvement in decision-making processes.

**2.4 Infrastructural Challenges**

Infrastructure in Lesotho is often inadequate to support the demands of adventure tourism. Poor road conditions limit access to remote areas where many adventure activities take place, such as pony trekking or hiking in national parks like Bokong Nature Reserve. Furthermore, there is a lack of essential services such as accommodation facilities and tourist information centers that can enhance the visitor experience.

Investment in infrastructure is crucial; however, financial constraints often prevent necessary developments from occurring. Without proper roads and facilities, attracting international tourists becomes increasingly difficult.

**2.5 Governance-Related Challenges**

Governance plays a critical role in shaping the landscape for adventure tourism development in Lesotho. There are often conflicts between government policies aimed at promoting tourism and local community interests (Saarinen & Wall-Reinius, 2019). The imposition of regulations without community consultation can lead to resentment among locals who feel excluded from the benefits of tourism initiatives.

Moreover, issues related to land ownership complicate matters further; as new tourist enclaves are developed on land traditionally used by local communities for agriculture or gathering resources (Kilipris, 2005). This creates tensions between preserving cultural practices and advancing economic interests through tourism.

**2.6 Strategies for Unlocking Potential**

**Infrastructure Development:** Investing in transportation networks and hospitality services is essential to improve access to remote regions where adventure activities are common. It is essential, as better road networks, adventure-friendly facilities, and digital access can enhance the overall experience for tourists. Upgrading routes to popular sites like Sani Pass and Sehlabathebe National Park, along with developing eco-lodges and adventure gear rental services, can significantly boost accessibility and comfort for visitors.

**Training and Capacity Building:** Creating training initiatives for local communities will improve their abilities as tour guides or service providers within the adventure tourism industry. **Marketing Campaigns:** Strategic marketing efforts aimed at international tourists can help establish Lesotho as a top destination for adventure tourism. Improving marketingefforts should position Lesotho as "Africa's Ultimate Adventure Destination" by leveraging international campaigns, strengthening its online presence, and collaborating with travel agencies to create specialized adventure tourism packages

**Service improvement:** Focusing on eco-friendly methods in adventure tourism can draw environmentally aware travelers while safeguarding natural resources. Training programs for


https://doi.org/10.52202/081568-0006

tour guides, hospitality staff, and adventure service providers can enhance professionalism and safety, while certification programs ensure high-quality services. Encouraging community involvement in tourism-related businesses can also create job opportunities and promote cultural exchange.

2.7 **Sustainable tourism and environmental conservation:** must be prioritized by implementing eco-tourism policies, improving waste management at key attractions, and introducing controlled tourism zones to prevent environmental degradation.
**Investment and policy support:**play a vital role in unlocking Lesotho's adventure tourism potential. The government can attract private sector investment through incentives like tax breaks and grants while forming strong public-private partnerships to develop adventure tourism hubs. By integrating these strategies, Lesotho can harness its unique landscapes and cultural heritage to establish itself as a premier adventure tourism destination, fostering economic growth and sustainable development.

## 3 Methodology

Research methodology refers to the structured approach and techniques used in conducting research, gathering data, analyzing information, and deriving conclusions (Creswell & Creswell, 2017). Yin (2017) further elaborates that research methodology encompasses the entire framework of a research study, including its design, structure, and the selection of research methods, data collection strategies, and data analysis techniques. A well-structured research methodology is essential to ensure the accuracy, reliability, and applicability of research findings.

For this study, the researcher adopted a mixed-methods approach, which integrates both quantitative and qualitative research methodologies within a single study (Bryman, 2016). This approach was chosen to leverage the benefits of both research paradigms, offering a more comprehensive understanding of the research problem. Quantitative research involves gathering numerical data through structured surveys, experiments, and statistical tools to identify trends, correlations, and statistical generalizations. On the other hand, qualitative research seeks to explore deeper meanings, perceptions, and contextual factors by using interviews, observations, and non-numerical data analysis. The combination of these methods allowed for a holistic exploration of adventure tourism in Lesotho.

One of the main advantages of mixed-methods research dominantly, secondary research is its ability to capitalize on the strengths of both quantitative and qualitative approaches. Quantitative data ensures statistical precision, allowing researchers to establish relationships between variables, identify patterns, and make generalized conclusions. Meanwhile, qualitative data provides in-depth insights into the socio-cultural and environmental aspects influencing adventure tourism in Lesotho. This dual approach enabled the researcher to develop a well- rounded understanding of the sector's potential and challenges.

The use of multiple data sources and methods In mixed-methods research enhances the validity and reliability of findings through a process called triangulation. This approach involves cross- verifying information by comparing qualitative insights with quantitative statistics. In this study, quantitative data helped establish broad trends in Lesotho's adventure tourism industry, while qualitative insights provided a deeper understanding of tourist experiences, perceptions, and challenges. By integrating both, the research achieved a more accurate and balanced representation of the topic. Quantitative data offered a macro-level perspective, identifying the scale, frequency, and statistical relationships within the adventure tourism sector in Lesotho. Meanwhile, qualitative data provided a micro-level understanding by capturing personal experiences, motivations, and barriers encountered by stakeholders in the industry. This holistic approach enriched the study's findings, ensuring both breadth and depth in the analysis.The integration of qualitative and quantitative data enhances the practical relevance of research findings. The statistical significance of quantitative insights, combined with the contextual depth provided by qualitative data, ensures that the study's outcomes are actionable. Decision- makers in Lesotho's tourism sector can use these findings to develop evidence-based policies, improvetourism infrastructure, and create targeted marketing strategies that promote adventure tourism in the region.

Every research methodology has inherent limitations, but combining quantitative and qualitative techniques helps mitigate these shortcomings. For instance, qualitative research was used to shape the design of quantitative surveys, ensuring that key variables and tourism trends were accurately captured. Conversely, statistical trends derived from quantitative data highlighted specific areas that required further qualitative exploration. This dynamic interplay between methods strengthened the overall research design, reducing potential biases and knowledge gaps.

## CONCLUSION

Lesotho has immense potential for adventure tourism, thanks to its breath-taking high-altitude landscapes, diverse outdoor activities, and close proximity to South Africa. With rugged terrain perfect for activities like skiing at Afriski and high-altitude sports training, the country could become a prime destination for adventure seekers and nature lovers. However, realizing this potential requires addressing key challenges, such as the lack of sufficient accommodation in key tourism areas and limited marketing aimed at adventure travellers. Investing in better infrastructureparticularly lodging and transportand forming partnerships with regional tour operators could help attract more visitors and keep tourism revenue within Lesotho rather than losing it to South Africa. Expanding adventure tourism beyond seasonal skiing to year-round activities like trekking, water sports, and cultural experiences would also make the industry more sustainable. Community involvement is essential to ensure that tourism benefits local people and supports poverty reduction efforts. While obstacles like climate change and limited funding remain, targeted policies and strategic promotion could help Lesotho position itself as a top adventure tourism destination, driving economic growth and creating jobs.

Lesotho possesses immense potential to become a leading adventure tourism destination, with its rugged landscapes, rich cultural heritage, and unique high-altitude environment offering diverse opportunities for outdoor recreation. However, despite these natural advantages, the sector remains underdeveloped due to infrastructure challenges, inadequate marketing, low service quality, and environmental mismanagement. To unlock this potential, strategic interventions are required, including improving infrastructure, enhancing digital marketing, investing in skills development, and implementing sustainable tourism policies. By fostering collaboration between the government, private sector, and local communities, Lesotho can transform adventure tourism into a key economic driver, creating jobs, boosting local businesses, and increasing its competitiveness in the global tourism market. If effectively developed and managed, adventure tourism can not only contribute to economic growth but also promote environmental conservation and cultural preservation, ensuring long-term sustainability for the industry and the nation as a whole.


https://doi.org/10.52202/081568-0006

## REFERENCES


1. Theobald, W. F. (1994). Global tourism: The next decade. Oxford: Butterworth-Heinemann.
2. Murphy, P. E. (1983). Tourism as a community industry: An ecological model of tourism development. Tourism Management, 4(3), 180–193. https://doi.org/10.1016/0261-5177(83)90044-0Lekaota, L. (2015). The importance of rural communities' participation in the management of tourism management: A case study from Lesotho. *Worldwide Hospitality and Tourism Themes, 7*(5), 453–462.
3. Ţîţu, M. A., & colleagues. (2016). Measuring service quality in tourism industry. *Procedia – Social and Behavioral Sciences, 221*, 294–301.
4. Rasethuntsa, B. C. (2021). Government collaborative efforts supporting tourism development in Lesotho. *International Tourism and Hospitality Journal, 4*(11). https://doi.org/10.37227/ITHJ-2021-09-348
5. Gill, A., Williams, P. W., & others. (1994). Managing growth in mountain tourism communities. *Tourism Management, 15*(3), 212–220.
6. Spencer, J. P., & Matela, L. (2015). Can small, medium and micro enterprises contribute to tourism development through economic activities in Butha-Buthe, Lesotho? *African Journal for Physical Activity and Health Sciences, 20*(3-2).
7. Mashinini, V. (2003). Tourism policies and strategies in Lesotho: A critical appraisal. *Africa Insight, 33*(1–2), 87–92.
8. Mutana, B., & Mukwada, G. (2021). Multidimensional challenges in the Maloti-Drakensberg: Tourism and sustainability (focus on Lesotho). *Sustainability, 13*, Article 8511.
9. Camillo, A. A. (2015). Lesotho, a tourism destination: Analysis of Lesotho's current tourism products and potential for growth. In A. Camillo (Ed.), *Handbook of Research on Global Hospitality and Tourism Management* (IGI Global).
10. Mapeshoane, T. J., & Pather, S. (2016). The adoption of e-Commerce in the Lesotho tourism industry. *Electronic Journal of Information Systems in Developing Countries, 75*(1), 1–24. http://dx.doi.org/10.1002/j.1681-4835.2016.tb00550.x
11. Makwindi, N., & Ndlovu, J. (2021). Prospects and challenges of community-based tourism as a livelihood diversification strategy at Sehlabathebe National Park in Lesotho. *African Journal of Hospitality, Tourism and Leisure, 10*(1), 333–348. http://dx.doi.org/10.46222/ajhtl.19770720-104
12. Cheng, M., Edwards, D., Darcy, S., & Redfern, K. (2016). A tri-method approach to a review of adventure tourism literature: Bibliometric analysis, content analysis, and a quantitative systematic literature review. *Journal of Hospitality & Tourism Research.*
13. Cloke, P., & Perkins, H. C. (1998). Cracking the canyon with the awesome foursome: Representations of adventure tourism in New Zealand. *Environment and Planning D: Society and Space, 16*, 185–218.
14. Cloke, P., & Perkins, H. C. (2002). Commodification and adventure in New Zealand tourism. *Current Issues in Tourism, 5*(6), 521–549.
15. Giddy, J. K., & Webb, N. L. (2016). The influence of the environment on motivations to participate in adventure tourism: The case of the Tsitsikamma. *South African Geographical Journal, 98*(2), 351–366.

# IMPACT OF CULTURAL TOURISM IN INDIA: CASE STUDY ON KERALA

Ashik.P[1*], Shikha Sharma[2]
[1&2]Chandigarh University, Mohali, India

*Correspondence author: shikhasharma.uithm@cumail.in

**Abstract.** Cultural tourism significantly impacts social progress, economic prosperity, and cultural preservation. This study examines the effects of cultural tourism in Kerala, India. Kerala effectively leverages its unique blend of vibrant traditions, rich history, and natural beauty to attract tourists. The study explores the positive impacts, including economic growth through job creation and income generation, community empowerment through local participation in tourism, and cultural preservation through the revival of traditional arts and crafts. However, it also acknowledges potential drawbacks such as urbanization, environmental damage, and the risk of cultural commodification. The findings emphasize the need for sustainable tourism practices to safeguard the environment, support the well-being of local communities, and authentically showcase Kerala's rich cultural heritage.

**Key Words:** Economic growth, community empowerment, cultural preservation.

## 1 Introduction

Cultural tourism is an important part of India's tourism industry, attracting visitors from all over the world who want to experience its rich traditions, historical landmarks, and diverse cultural heritage. India's long history and vibrant customs make it a unique destination for travelers. Among the many states known for cultural tourism, Kerala stands out for its deep- rooted traditions, art, festivals, and way of life. Often called "God's Own Country," Kerala offers visitors an authentic experience through its famous classical dance forms like Kathakali and Mohiniyattam, Ayurveda treatments, and grand festivals such as Onam, Thrissur Pooram, and Theyyam.
Apart from its beautiful landscapes and serene backwaters, Kerala's culture is reflected in its literature, cuisine, and traditional art forms. The state is home to ancient temples, folk music, martial arts like Kalaripayattu, and skilled artisans who create handloom textiles, coir products, and wooden crafts. Cultural tourism not only allows visitors to appreciate these traditions but also provides a source of income for local communities.
This study examines the impact of cultural tourism in India, with a focus on Kerala. It looks at how tourism contributes to the economy by creating jobs, supporting small businesses, and helping preserve cultural heritage. Additionally, it explores how tourism strengthens local identity and encourages communities to take pride in their history and art. However, cultural tourism also faces challenges, including commercialization, sustainability concerns, and the need for responsible tourism practices. This paper aims to highlight both the benefits and challenges of cultural tourism in Kerala while offering insights into how it can be developed sustainably for the future.

## 2 Literature Review

Cultural tourism has become a prominent sector within India's tourism landscape, providing an engaging insight into the nation's varied traditions, heritage, and artistic expressions (Smith, S. 2015). Among the Indian states, Kerala is particularly distinguished as a leading destination for cultural tourism, owing to its extensive history, distinctive customs, and profound cultural legacy. Numerous studies emphasize that cultural tourism in Kerala is instrumental in fostering economic development, Sebastian, L. M., & Rajagopalan, P. (2009) enhancing social progress, and safeguarding traditional practices, thereby serving as an essential element of the state's tourism framework. The cultural tapestry of Kerala is influenced by its long-standing history, which encompasses both indigenous customs and external influences (Ravichandran, G., & Nam, S. 2024). The state is celebrated for its classical dance


https://doi.org/10.52202/081568-0007

forms, such as Kathakali and Mohiniyattam, lively temple festivals, and the traditional martial art of Kalaripayattu. These facets of Kerala's cultural identity draw a significant number of tourists, both from within the country and abroad. Research indicates that the promotion of these cultural attributes has resulted in a rise in tourist numbers, thereby facilitating job creation and the expansion of local businesses. Furthermore, studies reveal that cultural tourism in Kerala encompasses not only performing arts but also heritage sites, architectural marvels, and culinary experiences. The state's backwaters, alongside its historic forts, temples, and churches, provide a distinctive cultural journey (Singh, R. P. 2015). Visitors in search of genuine experiences are attracted to traditional Kerala homes, ayurvedic wellness centers, and homestays that embody the local way of life. Many researchers have observed that cultural tourism has bolstered the local economy by creating income opportunities for artisans, performers, and craftsmen who depend on tourism for their livelihoods. Despite the positive impact of cultural tourism on Kerala's economy, some studies have also addressed the challenges it faces. Issues related to the commercialization of cultural practices, the erosion of authenticity, and the potential loss of cultural integrity have been raised.

## 3 Cultural Tourism: Positive and Negative Impacts

Cultural tourism in Kerala has brought both advantages and challenges, shaping the state's economy, society, and environment in different ways (Venugopalan, T., & Kumar, D. 2017).

On the positive side, it has significantly boosted the local economy by creating jobs and supporting traditional businesses. Many artisans, craftsmen, and performers have benefited from the growing interest in Kerala's rich cultural heritage. Visitors from different parts of the world seek authentic experiences, such as watching Kathakali performances, purchasing handmade crafts, and exploring Ayurveda-based wellness treatments. This increased demand has helped in preserving these cultural traditions, which might have faded over time. Additionally, tourism has provided many local communities with stable income sources, improving their quality of life. Another benefit is the cultural exchange that happens when people from different backgrounds interact, fostering mutual respect and understanding.

However, cultural tourism also has its downsides. One major concern is the commercialization of traditions (Zhao, K. 2024), where performances and rituals are sometimes altered to suit tourist preferences rather than staying true to their original form. This can lead to a loss of authenticity over time. Popular tourist destinations like Fort Kochi, Munnar, and Alappuzha often face overcrowding, putting pressure on local infrastructure and resources. Pollution, waste management issues, and environmental damage are common problems in such areas. The rise in tourism-driven businesses has also caused property prices to increase, making it difficult for local residents to afford housing in some places. Furthermore, exposure to different cultures sometimes creates social tensions, as rapid modernization can clash with traditional values and lifestyles.

To ensure that the benefits of cultural tourism outweigh the drawbacks, Kerala needs to adopt sustainable tourism practices (Chettiparamb, A., & Kokkranikal, J. 2012). Encouraging responsible tourism, preserving heritage sites, and involving local communities in decision-making can help maintain the balance between economic growth and cultural preservation. By doing so, Kerala can continue to thrive as a cultural hotspot while protecting its unique identity.

### 3.1 Environmental Impact : Cultural Tourism

Cultural tourism in Kerala has played a pivotal role in enhancing the state's economy and elevating its global profile; however, it has also introduced various environmental challenges. The growing number of tourists, particularly in culturally rich locations such as Fort Kochi, Kumarakom, and heritage sites like the Muziris region, has raised significant sustainability concerns (Munoth, N., Thomas, L., & Gehlot, S. 2025). Apple Academic Press.. A primary environmental issue is the generation of waste. Popular tourist spots experience a notable increase in plastic waste, food packaging, and other pollutants, which often overwhelm local waste management systems.

Numerous traditional festivals and cultural events draw large crowds, resulting in littering and improper waste disposal, thereby impacting both the natural environment and cultural heritage. The escalating demand for infrastructure to accommodate cultural tourism has prompted rapid urbanization in certain heritage areas. The construction of hotels, resorts, and commercial establishments can disrupt the ecological balance, particularly in coastal and backwater regions. Furthermore, the increased foot traffic in historically and culturally significant sites has led to the deterioration of monuments and traditional structures, necessitating extensive restoration efforts. Water consumption presents another critical challenge. Many cultural sites in Kerala, including temple towns and festival venues, witness a significant rise in water usage due to tourist activities, which adversely affects local communities dependent on limited water resources (Cosgrove, W. J., & Loucks, D. P. 2015). Additionally, the backwaters and rivers, integral to Kerala's cultural tourism, are increasingly contaminated by tourism- related activities, such as houseboat operations and unregulated sewage disposal. The transportation sector associated with tourism also raises environmental concerns. A higher influx of visitors results in more vehicles on the roads, contributing to increased carbon emissions. Areas of cultural significance often face traffic congestion, which disrupts local life and exacerbates air and noise pollution.

### 3.2 Challenges and Issues

Cultural tourism in Kerala has introduced both advantages and difficulties for the region. It has played a crucial role in bolstering the economy by generating employment, enhancing local crafts, and fostering traditional art forms. However, it has also given rise to various issues that impact the social, economic, and environmental landscape of the state. A primary concern is the over-commercialization of cultural practices. As tourism expands, there is a growing inclination to alter or embellish cultural traditions to meet tourist expectations. This often leads to a compromise in authenticity, where traditional art forms, festivals, and rituals are modified to create a more marketable image. The commercialization of cultural heritage can divert attention from its original meaning and significance, transforming it into a performance rather than a true representation of tradition. (Santa, E. D., & Tiatco, A. 2019). Another critical issue pertains to the effects on local communities. Although tourism generates revenue, it also contributes to increased living costs in certain regions, particularly in popular tourist spots such as Fort Kochi, Kumarakom, and Wayanad. Rising land prices, rental costs, and the price of goods can make it challenging for local residents to maintain their livelihoods. Furthermore, a growing cultural divide between locals and tourists can lead to social tensions and a decline in traditional lifestyles. Environmental degradation is yet another challenge associated with cultural tourism (Nair, V., Mohamed, B., Thomas, T. K., & Teare, R. 2016). Popular cultural sites draw large numbers of visitors, resulting in waste production, pollution, and pressure on natural resources. Historical sites, temples, and heritage buildings often experience deterioration due to heavy foot traffic, and without adequate conservation measures, these locations face the risk of long-term damage. Unregulated tourism in ecologically sensitive cultural areas can also result in deforestation, water pollution, and habitat destruction.

### 3.3 Cultural Festivals and Tourism

Cultural festivals are integral to the tourism framework of Kerala. These occasions not only highlight the state's diverse traditions but also draw thousands of visitors from both domestic and international locales, thereby making substantial contributions to the economy and the preservation of cultural heritage. Among the most renowned festivals is Thrissur Pooram, celebrated for its magnificent elephant parades, traditional drumming, and breathtaking fireworks displays. This festival, which is deeply embedded in Kerala's cultural legacy, attracts tourists eager to experience its lively festivities. The surge in visitors positively impacts local enterprises, including hotels, restaurants, and artisans who create decorative items for the elephants. Nonetheless, the environmental and logistical issues associated with managing large crowds and preserving the festival's authenticity pose significant challenges. Onam, the harvest festival of the state, serves as another prominent attraction. It is marked by


https://doi.org/10.52202/081568-0007

traditional dances such as Thiruvathira, intricate floral arrangements, and the renowned Vallamkali (boat races). Onam fosters community spirit and acts as a platform for cultural tourism, allowing visitors to indulge in Kerala's traditional cuisine, particularly the Onam Sadya, a lavish vegetarian meal served on banana leaves (Ozukum, L., & Aswathy, V. K. 2024). This festival is vital for maintaining Kerala's cultural identity while simultaneously appealing to tourists in search of genuine experiences. The Nehru Trophy Boat Race in Alappuzha stands out as a distinctive cultural event that captivates adventure and sports enthusiasts. The synchronized rowing of snake boats, accompanied by rhythmic chants and enthusiastic spectators, creates a mesmerizing spectacle. This event has significantly promoted water tourism in Kerala and has spurred the development of tourism-related infrastructure in the area.

### 3.4 Impact on Local Handicrafts and Cuisine

Cultural tourism in Kerala has profoundly influenced local handicrafts and cuisine, yielding both beneficial and adverse effects. The state's rich artistic heritage, encompassing coir products, mural paintings, Aranmula mirrors, and handwoven textiles, attracts numerous tourists. This heightened interest has resulted in increased demand for these crafts, thereby providing artisans with broader market opportunities and aiding in the preservation of traditional skills that have been handed down through generations. Many local craftsmen have modified their offerings to align with contemporary preferences while maintaining their authenticity (Suroyo, S., Soedjiwo, N. A. F., Entas, D., & Putra, B. M. 2023). Nonetheless, the commercialization of handicrafts has led to mass production, which can sometimes undermine quality and originality. In a similar vein, Kerala's varied cuisine has garnered international acclaim as a result of cultural tourism. Tourists are eager to sample traditional dishes such as Sadya, Malabar biryani, appam with stew, and an array of seafood specialties. This rising interest has prompted the growth of local restaurants, food festivals, and culinary tourism initiatives, including spice plantation tours and cooking classes. While tourism has contributed to the preservation of traditional recipes, it has also prompted alterations to cater to global tastes, occasionally diluting the authenticity of the flavors. Furthermore, the increased demand for unique local ingredients has influenced pricing, affecting accessibility for local communities.

### 3.5 Role of Government and Policy Framework

The government plays an essential role in influencing the effects of cultural tourism in Kerala through a variety of policies and initiatives. The tourism industry in Kerala is significantly fueled by its abundant cultural heritage, (Cherian, A. M., & Natarajamurthy, P. 2024). which encompasses traditional art forms, historical monuments, religious sites, and vibrant festivals. To safeguard and enhance these cultural treasures, the state government, in collaboration with the Kerala Tourism Development Corporation (KTDC), has enacted numerous strategies aimed at fostering sustainable development. A primary focus of government involvement is the preservation of heritage sites. Initiatives such as the Muziris Heritage Project have been pivotal in restoring and maintaining historical landmarks, ensuring their continued relevance as attractions for cultural tourists. By revitalizing ancient trade routes and structures, the government has not only safeguarded historical significance but also generated economic opportunities for local communities. To stimulate the growth of cultural tourism, financial assistance and policy frameworks have been established. Incentives for local artists, performers, and craftsmen are designed to support traditional art forms such as Kathakali, Theyyam, and Mohiniyattam, which may otherwise face decline due to modernization. Furthermore, the government organizes and funds cultural festivals to draw both domestic and international visitors, providing a platform for local artisans and performers to display their talents. Sustainability has been a central theme in policy formulation, as cultural tourism must reconcile economic advantages with the imperative of conservation. Rigorous guidelines are enforced on infrastructure development to prevent adverse effects on cultural heritage. Additionally, the promotion of homestays and community tourism initiatives has been encouraged, enabling visitors to engage with Kerala's culture in an authentic manner while benefiting local communities.

### 3.6 Sustainable Cultural Tourism

Cultural tourism is a vital component of Kerala's tourism sector, drawing visitors from around the world to engage with its diverse traditions, heritage, and artistic expressions (Sebastian, L. M., & Rajagopalan, P. 2009). As the appeal of cultural tourism grows, it becomes increasingly important to adopt sustainable practices that safeguard local traditions and heritage sites while simultaneously benefiting the surrounding communities. In Kerala, sustainable cultural tourism aims to strike a harmonious balance between economic development, environmental stewardship, and the preservation of cultural identity. A fundamental element of sustainable cultural tourism in Kerala is the commitment to safeguarding and promoting traditional art forms, including Kathakali, Mohiniyattam, Theyyam, and Kalaripayattu. The rise in tourism has led to greater recognition of these art forms, motivating local artists to continue their craft. However, the challenge of commercialization can lead to alterations in performances to cater to tourist expectations. Sustainable practices focus on maintaining the authenticity of these art forms while ensuring that artists receive equitable compensation and acknowledgment for their work. When sustainability is prioritized, local communities reap substantial benefits from cultural tourism. The emergence of homestays and locally-owned lodging options enables tourists to immerse themselves in Kerala's traditions, providing financial support to local families. Additionally, artisans who produce handicrafts such as coir products, mural paintings, and Aranmula mirrors benefit from heightened tourist interest. Initiatives that advocate for responsible tourism help ensure that these crafts are not mass-produced in environmentally harmful ways or exploitative to the artisans. Religious and spiritual tourism also plays a significant role in Kerala's cultural tourism landscape. Pilgrimage destinations like Sabarimala, Guruvayur, and Padmanabhaswamy Temple attract millions of devotees and tourists annually. While these sites bolster the local economy, issues such as overcrowding and pollution pose risks to their sustainability. Implementing measures like eco-friendly infrastructure and effective waste management is essential to address these challenges.

## Conclusion

Cultural tourism in Kerala has had a profound impact on the state's economy, society, and the preservation of its heritage. It has created employment opportunities for local artisans, performers, and small enterprises, while simultaneously aiding in the sustenance of traditional art forms, festivals, and customs. The arrival of tourists has fostered the promotion of Kerala's rich historical narrative, architectural marvels, and culinary diversity, establishing it as a prominent cultural destination on the global stage. Nonetheless, the swift expansion of tourism has introduced challenges such as commercialization, cultural erosion, and environmental issues. The overcrowding of heritage sites and places of worship, coupled with the excessive exploitation of natural resources, threatens the sustainability of cultural tourism. Although government initiatives and responsible tourism policies have made strides in reconciling economic development with cultural preservation, ongoing efforts are essential to uphold this balance. Looking forward, the future of cultural tourism in Kerala hinges on sustainable practices that honor local traditions while welcoming innovation. With a strategic approach, cultural tourism can continue to flourish, providing benefits to both local communities and visitors in search of genuine experiences.

## References


1. Smith, S. (2015). A sense of place: Place, culture and tourism. *Tourism Recreation Research*, *40*(2), 220-233.
2. Sebastian, L. M., & Rajagopalan, P. (2009). Socio-cultural transformations through tourism: a comparison of residents' perspectives at two destinations in Kerala, India. *Journal of Tourism and Cultural Change*, *7*(1), 5-21.
3. Ravichandran, G., & Nam, S. (2024). Interwoven Legacies: Exploring the Rich Textile Traditions of Korea and Tamil Nadu (India) through Cross-Cultural Influences. *International Journal*, *5*(5), 152-164.
4. Singh, R. P. (2015). Heritage value of religious sites and built archetypes: the scenario of Hinduism, and illustrating the riverfront Varanasi. *Aatmbodh, the Journal of*



https://doi.org/10.52202/081568-0007

*Knowledge of Self, 12*(1), 21-40.
5. Venugopalan, T., & Kumar, D. (2017). Sustainable development through sustainable tourism in India-A case study of Kerala tourism. *Asian Journal of Research in Business Economics and Management, 7*(12), 10-27.
6. Zhao, K. (2024). Commercialized Traditions: Costs and Benefits of Cultural Tourism of the Miao Ethnic Group. *Journal of World Economy, 3*(4), 53-62.
7. Chettiparamb, A., & Kokkranikal, J. (2012). Responsible tourism and sustainability: the case of Kumarakom in Kerala, India. *Journal of Policy Research in Tourism, Leisure and Events, 4*(3), 302-326.
8. Munoth, N., Thomas, L., & Gehlot, S. (2025). The Saga of Kochi: Cultural and Heritage Tourism Overview. In *Language and Cross-Cultural Communication in Travel and Tourism* (pp. 191-221). Apple Academic Press.
9. Cosgrove, W. J., & Loucks, D. P. (2015). Water management: Current and future challenges and research directions. *Water resources research, 51*(6), 4823-4839.
10. Santa, E. D., & Tiatco, A. (2019). Tourism, heritage and cultural performance: Developing a modality of heritage tourism. *Tourism Management Perspectives, 31*, 301-309.
11. Nair, V., Mohamed, B., Thomas, T. K., & Teare, R. (2016). How can the tourism industry respond to the global challenges arising from climate change and environmental degradation?. *Worldwide Hospitality and Tourism Themes, 8*(5), 611-616.
12. Ozukum, L., & Aswathy, V. K. (2024, June). Bridging Cultures: Fostering Sustainable Livelihoods and Tourism in Ao Naga Heritage. In *International Conference on Innovation and Regenerative Trends in Tourism and Hospitality Industry (IRTTHI 2024)* (pp. 295-307). Atlantis Press.
13. Suroyo, S., Soedjiwo, N. A. F., Entas, D., & Putra, B. M. (2023). Aghi Ghayo Onam: Religious and Costumery Tradition by Malay Kampar People in Islamic Perspective. *Analisa: Journal of Social Science and Religion, 8*(2), 203-219.
14. Cherian, A. M., & Natarajamurthy, P. (2024). TOURISM IN RURAL KERALA: AN OVERVIEW. *CAHIERS MAGELLANES-NS, 6*(2), 1704-1707.
15. Sebastian, L. M., & Rajagopalan, P. (2009). Socio-cultural transformations through tourism: a comparison of residents' perspectives at two destinations in Kerala, India. *Journal of Tourism and Cultural Change, 7*(1), 5-21.

# POTENTIAL AND CHALLENGES OF SUSTAINABLE VILLAGE TOURISM KHONOMA NAGALAND

[1*]Shruti Kumari, [2]Shikha Sharma,

[1&2]Chandigarh University, Mohali, India

*Correspondence author: 23BTT10002@cuchd.in

**ABSTRACT.** The beautiful village of Khonoma in the Indian state of Nagaland, known for its lush landscapes, diverse wildlife and strong cultural heritage, is becoming a popular example of how sustainable tourism can beDeveloped through community-based eco-friendly initiatives andConservation efforts. The village successfully combines traditional life. With modern ideas drawing visitors who want genuine cultural experiences and natural beauty. Khonoma's sustainable tourism works because it prioritizes protecting the environment and empowering theLocal community. Initiatives such as community-managed forestry organic farming and ecotourism ensure that tourism benefits both natureAnd the people who live there. Visitors are attracted to the village terraces, untouched nature and the unique Angami culture that offers locals an opportunity to earn money and share their culture. HoweverKhonoma still faces some obstacles. Poor infrastructure, a lack of touristAwareness about responsible travel, and the risk of too many tourists could harm the village's environment and culture. Finding a balanceBetween development and conservation requires careful planning and active participation of the local community. Promoting responsibleTourism, improving infrastructure in an environmentally friendly manner and highlighting the village's unique identity are essential for its futureSustainable growth.

**Keywords:** Sustainable Tourism, Community-Based Tourism, Ecotourism, Cultural Heritage, Responsible Travel

## 1 Introduction

Tourism has become a crucial factor in fostering economic growth, facilitating cultural exchange, and promoting environmental conservation in rural regions across the globe. Specifically, sustainable village tourism offers a unique opportunity to harmonize economic advancement with ecological integrity and socio-cultural sustainability. A notable example of this is Khonoma, a historic village located in Nagaland, India, celebrated for its rich biodiversity, distinctive traditions, and community-driven conservation initiatives.

Khonoma has earned the distinction of being India's first green village, thanks to its innovative conservation efforts, such as the Khonoma Nature Conservation and Tragopan Sanctuary (KNCTS), which exemplify the villagers' dedication to sustainable practices. The village's tourism appeal is rooted in its unspoiled natural surroundings, the indigenous Angami Naga culture, traditional farming methods, and eco-friendly homestay options. Nevertheless, while Khonoma serves as a model for sustainable rural tourism, it also encounters various challenges. These challenges include the need to maintain a delicate equilibrium between rising tourism and ecological preservation, addressing infrastructure constraints, ensuring fair economic benefits, and safeguarding its cultural heritage against external pressures.

This paper investigates the potential of sustainable village tourism in Khonoma, emphasizing its environmental, economic, and socio-cultural importance. It also addresses the obstacles that could impede its long-term viability and suggests strategies for fostering responsible tourism that aligns with the village's conservation principles. By examining Khonoma's experiences, this study seeks to offer insights into how rural communities can create sustainable tourism frameworks that benefit both the local populace and the environment.



https://doi.org/10.52202/081568-0008

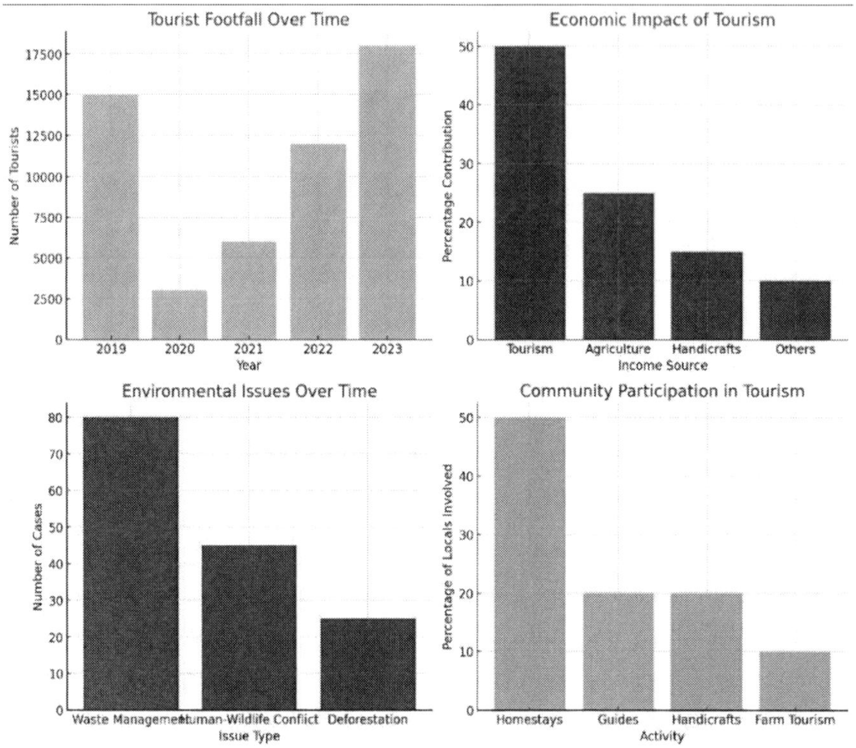


Figure 1. The data used for the graph is on based from the following Sources: Author

- Tourism footfall over time:-The Guardian (Pinto-Rodrigues, 2020) – "India's first 'green' village adapts to life without tourists".
- Economic impact of tourism:- Times of India (2024) – "Khonoma, Nagaland: The success story of eco-tourism".
- Environmental issues :- The Morung Express (2023)-"Asia's first 'Green village' faces waste management challenge".
- Community participation in tourism:-Sanctuary Nature Foundation – "Khonoma and the Angami".

## 2. OBJECTIVES

The key objectives of this research is paper is to analyze the sustainable village tourism in khonoma. To evaluate its impact on the environment, participation of local communities, environmental conservation, the economic development. The key objectives of the research are:

- To examine khonoma approach to sustainable tourism

- To know about the challenges and potential of sustainable village tourism of khonoma
- To evaluate the role of local community participation and to explore the economic benefits through sustainable tourism.
- To identify the key issue of environment such as waste management, human-wildlife conflict and economic dependency on tourism.

# 3 LITERATURE REVIEW

Khonoma, located in Nagaland, India, is frequently recognized as Asia's inaugural "Green Village" due to its commitment to sustainable village tourism, which harmonizes environmental conservation with the preservation of cultural heritage. In 1998, the village initiated the Khonoma Nature Conservation and Tragopan Sanctuary (KNCTS) to safeguard its diverse ecosystems, particularly focusing on the endangered Blyth's tragopan. This pivotal shift from hunting to conservation has significantly transformed community practices, establishing Khonoma as a benchmark for eco-tourism (Tripoto, n.d.). The sanctuary not only protects local wildlife but also draws ecotourists, thereby enhancing the local economy.
The tourism strategy in Khonoma is fundamentally community-driven. Residents actively engage in managing homestays, providing guided treks, and presenting cultural experiences, which ensures that tourism profits are fairly distributed and cultural authenticity is preserved (Times of India, 2024). These initiatives have economically empowered the community, enabling a variety of income sources while maintaining traditional agricultural methods (Department of Information & Public Relations, 2021). Furthermore, the community-based tourism model reinforces the village's social cohesion by promoting shared responsibility in resource management and decision-making processes.

Nevertheless, the surge in tourism has brought about significant challenges, particularly concerning waste management. Despite Khonoma's strong commitment to environmental stewardship, the influx of visitors has resulted in littering and inadequate waste disposal practices. Although the Khonoma Village Students' Union conducts cleanliness campaigns, the absence of effective waste segregation and disposal systems remains a concern, threatening the village's environmental sustainability (The Morung Express, 2023). Additionally, the successful conservation initiatives have led to increased human-wildlife conflicts, with wildlife encroaching on agricultural areas and causing crop damage, which has created tensions within the communityThe balance between the necessity for biodiversity conservation and the sustainability of agriculture has been emphasized (Pinto-Rodrigues, 2020).

The economic reliance on tourism has emerged as a significant challenge, particularly during the COVID-19 pandemic. Travel restrictions led to a notable decline in tourist arrivals, adversely affecting the livelihoods of the community and highlighting the urgent need for alternative economic avenues to lessen dependence on tourism (Pinto-Rodrigues, 2020). To ensure the long-term viability of Khonoma's tourism model, it is vital to tackle these issues through effective waste management practices, reducing human-wildlife conflicts, and investigating alternative sources of income.

## 3.1 Conservation and Ecotourism

The creation of the Khonoma Nature Conservation and Tragopan Sanctuary (KNCTS) in 1998 represented a key achievement in the village's conservation initiatives. The sanctuary was established to safeguard the endangered Blyth's tragopan and other indigenous species, focusing on the transition from hunting to sustainable wildlife conservation (Kikon, 2005). This change has inspired similar ecotourism projects driven by conservation in other areas, where


https://doi.org/10.52202/081568-0008

local inhabitants play an essential role in managing resources and protecting habitats (Dutta, 2012).

Khonoma's ecotourism approach is founded on community-based conservation, ensuring that locals actively engage in and gain advantages from tourism-related endeavors. Homestays, guided treks, and cultural exhibitions support a sustainable economic model in which profits are equitably shared (Times of India, 2024). Research by Jamir and Longkumer (2019) illustrates how participatory conservation fosters local dedication to sustainable tourism while maintaining indigenous knowledge systems.

### 3.2 Economic and Social Impacts

The transition to ecotourism has resulted in economic diversification, offering alternative livelihoods apart from subsistence farming. Research indicates that community-managed tourism promotes social unity by strengthening collaborative decision-making and sustainable resource management (Department of Information and Public Relations, 2021). The combination of traditional agricultural methods with tourism further ensures food security and economic resilience (Aier, 2018).

Nonetheless, the heavy dependence on tourism income creates economic risks, particularly evident during the COVID-19 crisis. Travel limitations caused a drastic reduction in tourist visits, greatly affecting local earnings (Pinto-Rodrigues, 2020). This situation highlights the necessity for alternative income-generating approaches, like agroforestry and handicraft creation, to alleviate the hazards linked to over-reliance on tourism (Lal and Singh, 2021).

### 3.3 Environmental Challenges

Despite Khonoma's robust environmental commitment, increasing visitor numbers have intensified waste management issues. The village does not have a thorough waste segregation and disposal system, leading to environmental harm despite ongoing cleanliness campaigns organized by the Khonoma Village Students' Union (The Morung Express, 2023). Comparative analysis of sustainable waste management in ecotourism locations suggests that incorporating circular economy practices, such as composting and recycling, could greatly improve waste reduction efforts (Sharma et al. , 2020).

Another urgent issue is human-wildlife conflict stemming from successful conservation initiatives. Growing wildlife populations have resulted in crop damage, impacting farmers and creating discord within the community. Research indicates that adaptive land-use planning and compensation programs can aid in reconciling conservation objectives with agricultural sustainability (Nair and Deka, 2017). Such strategies are vital for maintaining a balance between biodiversity conservation and rural livelihoods.

### 3.4 Policy and Future Directions

The sustainable tourism model of Khonoma's long-term success relies on integrated policy measures. Prioritizing effective waste management practices, diversifying income streams, and implementing conflict resolution strategies is essential. Researchers stress the importance of local governance in securing the sustainability of community-led tourism models (Das and Saikia, 2015). Policies that encourage eco-friendly infrastructure and responsible tourism practices could further enhance Khonoma's conservation legacy (Bhattacharya, 2019).

Khonoma exemplifies the potential of sustainable village tourism through its initiatives focused on community-led conservation and cultural heritage preservation. Nevertheless, it is imperative to address challenges such as waste management, human-wildlife interactions, and economic dependency to secure the village's ongoing success and sustainability.

https://doi.org/10.52202/081568-0008

# 4 METHODOLOGY

This report examines both the potential and challenges of sustainable village tourism in Khonoma, Nagaland. This report is based on mixed-method approach which is a combination of both qualitative and quantitative research techniques. This research aims to know the impact of sustainable tourism on the local communities, economy and on environment. This alsoresearch explore the role of community in eco-friendly tourism and also identify some key challenges such as waste management and human-wildlife conflict.

**4.1 Research design**- The analysis of sustainable and eco tourism approach of khonoma is done by case study for in-depth analysis of eco-tourism model. The study combines secondary data analysis to understand the dynamics of sustainable tourism in village.

### 4.1 Data collection method- Secondary data collection

- As per the literature review mentioned in my research on sustainable tourism that includes government report, academic research and media articles like The guardian, The times of India, The Morung express.
- Analysis of the official tourism policies reports from sources like department of information and public relations, Nagaland.

### 4.2 Limitations

- The dependence on secondary data might limit the opinion and perspectives of the local community.
- The research lack primary data which means the research do not include direct interviews and observations.

# 5 Case study

Khonoma, a village located in Nagaland, India, serves as a compelling example of sustainable village tourism due to its rich cultural heritage, dedication to environmental conservation, and breathtaking natural scenery. The village's cultural practices and eco-tourism projects, such as the Khonoma Nature Conservation and Tragopan Sanctuary, create opportunities for responsible tourism that uplifts local communities and protects the environment. Nonetheless, issues like environmental deterioration, commercialization of culture, inadequate infrastructure, and resistance from the community require careful attention. Successful strategies should include community-driven tourism initiatives, enhancing capacity, building eco-friendly infrastructure, encouraging low-impact activities, and fostering partnerships with governmental bodies and NGOs. With careful planning, Khonoma has the potential to set a precedent for sustainable tourism that benefits both the local population and the environment while maintaining its distinct cultural character.

# CONCLUSION

Khonoma serves as an outstanding model sustainable village tourism, effectively integrating conservation efforts with cultural preservation and economic advancement .initiatives led by the community in eco-tourism, such as the Khonoma nature conservation and TragopanSanctuary(KNCTS), have contributed to the protection of biodiversity while enhancing the local economy.

Nonetheless, issues like waste management, conflicts with wildlife, and over-reliance on tourism pose significant threats to its sustainability. To secure Lasting success, Khonoma needs to implement improved waste management system, address wildlife conflicts, and explore diverse income streams that extend beyond tourism

In summary, Khonoma illustrates how nature focused tourism driven by the community can promote both ecological sustainability and economic well-being, providing a model for other rural regions seeking sustainable development.


https://doi.org/10.52202/081568-0008

## REFERENCES


1. Department of Information & Public Relations, Nagaland. (2021, May 1). The Khonoma Experience Held.
2. Pinto-Rodrigues, M. (2020, May 28). India's first 'green' village adapts to life without tourists. The Guardian.
3. The Morung Express. (2023, May 18). Asia's first 'Green Village' faces waste management challenge.
4. Times of India. (2024, September 15). Khonoma, Nagaland: The success story of eco-tourism, sustainable travel, and living in India.
5. Tripoto. (n.d.). Khonoma Village: India's first Green Village and a model of eco-tourism.
6. Aier, T. (2018). Indigenous Agriculture and Ecotourism: A Study of Naga Villages. Journal of Rural Development, 37(2), 189-203.
7. Bhattacharya, P. (2019). Sustainable Tourism in Northeast India: Challenges and Opportunities. Tourism Management Perspectives, 30, 34-45.
8. Das, K., &Saikia, R. (2015). Community-Based Tourism in Northeast India: A Policy Perspective. International Journal of Tourism Research, 17(6), 495-505.
9. Dutta, S. (2012). Wildlife Conservation and Tourism: Lessons from Community Initiatives. Ecology and Society, 17(3), 28.
10. Jamir, A., &Longkumer, N. (2019). Indigenous Knowledge and Sustainable Tourism in Nagaland. Asian Journal of Tourism Studies, 9(1), 55-72.
11. Kikon, D. (2005). Conservation and Cultural Identity: The Case of Khonoma. South Asian Journal of Sustainability, 12(4), 112-129.
12. Lal, R., & Singh, A. (2021). Economic Diversification in Rural Ecotourism: A Study of Northeast India. Journal of Sustainable Development, 14(3), 76-91.
13. Nair, R., & Deka, J. (2017). Human-Wildlife Conflict and Conservation Strategies in Northeast India. Biodiversity Conservation Journal, 26(2), 289-310.
14. Pinto-Rodrigues, K. (2020). Tourism and Economic Resilience in India's Northeast: A Post-Pandemic Analysis. Current Issues in Tourism, 23(10), 1254-1268.
15. Sharma, P., et al. (2020). Waste Management in Ecotourism: Circular Economy Perspectives. Environmental Research Letters, 15(4), 045007.

# Developing Religious Tourism Circuits in Jammu: Linking Vaishno Devi, Shiv Khori and Raghunath Temple

Dr. Pankaj Sharma [1], Rakesh Sharma [2], Dr. Trilochan Kumar* [3], Virender Mehta [4]

[1,3]University of Jammu, Jammu, India
[2,4]Department of Tourism, Chandigarh University, India

* Corresponding author- trilochankumar22@gmail.com

**Abstract**- Religious tourism possesses considerable potential to stimulate economic growth, enhance cultural legacy, and encourage regional development in the Jammu region. This research examines the establishment of a religious tourist circuit connecting three significant spiritual sites: Vaishno Devi, Raghunath Temple, and Shiv Khori. Each location possesses profound religious importance and draws millions of adherents each year. The study analyses the existing infrastructure, visitor experience, and connection among these locations, while highlighting issues such as overpopulation, environmental deterioration, and insufficient facilities. Furthermore, it assesses the impact of technology, marketing, and community engagement on improving the tourist experience. The report offers measures to enhance accessibility, promote sustainability, and stimulate the area economy by combining these locations into a unified tourism circuit. The findings seek to furnish actionable insights for policymakers, tourist authorities, and local stakeholders to establish Jammu as a leading religious tourism destination while promoting sustainable development.

**Keywords** – Religious Tourism, Tourism Circuit, Sustainable Development, Pilgrimage Tourism, Visitor Experience.

## 1. Introduction

Religious tourism, often termed pilgrimage tourism, refers to the journeys undertaken by individuals or groups for spiritual or religious purposes. This form of tourism encompasses a wide range of activities, including visits to sacred sites, participation in religious rituals, and attendance at festivals or ceremonies. The World Tourism Organization (UNWTO) defines religious tourism as "travel undertaken for religious purposes, whether for pilgrimage, worship, or spiritual renewal" (UNWTO, 2019). In the Indian context, religious tourism encompasses visits to a variety of sacred destinations associated with multiple faiths, including Hinduism, Buddhism, Jainism, Sikhism, Christianity, and Islam. The scope of religious tourism in India is vast, covering pilgrimage sites like Varanasi and Bodh Gaya, as well as major festivals like Kumbh Mela, which attract millions of devotees.

The significance of religious tourism is multifaceted. Economically, it contributes to local economies through tourism-related spending, creating jobs and supporting local businesses. Socially, it fosters a sense of community and shared identity among pilgrims. Culturally, religious tourism plays a critical role in the preservation of traditions and heritage, ensuring that the customs

 https://doi.org/10.52202/081568-0009

and practices associated with various faiths continue to thrive in a rapidly changing world (Sharma, 2018). Religious tourism also offers opportunities for intercultural dialogue and understanding. As people from diverse backgrounds come together in shared spiritual spaces, they can engage in meaningful exchanges that enhance mutual respect and appreciation for different cultures and beliefs. This aspect of religious tourism can promote social cohesion and reduce cultural misunderstandings, making it a significant force for peace and unity in a diverse nation like India.

Religious tourism in Jammu has been fundamental to the region's cultural and spiritual identity for centuries. Renowned for its hallowed temples and shrines, Jammu draws millions of devotees each year, especially to locations such as Vaishno Devi, Shiv Khori, and Raghunath Temple. The region's abundant religious heritage, coupled with enhanced infrastructure and accessibility, has established it as one of India's foremost pilgrimage sites. Religious tourism in Jammu has transformed from conventional pilgrimage activities into a structured and prosperous industry, substantially benefiting the local economy.

### 1.1 Linking Vaishno Devi, Shiv Khori and Raghunath Temple

Jammu, renowned for its sacred and culturally significant temples, has historically served as a center for religious tourism. The Vaishno Devi Temple, Shiv Khori, and the Raghunath Temple are among the most renowned pilgrimage destinations. These temples, each possessing unique spiritual importance, draw millions of devotees annually. Nevertheless, despite their significant popularity, the opportunity to link these sites via meticulously crafted religious tourism circuits has not been fully actualized. This paper examines the establishment of a tourism circuit connecting three significant religious sites to improve the pilgrimage experience, stimulate the local economy, and foster sustainable tourism in Jammu.

### 1.2 Principal Religious Locations

**1.2.1Vaishno Devi Temple** - The Vaishno Devi Temple, situated on Trikuta Mountain near Katra, is among the most frequented religious sites in India. The site is devoted to Goddess Vaishno Devi, drawing millions of devotees each year. Pilgrims undertake a 13 km pilgrimage from Katra to the shrine, rendering it a physically arduous yet spiritually fulfilling endeavor.

**1.2.2 Shiv Khori**- Shiv Khori, a cavernous sanctuary situated in the Reasi district, is devoted to Lord Shiva. The temple contains a naturally occurring Shiv Lingam and is enveloped in legends and folklore. It provides a more tranquil and isolated pilgrimage experience in contrast to the crowded Vaishno Devi shrine. The cave's inherent beauty and spiritual ambiance render it a distinctive location for worshippers.

**1.2.3 Raghunath Temple**- The Raghunath Temple, situated in the center of Jammu city, is a significant Hindu sanctuary devoted to Lord Rama. It is among the largest temple complexes in North India, featuring numerous shrines devoted to various deities from the Ramayana and other Hindu texts. The temple's historical importance and its magnificent architecture render it a vital destination for religious tourists.

## 2. Objectives of Linking Vaishno Devi, Shiv Khori, and Raghunath Temple

### 2.1 Enhancing Pilgrimage Experience

By linking these sacred sites into a cohesive circuit, pilgrims can experience a more comprehensive spiritual journey. Instead of visiting these sites in isolation, travelers can explore a range of religious, cultural, and natural offerings, enhancing their overall experience.

### 2.2 Economic Boost for Local Communities

Creating a religious tourism circuit will encourage pilgrims to visit multiple sites in one trip, thereby increasing tourism revenue for local businesses, including accommodations, transport services, and retail outlets.

### 2.3 Promoting Sustainable Tourism

A well-managed tourism circuit can ensure that the natural and cultural heritage of the region is preserved, while accommodating the needs of a growing number of visitors. Sustainable practices in transportation, waste management, and conservation will be integral to this development.

## 3.Designing the Tourism Circuit: Vaishno Devi, Shiv Khori, and Raghunath temple

The tourism circuit should be designed with convenience, accessibility, and spiritual value in mind. The following are essential aspects to consider:

### 3.1 Circuit Route and Accessibility

The journey should connect the three temples in a logical sequence, making it easier for pilgrims to travel between the sites.

**Proposed Circuit Design:**

**1. Vaishno Devi to Shiv Khori:** After completing the pilgrimage to Vaishno Devi, pilgrims can travel by road to Shiv Khori, which is approximately 75 km away. A well-maintained road with transport options such as buses, taxis, and private vehicles can connect these two sites.

**2.Shiv Khori to Raghunath Temple:** After visiting Shiv Khori, pilgrims can head towards Jammu city to visit the Raghunath Temple, which is approximately 70 km from Shiv Khori. This leg of the journey can be facilitated through a combination of road travel and dedicated pilgrimage transport services.

**3.Circular Itinerary:** A circular itinerary can be designed, allowing pilgrims to either begin their journey at Vaishno Devi or Raghunath Temple, depending on their preference. This flexibility will help maximize convenience for visitors.

## 4. Marketing and Promotion

Promoting the tourism circuit effectively will ensure that it reaches a wider audience. The following strategies can be adopted:

### 4.1 Digital Promotion

- Website and App: Develop a dedicated website and mobile app for the tourism circuit that provides information on the temples, itinerary, transport options, accommodations, and online booking services.
- Social Media Campaigns: Utilize platforms like Instagram, Facebook, and Twitter to share the beauty, history, and spiritual significance of these temples. Video content showcasing the serene atmosphere at Shiv Khori or the majestic Vaishno Devi trek can attract pilgrims.
- Influencer Collaborations: Collaborate with religious and travel influencers to promote the circuit, helping to create awareness among a global audience.

## 5. Historical Context and Evolution of Religious Tourism in India


https://doi.org/10.52202/081568-0009

The roots of religious tourism in India can be traced back thousands of years. Ancient texts, including the Vedas, Mahabharata, and Ramayana, reference pilgrimage practices and sacred journeys undertaken by devotees seeking spiritual fulfillment. These texts outline the significance of various sacred sites and emphasize the importance of pilgrimage as a means of attaining spiritual merit (Vikram & Gupta, 2020).

**5.1 Historical Context and Development of Religious Tourism in Jammu**- Jammu, referred to as the "City of Temples," possesses a lengthy and illustrious history that has established it as one of India's foremost religious tourism centers. The site's development into a prominent pilgrimage destination is fundamentally anchored in ancient religious traditions, chiefly Hinduism, while also shaped by Sikh, Muslim, and Buddhist influences. Throughout the centuries, the region has evolved from a local religious hub to a globally acknowledged site, drawing millions of pilgrims annually.

**5.2 Antiquity and the Middle Ages**- The religious significance of Jammu extends back thousands of years. The temples, shrines, and sacred sites are integral to its spiritual identity. The Vaishno Devi Temple, honoring Goddess Vaishno Devi, is among the oldest and most venerated pilgrimage sites in India. The Vaishno Devi legend, documented in the Skanda Purana, originates from antiquity when the goddess meditated in the Trikuta Mountains. The temple's historical importance commenced in the 8th century, subsequently evolving into a vital pilgrimage destination for Hindus.

In addition to Vaishno Devi, the area witnessed the establishment of other significant temples, including the Raghunath Temple, constructed in the 19th century by Maharaja Ranbir Singh. This temple emerged as one of the largest and most important centers of worship in North India. During the medieval era, locations such as Shiv Khori, a cave temple devoted to Lord Shiva, emerged, thereby enhancing Jammu's religious landscape.

**5.3 Colonial and Post-Colonial Period**- During British colonial rule, Jammu's religious tourism commenced its development due to the establishment of infrastructure, including roads and railways, which enhanced accessibility to pilgrimage sites. The Vaishno Devi temple has become more accessible to pilgrims due to enhanced transportation connections, resulting in an increase in visitor numbers. The colonial era witnessed the systematic development of pilgrimage services, including the creation of inns and rest houses for travelers.

Following independence, Jammu's religious tourism sector underwent substantial expansion. The Vaishno Devi Shrine Board, founded in 1986, significantly improved pilgrimage amenities, encompassing the development of superior pathways, lodging, and transportation services. The formalization of pilgrimage management signified a transition to a more systematic method of religious tourism, facilitating more efficient travel for millions of pilgrims.

**5.4 Contemporary Period and Expansion of Pilgrimage Tourism**- Since the late 20th century, religious tourism in Jammu has experienced significant growth, with annual visitor numbers rising consistently. The religious sites of Jammu, particularly Vaishno Devi, draw pilgrims from throughout India and overseas. The establishment of supplementary pilgrimage infrastructure, such as Shiv Khori and Raghunath Temple, enhanced the variety of the region's religious tourism options. Technological innovations significantly contributed to the modernization of the pilgrimage experience. The implementation of online booking systems for Vaishno Devi enabled pilgrims to organize their visits with greater efficiency. The promotion of these sites via digital platforms and social media facilitated global awareness.

**5.5 Obstacles and Future Opportunities**- Notwithstanding its expansion, religious tourism in Jammu encounters numerous challenges, including overcrowding, environmental issues, and pressure on local infrastructure. To mitigate these concerns, there is a growing emphasis on sustainable tourism practices, encompassing waste management and environmentally friendly initiatives. Jammu's religious tourism sector possesses considerable potential for growth. Establishing interconnected pilgrimage routes, improving amenities, and adopting sustainable practices can guarantee the region's enduring viability as a spiritual and cultural destination. Jammu, by upholding its historical and spiritual heritage, can offer a profound and conscientious pilgrimage experience for future generations.

## 6.Findings

- **Potential for Economic Growth:** Religious tourism in Jammu, facilitated by the establishment of a cohesive circuit connecting Vaishno Devi, Raghunath Temple, and Shiv Khori, possesses considerable economic potential. Enhancing accessibility and marketing these sacred sites as a cohesive tourism route could augment tourism revenue in Jammu, thereby benefiting local enterprises, hospitality sectors, and associated industries.
- **Cultural and Religious Heritage Preservation:** Culturally and historically significant, the holy places of Vaishno Devi, Raghunath Temple, and Shiv Khori are also important spiritual centers for millions of worshippers. The planned tourist circuit has the potential to both conserve these traditions and bring in visitors interested in Jammu's religious and cultural history.
- **Infrastructure Challenges and Opportunities:** Even though it has been upgraded throughout the years, the current infrastructure still encounters problems including congestion, limited housing, and inadequate facilities, particularly during the busiest seasons for pilgrimages. To handle the increasing number of tourists and make sure the pilgrimage goes smoothly, it is vital to improve transportation networks, lodgings, and restroom facilities.
- **Role of Technology and Marketing:** Streamlining visitor management and improving the overall tourist experience could be achieved through technological improvements like digital promotions, smartphone apps, and online booking systems. With better advertising, Jammu may be promoted as a major religious tourist site, drawing in visitors from all over the world who are seeking spiritual solace, and all the while bringing attention to the importance of ecotourism.
- **Community Engagement and Local Involvement:** Local communities that are actively involved in religious tourism can improve the experience for visitors and make sure that the economic benefits are shared fairly. One way to foster community development and cultural preservation is to provide local stakeholders with job opportunities. Another is to support traditional crafts and businesses.

## 7. Recommendations

Jammu can position itself as a premier religious tourism destination in India by creating a circuit that links Vaishno Devi, Shiv Khori and Raghunath Temple. This circuit can lead to long-term growth and prosperity for the region if it is well-coordinated and incorporates accessibility improvements, sustainable practices, and community engagement.

## 8. Discussion and Conclusion

The exploration of religious tourism in India reveals its profound significance as a multidimensional phenomenon that intertwines faith, culture, and community. The definition and scope of religious tourism encompass a vast array of practices and experiences that attract millions


https://doi.org/10.52202/081568-0009

of pilgrims and tourists to various sacred sites across the country. The historical context of religious tourism in India showcases its evolution from ancient pilgrimage practices to contemporary tourism trends. This evolution reflects not only the resilience of religious traditions but also the dynamic nature of societal values and beliefs. As religious tourism continues to grow in contemporary society, it holds significant economic potential for local communities while fostering a deeper understanding of diverse cultural and religious practices.

The establishment of a religious tourism circuit connecting Vaishno Devi, Shiv Khori, and Raghunath Temple offers a distinctive opportunity to enrich the spiritual and cultural experiences of pilgrims in Jammu. By strategically linking these sacred sites, the circuit can provide a cohesive, enriching journey that enhances the pilgrimage experience and significantly benefits the local economy and community development.

The circuit features a meticulously designed infrastructure that facilitates easy access, comfortable lodgings, and essential amenities, thereby catering to the increasing influx of religious tourists and offering them both convenience and spiritual enrichment. Efficient marketing strategies and sustainable tourism practices will guarantee the enduring success of this initiative while safeguarding the region's natural and cultural heritage.

Ultimately, connecting these three iconic temples will enhance Jammu's status as a prominent religious tourism destination and foster a more integrated and comprehensive spiritual experience for pilgrims, motivating them to delve into the region's rich spiritual traditions. Through meticulous planning and cooperation among the government, local communities, and tourism operators, the religious tourism circuit can serve as a paradigm of sustainable, culturally attuned tourism that advantages both visitors and residents.

Moreover, the overview of major religious traditions in India, including Hinduism, Buddhism, Jainism, Sikhism, and Islam, highlights the rich tapestry of beliefs and practices that coexist within the nation. Each tradition contributes uniquely to the landscape of religious tourism, offering distinct experiences that resonate with pilgrims and tourists alike. As religious tourism evolves, it is essential to balance the spiritual, cultural, and economic dimensions to ensure that it remains sustainable and respectful of the sacredness of these sites, the significance of religious tourism in India extends beyond mere visitation; it plays a crucial role in shaping identities, fostering community bonds, and promoting intercultural dialogue. As we delve deeper into the subsequent chapters, we will explore the various facets of religious tourism, including pilgrimage sites, festivals, community engagement, and the challenges and opportunities that lie ahead. By understanding and appreciating the complexities of religious tourism, stakeholders can work together to create a more sustainable and enriching experience for all involved.

## References


1. Chaudhary, S. (2019). Religious Tourism in India: A Comprehensive Overview. *Journal of Tourism and Cultural Change*, 17(2), 1-15. https://doi.org/10.1080/14766825.2018.1483461
2. Cohen, E. (2016). The Symbolic Construction of Pilgrimage. *Annals of Tourism Research*, 58, 1-13. https://doi.org/10.1016/j.annals.2015.12.003
3. Ghosh, A. (2014). Pilgrimage in India: The Historical and Cultural Context. *Journal of Religious Tourism and Pilgrimage*, 1(1), 5-19. https://doi.org/10.1080/23288227.2014.946775
4. Mishra, R. (2017). The Impact of Mughal Architecture on Indian Pilgrimage Sites. *International Journal of Heritage Studies*, 23(3), 223-237. https://doi.org/10.1080/13527258.2017.1297562
5. Ministry of Tourism, Government of India. (2021). *Annual Report on Pilgrimage Tourism in India*. New Delhi: Ministry of Tourism. Link
6. Sinha, S. (2020). Sustainable Religious Tourism: Challenges and Strategies in India. *Journal of Sustainable Tourism*, 28(1), 113-130. https://doi.org/10.1080/09669582.2019.1624168
7. Smith, V. L. (1992). Alternative Tourism: A New Paradigm for Tourism Studies. *Annals of Tourism Research*,

19(1), 5-18. https://doi.org/10.1016/0160-7383(92)90003-F
8. Vikram, S., & Gupta, R. (2020). Kumbh Mela: A Model for Pilgrimage Tourism Management. *Journal of Tourism Management Research*, 12(2), 77-90. https://doi.org/10.51783/jtmr.v12i2.64
9. World Tourism Organization (UNWTO). (2019). *Religious Tourism: A Global Perspective*. Madrid: UNWTO. Link

 https://doi.org/10.52202/081568-0009

# Smart City and Migration to its Population: A Sustainable Bibliometric Approach

Muskan[1*], Karampreet Kaur[2], and Ambar Mani Mishra[3]

[1]University School of Business, Chandigarh University, Mohali, India
[2,3]Department of Airlines, Chandigarh University, Mohali, India

*Corresponding author: muskandhiman3282@gmail.com

**Abstract.** Smart cities have become a growing concept, as it provides several new facilities as well as opportunities like Quality education, Health facilities and many more to the people living in the region and it contributes a lot in transformation of the lives of individuals as well as leads to development of the Nation. On the flip side, the increase in the population is resulting in various troubles. There are numerous challenges that can be faced due to increased urbanisation and one of the major challenges is the shortage of resources in the urban areas. In this research, Bibliometric analysis is utilised. Total Five Hundred and Eleven (511) documents are chosen from the Scopus (database). An analysis is done through various criteria like authors keywords, yearly publication, analysis of the subject area, local impact of the author with the help of h-index and production of author over time. For the purpose of visualization, VOS viewer and R studio are utilised. The cluster analysis assisted in providing more clarity about the links and relations between various aspects. The results after overall analysis and understanding indicates that in future there would be more publications in this field as the scope is emerging with the passage of time. Also, the aviation sector is highly impacted due to migration.

**Keywords:** Sustainable Tourist, SDG, Smart City, Population, Career Development, Talent Management, Aviation.

## 1. Introduction

Smart cities have been in the significant attention for the past few decades, due to rapid urbanization all over the world (Silva et al.,2018). It has gained interest of several researchers who concentrate on the management of urban or construction technologies and observe the exponential research extension in the area. Under such a situation, acquiring a properly organized outline from the affluence of information is a main challenge in discovering possible parts in smart city investigates (Jia et al., 2014). What is a smart city? An unsophisticated elucidation is that a smart city is a place where traditional networks as well as facilities are made extra flexible, well-organized, and sustainable with the usage of information, digital, as well as technologies related to telecommunication in order to advance the operations in the city for the wellness of its residents (Mohanty et al., 2016). Smart cities are extensively understood as areas that vigorously hold novel technologies in order to attain anticipated city results (Yigitcanlar et al., 2019). Though, the idea of a smart city is not innovative. Diverse opinion originates in literature concerning the beginning of the notion of a 'smart city'. Some

https://doi.org/10.52202/081568-0010

reflect the origins of the smart city date back to the 1960s under the 'cybernetically planned cities', whereas other people thought the smart city guessed in suggestions for interacted cities since the 1980s (Bibri et al., 2017). According to Bibri and Krogstie, the term 'SMART CITY' was initially invented in the mid-1800s to define the novel cities of the American West that were well-organized as well as self-governed (Bibri et al., 2017). On the contrary, urbanization is a characteristic of contemporary civilization, and the method of this procedure is formed by globalization and technological advancement (Sikora-Fernandez, D 2013). For years, rural areas have been a point of consideration for scientific as well as political deliberations. The progress and organization of rural regions pose complex challenges, assigned to multiple factors. One crucial aspect is their examination across various scientific matters, leading to incorporative research attempts (Bruckmeier, K. 2011). Also, the definitions of "rurality" and the moralities and execution of rural arrangement vary significantly between nations and even within more localized geographical areas (Gkartzios et al., 2022a). Inspecting migration patterns is important for understanding transfers in people's actions within a country. It reflects the most energetic aspect of populace change and is highly amenable to economic, political, and cultural.

The objective of the research is:

1. Which are the most productive years?
2. Which are the most Productive and Cited keywords?
3. Which are the most co-occurred keywords?

## 2. Review of Literature

The tag "smart city" was an uncertain idea and utilised in conducts that were not constantly steady (Albino et al.,2015). In the 1950s and 1960s, ecological behaviour proposal researchers incorporated the idea of liveability in order to discover in what way individuals observe and utilise cities and then established design rules and references in order to incorporate individual's opinions (Ahmed et al., 2019; Kaal, 2011). Meanwhile then the term has had continuing petition for social scientists, human geographers, development researchers as well as urban organizers. In the 1990s, liveability developed as a widespread perception amongst municipal planners in cities in High Income Countries (HICs) (Pacione, 1990). Humanoid settlements as well as urban progress have reflective impacts on the natural environments (Grimm et al., 2008). Urban growth outcomes in loss, deprivation as well as shattering of natural surroundings, surge in resistant surfaces and in conservational impacts related with the heat-island effect, water, air, noise as well as light pollution, and the origin of non-native species (Grimm et al., 2008). Eventually, urbanisation resulted in the formation of new ecosystems that display different abiotic as well as biotic features and pose a problem to several creatures, together with insects (McKinney et al., 2008). With growing urbanization, population growth, as well as growing standard of living, sustainability—concerning utilisation of energy, buildings, as well as the environment—is becoming more extensively debated (Mazraeh et al., 2018). In an age of incredible variations in the scale of civilization as well as technology and sustainability can support in re-establishing significant local resources and reserve traditional characteristics (Hărmănescu et al., 2016). Additionally, the primary determination of sustainability is to bind the ecological ability to deliver ease to inhabitants (Motealleh et al., 2018). Though, environmental problems are not exact to single countries, and discovering sustainable remedies for current buildings remains stimulating in the terms of energy effectiveness (Michael et al., 2017). The preceding


https://doi.org/10.52202/081568-0010

literature discloses that researchers decide that vernacular construction, as a creation of adjustment to the local circumstances, attaches the historical and is evidence of identity related to culture. Through the usage of vernacular substances as well as resources that are available, sustainability for the environment can be attained (Philokyprou, et al., 2021). Cities are multifaceted organizations as well as a group of essentials such as interrelated actors (citizens as well as businesses), utilities, transportation and networks, too (Mora et al. 2017). Cities play a vigorous character in economic, social, as well as environmental advancement. An ever-swelling populace as well as urban growth describe today's biosphere. The urban population of the World is anticipated to be 68% by 2050 (UN Department of Economic and Social Affairs 2018). This, in line, resulted in new difficulties, together with pollution as well as waste management, parking lots, traffic, availability of resources, social as well as physical infrastructure, and access to education, healthcare, justice, too (Camero and Alba 2019). The smart city thought had been attaining attention (Zhou et al. 2020) meanwhile its arrival many decades before. While some Researchers have annoyed to response what makes up a smart town (Dameri, 2017), later more than 20 years, there is no commonly acknowledged meaning for the perception (Ruhlandt, 2018). The reason behind this is the overlay of the smart city thought with its relatives, together with digital town, sustainable city, as well as intelligent city (Nam and Pardo,2011) and the complex nature of the notion. We describe a smart town in general terms to retain our complex viewpoint as well as evade a slight opinion. A smart urban is a progressive contemporary environment of urban region joining ICT as well as other technologies to attain economic, social, together with environmental targets. According to several Researchers (Ahad et al. 2020), a smart city is a multifaceted system that contains of keen on-demand transportation, smart energy, smart grids, smart healthcare, smart environment, smart education, smart industry, smart roads, smart houses, smart flats, and many more. Preceding studies have discovered progress of China in association to removing poverty. By Following the cultural Revolution, China applied a sequence of 'reform and opening up' guidelines in 1978 with the purpose of improving economic growth (Lu et al. 2019). The swiftly growing population as well as growing industrialization and urbanization caused in rapid economic development and the elimination of poverty of hundreds of millions of people. Though, this was escorted by environmental deprivation and a series of natural tragedies (Lu et al. 2019). Also, during China's urbanization as well as economic development, the provision of public facilities in the comparatively underdeveloped western areas has been moderately insufficient as compared to the more developed eastern Zones (Zhao et al. 2022b). Regional inequalities in the expansion of public amenities delay the sustainable development of the overall country as well as aggravate the unfairness among poverty and non-poverty regions. In a current meta-analysis review of urbanization studies, Liu and Niyogi (2019) achieve that there is a continuing requirement for experimental valuations associated to the influence of urban lands on dissimilar weather constraints. Utilising the mixture of modelling as well as observational datasets, researches have resolved that the variations in rainfall over a city are connected with a surge in development of the urban (Zhang et al., 2019). The urban-rural collection reinforces the mesoscale convection. Urbanization may overpower rainfall over the town as well as generate favourable circumstances for descendant development. With the growth in developed zones, urbanization usually changes the local atmospheric motions, which outcomes in more powerful rainfall (Kumar et al. 2013). Researches like (Yang et al., 2017) indicates that short thrilling rainfall actions are powerfully related with urbanization. Study (Niyogi et al., 2020) also recommend a probable postponement in late afternoon rainfall because of urbanization. Investigation of spatio-temporal dissimilarity of rainfall as well as urbanization discloses that urban regions with the maximum population zones are conquered by the somewhat extended rainfall period and higher rainfall concentration (Zhu et al., 2019). There is also an important association between the spatial distribution of substantial rainfall amount as well as a growth in urbanization over India, the United States, and China (Zhi-Hong et al., 2016). Employee flexibility is a vigorous characteristic in today's active as well as frequently thought-provoking work situations.

## 3. Research Methodology

A proper literature review was directed for the purpose of this study and PRISMA framework was utilised for the purpose of analysis.

The Scopus database was selected for the purpose of this bibliometric study, because it is examined as the most substantial broad content and citation database (Oliveira et.al ,2018) consisting publications from scientific publishers around the world, it is now utilised all over world for the purpose of many large-scale analyses (Baas et. al ,2020).

SCOPUS was nominated amongst various academic databases because of its comprehensive handling of quality peer-reviewed literature associated to the topic. We can purify the data on the basis of Subject Area, Document type, Publication Stage, Key-words, Source type, Language through SCOPUS.

**PRISMA FRAMEWORK:**

Table 1. Filtering Criteria

| FILTERING CRITERIA | ACCEPT |
|---|---|
| **Searching Criteria** | |
| Search Engine: SCOPUS | |
| Search Date: 20 December 2024 | |
| Search Term: "SMART CITY" AND "POPULATION" | 3118 |
| Subject Area: "SOCIAL SCIENCE", "BUSINESS MANAGEMENT AND ACCOUNTANCY", "ENVIRONMENTAL SCIENCE" | 1345 |
| Document Type: "ARTICLE" And "REVIEW" | 662 |
| Article Selection | |
| Language screening: ENGLISH | 511 |
| Year (UPTO 20 December 2024) | 511 |


https://doi.org/10.52202/081568-0010

| | |
|---|---|
| **CONTENT SCREENING:** includes articles if "TITLES, ABSTRACT and KEYWORDS" indicate relevance to scope of study only | 511 |

*(Source: AuthorS' Compliance)*

This study used VOS viewer to construct the cluster analysis and recognize the present state of study in this field. The overall procedure to create such a network is as follows:

- Introduce VOS viewer and import Scopus csv files;
- Nominate keywords Co-Occurrence from the files;
- apply the full counting method to produce the co-occurrence network; and select the minimum number of occurrences. Succeeding the keyword co-occurrence network analysis, we conducted a co-citation analysis to identify the key subfields in smart city and migration research and then applied the research potential evaluation model to examine each subfield's prospective growth and future directions.

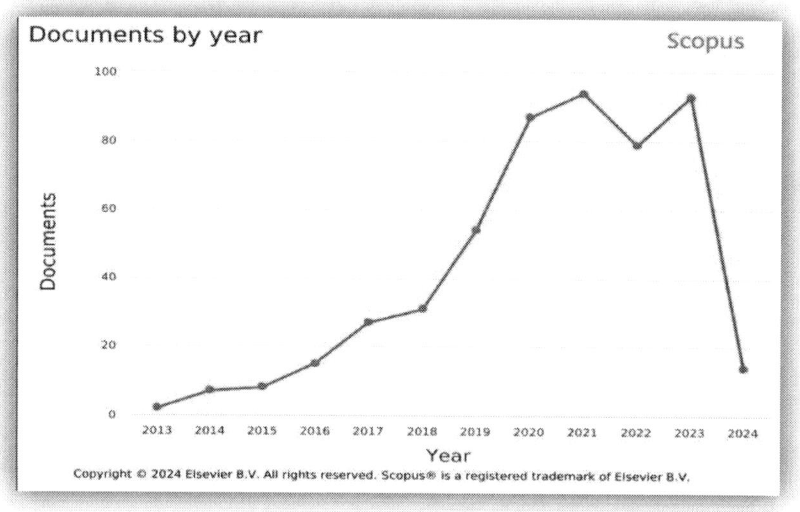


Figure 1: Publication of Documents in Different Years (Source: Scopus)

## 4. Bibliometric Data Analysis

A comprehensive analysis of the selected articles was undertaken in a systematic manner. Various Criteria were chosen for the purpose of analysis like:

- *"Yearly Production", "Productive and Cited Keywords","Co-occurrence of Keywords" and an overall analysis, and so on.*

Different software was utilised for the analysis purpose like VOS viewer (v.1.6.18), RStudio, MS excel and Scopus(database).

### 4.1 Yearly Production

The figure (1) and Table (2) show the publication of documents in different years from 2013 to 2024. There were several articles published in various years but the number of articles varies from year to year as in some years more articles were produced whereas, in some years less articles were published. The figure (1) is the graphical representation of the articles' publication year-wise. It is observed that in the initial year which is 2013, least number of articles were there which were only 2 documents. The number of articles is increasing with an increasing year i.e. 2014, 7 articles, 2015- 8 documents and so on. It is observed that the highest number of articles were published in 2021 which numbered to 94 and second highest were in 2023 i.e. 93. But it is analysed that there was a sudden downfall in the production of articles in 2022 which were decreased from 94 articles to 79. And it is depicted in the figure that in 2024, the number of articles decline to 14 in number which is an expected downfall.

Table 2: Yearly Publication of Documents (Source: Authors Compliance)

| *YEAR* | *NO. OF ARTICLES* |
|---|---|
| **2024** | **14** |
| **2023** | **93** |
| **2022** | **79** |
| **2021** | **94** |
| **2020** | **87** |
| **2019** | **54** |
| **2018** | **31** |
| **2017** | **27** |
| **2016** | **15** |
| **2015** | **8** |
| **2014** | **7** |
| **2013** | **2** |

Table (2) is the analysis of documents publication in different years starting from 2013 to 2024. It depicts that highest number of documents are in 2021 which are 94 articles and least are published in 2013 which is 2 documents and there are several many years also with different number of articles produced.


https://doi.org/10.52202/081568-0010

### 4.2 Subject Area

Figure (2) and Table (3) shows the documents published in different subject areas with the help of a pie chart. There are many subjects which were chosen by different authors as per their compliance. Different subject areas are- social sciences, environmental sciences, engineering, energy, computer science, earth and planetary sciences, decision sciences, psychology, medicine, business, management and accounting,*and many more.*

It is analysed from the pictorial representation that the highest number of authors selected ***Social Sciences*** as their preferred subject area and total documents published are 348 (27.0%), followed by ***Environmental Sciences*** with 241 (18.7%) articles and 158

(12.3%) documents from ***Engineering.*** On the other hand, the least number of documents are from ***Health Professions*** which is only 1 document, and second lowest documents are from ***Physics and Astronomy***, ***Pharmacology, Toxicology and Pharmaceutics*** with 2 documents each. There are many other Subjects like ***Multidisciplinary*** (8), ***Computer Science*** (118) and so on.

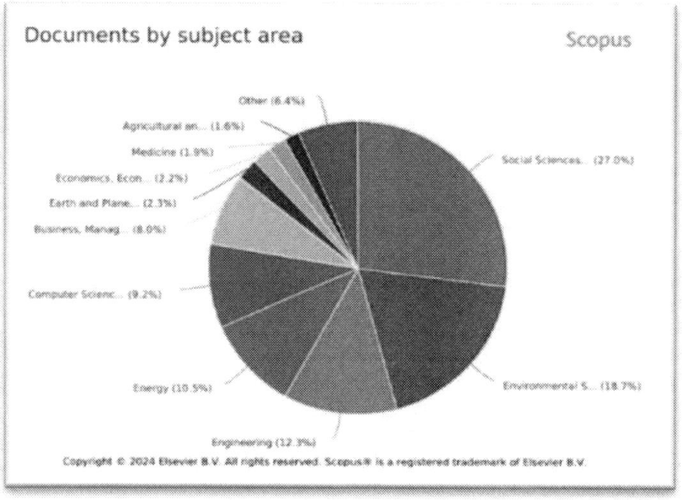


*Figure2: Documents by Different Subject Area (Source: Scopus)*

Table 3: Publication of Documents in Different Subject Area (Source: Authors Compliance)

| Subject Area | No. of Documents |
|---|---|
| SOCIAL SCIENCE | 348 |

| | |
|---|---|
| ENVIRONMENTAL SCIENCE | 241 |
| ENGINEERING | 158 |
| ENERGY | 135 |
| COMPUTER SCIENCE | 118 |
| BUSINESS, MANAGEMENT AND ACCOUNTING | 103 |
| EARTH AND PLANETARY SCIENCES | 30 |
| ECONOMICS, ECONOMETRICS AND FINANCE | 28 |
| MEDICINE | 25 |
| AGRICULTURAL AND BIOLOGICAL SCIENCES | 20 |
| DECISION SCIENCES | 20 |
| PSYCHOLOGY | 12 |
| ARTS AND HUMANITIES | 8 |
| BIOCHEMISTRY, GENETICS AND MOLECULAR BIOLOGY | 8 |
| CHEMICAL ENGINEERING | 8 |
| MATHEMATICS | 8 |
| MULTIDISCIPLINARY | 8 |
| MATERIALS SCIENCE | 4 |
| NURSING | 2 |
| PHARMACOLOGY, TOXICOLOGY AND PHARMACEUTICS | 2 |
| PHYSICS AND ASTRONOMY | 2 |


https://doi.org/10.52202/081568-0010

| | |
|---|---|
| HEALTH PROFESSIONS | 1 |

Table (3) represents the number of documents in different subject area and the table arranges the documents in a series from highest number of documents to lowest number of documents on the basis of preference of the subject area.

**4.3 Keywords**

It is important to perform an analysis on the basis of the keywords and for this purpose VOS viewer is utilised and a cluster analysis is made which represents the network between different keywords and this represented in the Figure (4). The figure is depicting that there are several keywords which are closely connected with each other. There is total 7 colours (cluster) that are used in the Network analysis of the keywords. Various Keywords are Urban infrastructure, Efficiency, Bibliometric analysis, machine learning, conceptual framework, smart city, Sustainable development, planning, water management and other are also there.

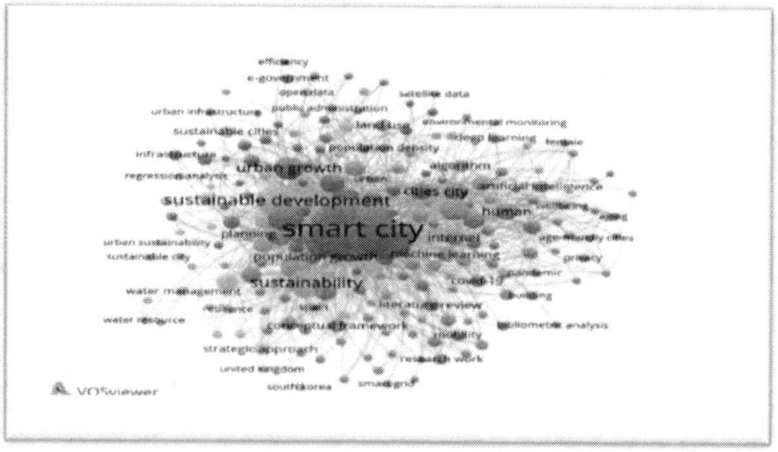


Figure 3: Cluster Analysis of the Keywords (Source: VOS viewer)

From Table (4), it is understood that there is total 7 clusters in different colours (red, green, blue, yellow, purple, sky blue and orange). The detailed analysis of the clusters is as follows:

Table 4: Different Clusters with Sub Keywords (Source: Authors Compliance)

| COLOUR | CLUSTER | SUB KEYWORDS |
|---|---|---|
| **RED** | URBAN GROWTH | ENERGY, EFFICIENCY, ECONOMIC GROWTH, ECONOMIC ANALYSIS |
| **GREEN** | SMART CITY | POLICY MAKING, PUBLIC TRANSPORT, ECOSYSTEM |
| **BLUE** | HUMAN | DEMOGRAPHY, EDUCATION, |

| | | |
|---|---|---|
| | | BIBLIOMETRIC ANALYSIS |
| **YELLOW** | SUSTAINABLE DEVELOPMENT | PLANNING, STRATEGIC APPROACH, INFORMATION AND COMMUNICATION |
| **PURPLE** | POPULATION GROWTH | POPULATION DENSITY, LAND USE, BIG DATA |
| **SKY BLUE** | INTERNET | WATER QUALITY, AIR QUALITY |
| **ORANGE** | URBAN POPULATION | DEMAND, GOVERNANCE APPROACH |

1. **CLUSTER (RED):** ***[46 items]*** The cluster 1 is in red colour and the main keyword is URBAN GROWTH with several sub keywords making a network and some of the keywords are ENERGY, EFFICIENCY, ECONOMIC GROWTH, ECONOMIC ANALYSIS and many more. These are basically the factors which are contributing towards the urban growth.

2. **CLUSTER (GREEN):** ***[44 items]*** The cluster 2 is represented in green colour with the main keyword SMART CITY and sub keywords or factors are POLICY MAKING, PUBLIC TANSPORT, ECOSYSTEM and so on, which play an essential role in the creation of a smart city.

3. **CLUSTER (BLUE):** ***[42 items]*** Green colour is used to represent the network made with the main keyword HUMAN and various sub keywords of this cluster are DEMOGRAPHY, EDUCATION and these are determinants that identifies the human in a society.

4. **CLUSTER (YELLOW):** ***[27 items]*** Yellow coloured cluster represents the network generated with SUSTAINABLE DEVELOPMENT and various sub keywords or determinants that support Sustainable Development are PLANNING, STRATEGIC APPROACH, INFORMATION AND COMMUNICATION and many other also.

5. **CLUSTER (PURPLE):** ***[26 items]*** The purple cluster shows the network of the POPULATION GROWTH with other sub keywords and showing connections and inter relationships between several factors.

6. **CLUSTER (SKY BLUE):** ***[17 items]*** The cluster 6 is of Sky blue colour which is making connections with the keyword INTERNET and with several other keywords like WATER QUALITY, AIR QUALITY and many more.

7. **CLUSTER (ORANGE):** ***[8 items]*** Cluster 7 is in orange and it has connections with main word which is URBAN POPULATION and several other sub keywords like DEMAND, GOVERNANCE APPROACH, and these factors assist in the


https://doi.org/10.52202/081568-0010

management or analysis of urban population.

### 4.4 Country-Wise Production

Figure (5) represents the country wise production of the documents in the form of line graph in different years from 2013 to 2024. It is shown that there are main five countries in which the production of documents considered and countries are: *CHINA, INDIA, SPAIN, UNITED KINGDOM, USA*

It is understood that the production of the documents with year is increasing. INDIA amongst the all five countries is producing the highest number of documents but in initial years like 2013, INDIA was not even producing a single article. Orange coloured line represents the production of documents by CHINA. China is producing second highest number of articles which is more than 150 documents in 2024. USA and SPAIN are contributing almost same number of articles in the recent years i.e. 2024 which is nearby 100 documents.

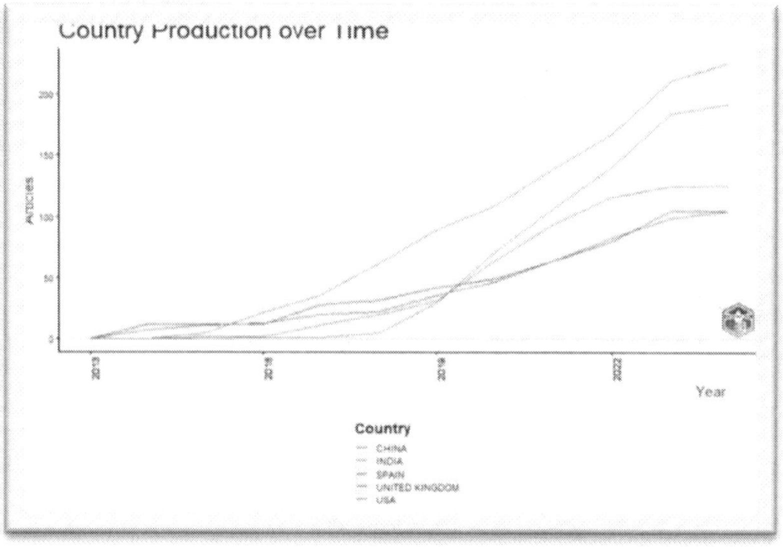


*Figure 4: Production of Documents Country Wise (Source: RStudio)*

### 4.5 Country-Wise Citation

Figure (4) represents the country wise citation of documents. Figure clears it with the help of a graphical representation. It shows that the highest number of citations is given to ITALY (1908 citations), followed by KOREA, with 1239 citations. It is observed that JAPAN is having least number of citations. And second lowest is PORTUGAL. Each country is having ample number of citations but the most cited ones is ITALY and least is JAPAN.

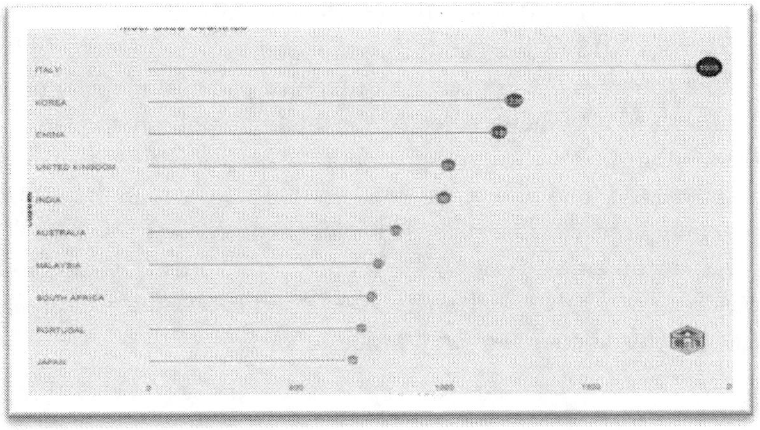


*Figure 5: Country wise Citation (Source: RStudio)*

## 5. Results and Discussion

The various results that are derived from the data analysis and the results are as follows:

- Which are the most productive years?
- Which are the most Productive and Cited keywords?
- Which are the most co-occurred keywords?

These were the three RESEARCH QUESTIONS which were to be analysed through this study. Hence, Results are discussed as under:

- ***MOST PRODUCTIVE YEARS:*** The most productive years as per the analysis are 2021, 2022 and 2023. In these three years, the highest number of documents were published with highest articles in 2021 with 94 documents, followed by 2023 with 93 documents. And third rank is attained by 2022 with 79 documents.

- ***MOST PRODUCTIVE AND CITED KEYWORDS:*** The most productive and cited keywords is SMART CITY, SUSTAINABLE DEVELOPMENT and URBAN GROWTH.

- ***MOST CO-*OCCURRED KEYWORDS:** If the most co-occurred keyword is analyzed, then, there are many keywords which have occurred highest number of times is URBAN GROWTH, SUSTAINABLE DEVELOPMENT.

It has been resulted from the analysis that there are several publications on these keywords as these are an emerging topics or issues. Therefore, it is expected that more publications will be there in this field of study. As total 511 documents were selected for the bibliometric study and it is observed that many publications are undertaken till now and more emerging and growing keywords are being noticed in this study. Hence, the highest number of publications were in 2021 and on the other side, the most cited and productive keywords were Smart city, Sustainable development.

## 6. Overview

Figure (6) provides a detailed overview of the overall searching criteria and several


https://doi.org/10.52202/081568-0010

other bases. The TIMESPAN for the searching criteria is 2013 to 2024 i.e. documents are selected from 2013 to 2024 and from the total sources, documents are 247 in which there are many sources like Journals, Conference papers and many more. Furthermore, the total number of documents selected for the final bibliometric study were 511 which are finalized after proper filtration procedure. There are 1687 Authors in the all whose study is analysed. If total references is taken in consideration then, from the figure it is observed that references are 29178. The average age of document is 3.54. The Average citations per document is 31.35. The Authors' keywords (DE) are 1796. On the basis of various criteria, the overall analysis is conducted and also it provided a lot of significant insights about the overall process.

*Figure 6: Overview (Source: RStudio)*

## 7. Future Scope

The constant and growing urbanisation emerged in a situation where a large number of global Population reside in urban regions (Ringerson et al., 2017). As nowadays more and more people are migrating from rural areas to urban areas in search of several better opportunities and facilities like healthcare, education, and many more. Basically, due to this movement of people from one place to another has led to the increase in the population of the urban areas which is directly and indirectly impacting the resources of urban areas as well as rural regions. In case of urban regions, there is shortage of resources due to an increment in the population because previously, in urban regions only limited population was residing, as the concept of smart cities and urbanisation is emerging, the population shift is increasing and therefore, it is resulting in the exploitation or shortage of resources as, large number of people are using those resources. Not only resources, but same is with the facilities and environment. Before Urbanisation, there were limited industries and environment was clean and not polluted, but with time, the condition is becoming worst which is not good for health as well as for the Environment. Therefore, rise in population in the cities has resulted in several challenges which must be taken care. Another adverse impact of migration to urban is on the aviation sector. As the resources of the urban regions are used in excess which is resulting in the over exploitation of the resources of the aviation sector. For instance, if there is one airport in the urban region and the population is increasing then, the ultimate impact will be overburden on the airports of the urban area as most of the population will travel through that particular airport, as to sustain such population the expansion of the airports will take place due to which several trees and natural habitat will get destroyed which will result in the imbalance in the environment. Hence, migration of people from rural to urban areas have shown several more adverse effects in different sectors. And also in urban regions more attention is paid to the career development and talent management.

***Following is a MODEL which is clearly depicting the "Environmental problems" and "Environmental safety" in the URBAN REGIONS.***

It is analysed that in urban areas, there are several challenges which are emerging due to constant increase in the population due migration of individuals from Rural areas to urban regions. Some problems are- lack of clean drinking water, problem of hygiene, management of wastage, air pollution. So, this model basically shows that in urban zones, Environmental Cleanliness must be initiated in order to make the environment a safer place to reside.

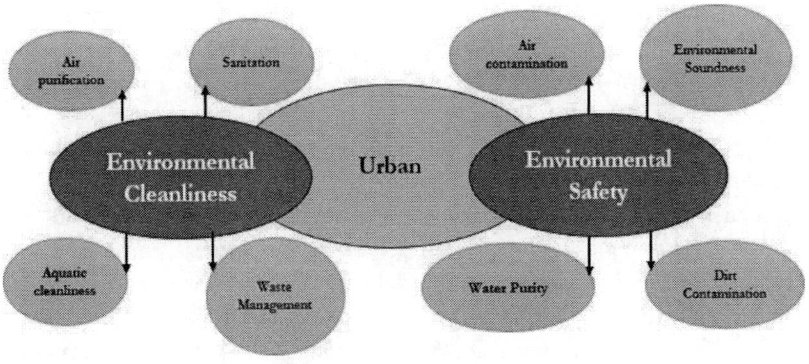


*Figure 7: Model Showing Environmental Safety and Environmental Cleanliness*

As due to Population growth in urban areas, there is uneven distribution of resources, which is resulting in the IMBALANCE in the society and also in a whole economy. This is because the resources with the urban areas are more and less with the rural regions and it is creating a situation of inequality and a sense of deprivation amongst the people living in the countryside. As this problem is growing and one day it may become a scenario which will be difficult to handle. So, it is essential to distribute the resources in an equitable manner so that the balance can be created.

## References


1. Oliveira, A. S., de Barros, M. D., de Carvalho Pereira, F., Gomes, C. F. S., & Da Costa, H. G. (2018). Prospective scenarios: A literature review on the Scopus database. *Futures*, *100*, 20-33.
2. Baas, J., Schotten, M., Plume, A., Côté, G., & Karimi, R. (2020). Scopus as a curated,



https://doi.org/10.52202/081568-0010

high-quality bibliometric data source for academic research in quantitative science studies. *Quantitative science studies*, *1*(1), 377-386.

3. Albino, V., Berardi, U., & Dangelico, R. M. (2015). Smart cities: Definitions, dimensions, performance, and initiatives. *Journal of urban technology*, *22*(1), 3-21.
4. Ahmed, N. O., El-Halafawy, A. M., & Amin, A. M. (2019). A critical review of urban livability. *European Journal of Sustainable Development*, *8*(1), 165-165.
5. Kaal, H. (2011). A conceptual history of livability: Dutch scientists, politicians, policy makers and citizens and the quest for a livable city. *City*, *15*(5), 532-547.
6. Pacione, M. (1990). Urban liveability: a review. *Urban geography*, *11*(1), 1-30.
7. Silva, B. N., Khan, M., & Han, K. (2018). Towards sustainable smart cities: A review of trends, architectures, components, and open challenges in smart cities. *Sustainable cities and society*, *38*, 697-713.
8. Jia, X., Dai, T., & Guo, X. (2014). Comprehensive exploration of urban health by bibliometric analysis: 35 years and 11,299 articles. *Scientometrics*, *99*, 881-894.
9. Mohanty, S. P., Choppali, U., & Kougianos, E. (2016). Everything you wanted to know about smart cities: The Internet of things is the backbone. *IEEE consumer electronics magazine*, *5*(3), 60-70.
10. Yigitcanlar, T., Kamruzzaman, M., Foth, M., Sabatini-Marques, J., Da Costa, E., & Ioppolo, G. (2019). Can cities become smart without being sustainable? A systematic review of the literature. *Sustainable cities and society*, *45*, 348-365.
11. Bibri, S. E., & Krogstie, J. (2017). Smart sustainable cities of the future: An extensive interdisciplinary literature review. *Sustainable cities and society*, *31*, 183-212.
12. Sikora-Fernandez, D. (2013). Koncepcja „smart city" w założeniach polityki rozwoju miasta–polska perspektywa.
13. Grimm, N. B., Faeth, S. H., Golubiewski, N. E., Redman, C. L., Wu, J., Bai, X., & Briggs, J. M. (2008). Global change and the ecology of cities. *science*, *319*(5864), 756-760.
14. McKinney, M. L. (2008). Effects of urbanization on species richness: a review of plants and animals. *Urban ecosystems*, *11*, 161-176.
15. Hărmănescu, M., & Enache, C. (2016). Vernacular and Technology. InBetween. *Procedia Environmental Sciences*, *32*, 412-419.
16. Motealleh, P., Zolfaghari, M., & Parsaee, M. (2018). Investigating climate responsive solutions in vernacular architecture of Bushehr city. *HBRC journal*, *14*(2), 215-223.
17. Michael, A., Demosthenous, D., & Philokyprou, M. (2017). Natural ventilation for cooling in mediterranean climate: A case study in vernacular architecture of Cyprus. *Energy and Buildings*, *144*, 333-345.
18. Philokyprou, M., & Michael, A. (2021). Environmental sustainability in the conservation of vernacular architecture. The case of rural and urban traditional settlements in Cyprus. *International Journal of Architectural Heritage*, *15*(11), 1741-1763.
19. Mora, L., Bolici, R., & Deakin, M. (2017). The first two decades of smart-city research: A bibliometric analysis. *Journal of Urban Technology*, *24*(1), 3-27.
20. Nam, T., & Pardo, T. A. (2011, June). Conceptualizing smart city with dimensions of technology, people, and institutions. In *Proceedings of the 12th annual international digital government research conference: digital government innovation in challenging times* (pp. 282-291).
21. Zhou, S., Zhang, X., Liu, J., Zhang, K., & Zhao, Y. (2020). Exploring development of smart city research through perspectives of governance and information systems: a scientometric analysis using CiteSpace. *Journal of Science and Technology Policy Management*, *11*(4), 431-454..
22. Dameri, R. P. (2017). Smart city implementation. *Progress in IS; Springer: Genoa, Italy*.
23. Ruhlandt, R. W. S. (2018). The governance of smart cities: A systematic literature review. *Cities*, *81*, 1-23.

24. Ahad, M. A., Paiva, S., Tripathi, G., & Feroz, N. (2020). Enabling technologies and sustainable smart cities. *Sustainable cities and society*, *61*, 102301.
25. Ahmed, S., Shah, M. A., & Wakil, K. (2020). Blockchain as a trust builder in the smart city domain: a systematic literature review. *IEEE Access*, *8*, 92977-92985.
26. Zhao, Z., Pan, Y., Zhu, J., Wu, J., & Zhu, R. (2022). The impact of urbanization on the delivery of public service–related SDGs in China. *Sustainable Cities and Society*, *80*, 103776.
27. Liu, J., & Niyogi, D. (2019). Meta-analysis of urbanization impact on rainfall modification. *Scientific reports*, *9*(1), 7301.
28. Chen, T. C., Wang, S. Y., & Yen, M. C. (2007). Enhancement of afternoon thunderstorm activity by urbanization in a valley: Taipei. *Journal of Applied Meteorology and Climatology*, *46*(9), 1324-1340.
29. Yang, P., Ren, G., & Yan, P. (2017). Evidence for a strong association of short-duration intense rainfall with urbanization in the Beijing urban area. *Journal of Climate*, *30*(15), 5851-5870.
30. Niyogi, D., Osuri, K. K., Busireddy, N. K. R., & Nadimpalli, R. (2020). Timing of rainfall occurrence altered by urban sprawl. *Urban Climate*, *33*, 100643.
31. Zhu, X., Zhang, Q., Sun, P., Singh, V. P., Shi, P., & Song, C. (2019). Impact of urbanization on hourly precipitation in Beijing, China: Spatiotemporal patterns and causes. *Global and Planetary Change*, *172*, 307-324.
32. Zhi-Hong, J., Yang, L., & Dan-Lian, H. (2016). Impact of urbanization in different regions of eastern China on precipitation and its uncertainty. *Journal of Tropical meteorology*, *22*(3), 382.
33. Bruckmeier, K. (2011). Towards interdisciplinary rural research–theorizing nature-society relations. *Natures Sciences Sociétés*, *19*(1), 3-13.
34. Gkartzios, M., Gallent, N., & Scott, M. (2022). A capitals framework for rural areas:'Place-planning'the global countryside. *Habitat International*, *127*, 102625.


https://doi.org/10.52202/081568-0010

# Exploring the Role of Digital Media in Advancing Cultural Tourism: Opportunities, Ethical Challenges, and Sustainable Development Strategies

Shagufta Pandit[1]* and Dr. Kirti Sharma[2]

[1&2]Department of Tourism and Hospitality Management, University School of Business, Chandigarh University, Mohali2

*Corresponding author: panditshagufta33@gmail.com

**Abstract.** This Study seeks to understand how digital media in the promotion of cultural tourism change the nature of heritage, access, and world participation. It discusses prospects presented by social media, virtual reality, and digital storytelling, while citing ethical dilemmas of fraudulent representation, commodification, and loss of authenticity. Different methods for data collection from industry practitioners, cultural experts, and tourists are applied in order to propose sustainable solutions that create a balance between technological advancements and cultural conservation. Policymakers, stakeholders, and researchers will use this study as a reference while developing digital content supporting the growth of sustainable cultural tourism.

Keywords: Digital Media, Cultural Tourism, Sustainable Development, Ethical Challenges, Tourism Promotion

## 1 Introduction

The rapid evolution of digital media has profoundly transformed various sectors, including tourism, by redefining how individuals interact with cultural heritage and travel experiences. Cultural tourism, which revolves around the exploration of cultural attractions, traditions, and historical sites, has particularly benefited from the integration of digital technologies. Digital media, encompassing social media platforms, virtual reality (VR), augmented reality (AR), mobile applications, and interactive websites, has emerged as a powerful tool for promoting cultural tourism, enhancing visitor engagement, and preserving cultural narratives. This digital transformation has not only expanded the reach of cultural tourism but also created new opportunities for destinations to showcase their unique heritage to a global audience. However, this shift is accompanied by significant ethical challenges and sustainability concerns that demand careful consideration. This research paper delves into the multifaceted role of digital media in advancing cultural tourism, exploring its opportunities, ethical dilemmas, and strategies for sustainable development.

The integration of digital media into cultural tourism has opened up a world of possibilities. Social media platforms such as Instagram, Facebook, and TikTok have become indispensable tools for promoting cultural destinations, allowing users to share

https://doi.org/10.52202/081568-0011

their experiences and inspire others to visit. Virtual and augmented reality technologies have revolutionized the way cultural heritage is experienced, enabling individuals to explore historical sites and artifacts from the comfort of their homes. For instance, VR tours of the Louvre Museum or the ancient city of Petra have made cultural tourism more accessible to people who may not have the means or ability to travel. Additionally, digital storytelling through interactive websites and mobile applications has enriched the visitor experience by providing immersive and educational content. These advancements have not only boosted tourism revenues but also fostered a deeper appreciation for cultural diversity and heritage.

Despite these opportunities, the use of digital media in cultural tourism raises several ethical concerns. One of the primary issues is the commodification of culture, where cultural heritage is reduced to a marketable product for tourist consumption. This can lead to the dilution or misrepresentation of cultural narratives, as destinations may prioritize profit over authenticity. For example, the portrayal of indigenous cultures in a stereotypical or superficial manner to attract tourists can perpetuate harmful stereotypes and undermine the dignity of local communities. Furthermore, the over-reliance on digital media risks eroding the authenticity of cultural experiences, as visitors may become more focused on capturing the perfect photo or video for social media rather than engaging meaningfully with the culture. The digital divide also poses a challenge, as not all communities have equal access to the resources needed to leverage digital media effectively, potentially exacerbating inequalities in the tourism sector.

Sustainability is another critical dimension that must be addressed in the context of digital media and cultural tourism. While digital technologies can reduce the environmental impact of tourism by minimizing physical travel through virtual experiences, they also contribute to electronic waste and energy consumption. The production and disposal of digital devices, as well as the energy required to power data centers and online platforms, have significant environmental implications. Moreover, the rapid digitization of cultural tourism must be balanced with the need to preserve the integrity of cultural heritage. Over-commercialization and the excessive use of digital tools can lead to the degradation of cultural sites and practices, threatening their long-term sustainability. Therefore, it is essential to develop strategies that promote the responsible use of digital media in cultural tourism, ensuring that technological advancements align with the principles of cultural preservation and environmental stewardship.

This research paper aims to provide a comprehensive exploration of the role of digital media in advancing cultural tourism, with a focus on three key areas: opportunities, ethical challenges, and sustainable development strategies. By examining case studies and analyzing current trends, the study seeks to highlight the transformative potential of digital media in enhancing cultural tourism experiences. At the same time, it will critically assess the ethical dilemmas associated with the digitalization of cultural heritage and propose strategies to address these challenges. The ultimate goal is to contribute to a more nuanced understanding of how digital media can be harnessed responsibly to promote cultural tourism while safeguarding its core values and ensuring its sustainability for future generations.

In the following sections, the paper will first explore the opportunities presented by digital media in cultural tourism, including its role in promoting destinations, enhancing visitor engagement, and preserving cultural narratives. It will then delve into the ethical


https://doi.org/10.52202/081568-0011

challenges, such as the commodification of culture, the risk of misrepresentation, and the digital divide. Finally, the paper will discuss sustainable development strategies, emphasizing the need for a balanced approach that integrates technological innovation with cultural preservation and environmental responsibility. Through this multifaceted analysis, the research aims to provide valuable insights for policymakers, tourism professionals, and cultural heritage managers, offering a roadmap for leveraging digital media to advance cultural tourism in an ethical and sustainable manner.

In conclusion, the intersection of digital media and cultural tourism represents a dynamic and evolving field with immense potential to transform how we experience and appreciate cultural heritage. However, this transformation must be guided by a commitment to ethical practices and sustainability to ensure that the benefits of digital media are realized without compromising the integrity of cultural tourism. By addressing the opportunities, challenges, and strategies outlined in this paper, stakeholders can work towards a future where digital media serves as a force for positive change in the cultural tourism sector.

## 2. Literature Review

The integration of digital media into cultural tourism has fundamentally transformed how cultural heritage is promoted, experienced, and preserved. Over the past two decades, digital technologies—ranging from social media platforms to virtual reality (VR) and augmented reality (AR)—have revolutionized the cultural tourism landscape. This review traces the evolution of this field chronologically, exploring the opportunities, ethical challenges, and sustainable development strategies associated with digital cultural tourism. By synthesizing insights from seminal and contemporary studies, this review provides a comprehensive understanding of how digital media has shaped cultural tourism and what the future holds for this dynamic field.

The early 2000s marked the beginning of digital media's influence on cultural tourism. During this period, the internet and early social media platforms began to reshape how destinations marketed themselves and how tourists shared their experiences. Seminal works by MacCannell (1973) and Urry (1990) laid the groundwork for understanding the relationship between tourism and cultural representation. MacCannell's concept of "staged authenticity" highlighted how cultural sites were often presented in a way that catered to tourist expectations, while Urry's "tourist gaze" emphasized the role of visual media in shaping tourist experiences.

The introduction of virtual reality (VR) and augmented reality (AR) technologies during this period also sparked interest in their potential applications for cultural tourism. Guttentag (2010) was among the first to examine how VR could enhance cultural tourism by providing immersive experiences of heritage sites, making them accessible to individuals who could not visit in person. These early developments set the stage for the rapid digital transformation of cultural tourism in the following decade.

The 2010s witnessed the rise of social media as a dominant force in cultural tourism. Platforms like Instagram, Facebook, and Twitter became essential tools for destination marketing, enabling tourists to share their experiences in real-time and influencing the travel decisions of their peers. Gretzel et al. (2012) and Munar and Jacobsen (2014) highlighted the role of user-generated content in shaping perceptions of cultural destinations. For example, Instagram's visual-centric platform allowed tourists to share

https://doi.org/10.52202/081568-0011

captivating images of cultural sites, inspiring others to visit.

During this period, concerns about the commodification of culture and the loss of authenticity began to emerge. Scholars like Shepherd (2002) and Higgins-Desbiolles (2010) warned that the digital promotion of cultural tourism could lead to the oversimplification or misrepresentation of cultural narratives. For instance, the portrayal of indigenous cultures in social media campaigns often perpetuated stereotypes, undermining the dignity of local communities.

The late 2010s saw the widespread adoption of immersive technologies like VR and AR in cultural tourism. Han et al. (2019) and tom Dieck et al. (2018) explored how these technologies could enhance visitor experiences by providing interactive and educational content. For example, VR tours of the Louvre Museum and AR apps like "Google Arts & Culture" allowed users to explore cultural heritage sites in unprecedented detail. These technologies not only enriched the visitor experience but also contributed to the preservation of cultural narratives by digitizing and archiving heritage sites.

However, this period also brought to light the ethical challenges associated with digital cultural tourism. Studies by Whitford and Ruhanen (2016) and Picard and Robinson (2019) highlighted the risks of exploiting indigenous cultures for commercial gain. Bandyopadhyay and Morais (2018) examined the privacy concerns arising from the use of drones and 360-degree cameras to capture cultural sites. The digital divide also became a pressing issue, as many communities lacked the resources to participate in the digital tourism economy (Scheyvens & Hughes, 2019).

The COVID-19 pandemic accelerated the adoption of digital media in cultural tourism, as travel restrictions forced destinations to rely on virtual experiences to engage tourists. Studies by Gretzel et al. (2020) and tom Dieck et al. (2021) documented the surge in virtual tours, online exhibitions, and digital storytelling initiatives during this period. For example, the British Museum and the Vatican Museums launched virtual tours that attracted millions of visitors worldwide.

The pandemic also underscored the importance of sustainability in digital tourism. Research by Gössling et al. (2020) and Dwyer (2021) highlighted the environmental benefits of virtual tourism, such as reduced carbon emissions from travel. However, concerns about electronic waste and energy consumption persisted, prompting calls for eco-friendly digital practices (Dwyer et al., 2021).

The years 2023 and 2024 have seen significant advancements in the use of digital media for cultural tourism, as well as increased attention to ethical and sustainability issues. Recent studies by Ketter (2023) and Mariani et al. (2023) have explored the growing influence of TikTok and other short-form video platforms on cultural tourism, particularly among younger generations. These platforms have enabled destinations to reach global audiences through viral content, but they have also raised concerns about the superficial portrayal of cultural heritage.

Ethical challenges remain a central focus of recent research. Higgins-Desbiolles et al. (2023) and Duxbury et al. (2023) have called for greater community involvement in the digital promotion of cultural tourism, emphasizing the need for participatory approaches that respect local voices and traditions. Meanwhile, Bandyopadhyay and


https://doi.org/10.52202/081568-0011

Morais (2023) have proposed ethical guidelines for the use of digital media in cultural tourism, addressing issues such as privacy, consent, and cultural representation.

Sustainability has also emerged as a key theme in recent studies. Dwyer et al. (2023) have examined the potential of blockchain technology to enhance transparency and sustainability in digital tourism, while UNESCO (2024) has launched initiatives to promote the digital preservation of cultural heritage in an environmentally responsible manner. These efforts reflect a growing recognition of the need to balance technological innovation with cultural preservation and environmental stewardship.

## 3. Gap of the study

Most studies emphasize digital media's role in general tourism marketing, with insufficient exploration of its impact on heritage conservation and cultural engagement.

The risks of cultural appropriation, heritage commercialization, and misinformation remain underexplored, raising concerns about the authenticity of digitally mediated cultural experiences.

While digital media enhances destination promotion, its potential mitigating over-tourism, cultural degradation, and environmental strain lacks in-depth analysis.

Limited research examines the role of AI, AR, VR, and block chain in fostering sustainable and ethical cultural tourism experiences.

Studies predominantly focus on Western contexts, neglecting the digital divide, accessibility, and cultural sensitivities in developing regions.

There is a lack of regulatory frameworks and policy-driven research addressing digital ethics, data privacy, and inclusive representation in cultural tourism.

## 4. Objectives of the Study

1. To analyze the role of digital media in promoting cultural tourism, focusing on its impact on engagement, accessibility, and heritage interpretation.

2. To examine the ethical challenges associated with digital media in cultural tourism, including issues of authenticity, cultural appropriation, and misinformation.

3. To evaluate the contribution of digital media to sustainable cultural tourism, particularly in heritage conservation, visitor management, and responsible tourism practices.

## 5. Research Methodology

This study adopts a mixed-methods research approach, incorporating both primary and secondary data sources to explore the role of digital media in advancing cultural tourism. The research methodology is designed to achieve the outlined objectives through systematic data collection, analysis, and interpretation.

### 5.1 Research Design

The study follows a descriptive and analytical research design. Primary data are gathered through surveys and interviews, while secondary data are collected from existing literature, reports, and digital resources. This combination allows for a comprehensive understanding of how digital media influences cultural tourism.

### 5.2 Data Collection

**Primary Data:**

Primary data are collected through:

- Surveys: Structured questionnaires distributed to tourists, tourism professionals, and cultural heritage site managers to gather quantitative and qualitative insights.
- Interviews: Semi-structured interviews with stakeholders such as tourism policymakers, cultural organization representatives, and digital media content creators.

**Secondary Data:**

Secondary data are gathered from credible and relevant sources, including:

- Academic journals and research papers
- Industry reports and publications
- Government and NGO reports on cultural tourism
- Digital media content analysis (e.g., social media campaigns, virtual tours)
- Case studies of successful digital media applications in cultural tourism
- Ethical guidelines and regulatory reports related to cultural heritage preservation

## 6. Data Analysis

A combination of quantitative and qualitative analysis methods is applied:

- Quantitative Analysis: Survey data are analyzed using statistical tools to identify patterns and correlations related to digital media's impact on cultural tourism.

- Qualitative Analysis: Content from interviews and secondary sources is analyzed thematically to explore perceptions, experiences, and ethical considerations.

- Comparative Analysis: Primary data findings are compared with insights from secondary sources to validate and enrich the results.

## 7. Analysis and Findings

Based on the data collected and analyzed, the following key findings are identified:

- Impact on Engagement and Accessibility: Digital media significantly enhances tourist engagement by providing immersive experiences through virtual tours, interactive content, and social media promotions. Accessibility to cultural heritage is increased for global audiences, particularly for those unable to visit sites physically.
- Ethical Challenges: Analysis highlights concerns related to the authenticity of cultural representations, risks of cultural appropriation, and the spread of misinformation. Stakeholders emphasize the need for transparent content creation and responsible digital storytelling.
- Sustainable Tourism Practices: Digital media supports heritage conservation through digital archives and virtual restoration projects. It also aids in visitor management by providing real-time information and promoting responsible


https://doi.org/10.52202/081568-0011

tourism behaviors.

- Stakeholder Perspectives: Policymakers and cultural organizations acknowledge the role of digital media in tourism promotion but stress the importance of regulatory frameworks to mitigate ethical concerns.

## 8. Scope of the study

This study investigates the role of digital media in promoting cultural tourism with a particular focus on engagement, accessibility, and heritage interpretation. Furthermore, it examines the ethical challenges posed by digital platforms in the context of cultural representation, misinformation, and cultural appropriation. The research also evaluates the contribution of digital media to sustainable tourism development, addressing aspects such as heritage conservation, visitor management, and responsible tourism practices.

The geographical scope of the primary data collection is limited to specific regions, providing insights from key stakeholders, including tourists, tourism professionals, and policymakers. Additionally, the study covers a range of digital media platforms, including social media channels, virtual tours, and multimedia content.

## 9. Limitations of the Study

1. The digital media environment is dynamic, with rapidly emerging platforms and technologies. This may limit the long-term applicability of findings.

2. The study captures a snapshot in time rather than tracking changes over a longer period, restricting insights into long-term trends and impacts.

3. Analyzing digital media's impact on cultural tourism may involve subjective interpretations influenced by cultural perspectives and biases.

4. While stakeholders from tourism and cultural heritage sectors are engaged, the absence of direct input from digital media developers may limit insights into technological innovations.

5. The study highlights ethical concerns and regulatory challenges, but in-depth legal analysis is beyond its scope.

## 10. Conclusion

This research provides a comprehensive examination of how digital media contributes to the advancement of cultural tourism. Through a combination of primary and secondary data, the study has highlighted the significant role digital platforms play in enhancing tourist engagement, improving accessibility, and preserving cultural heritage.

While digital media offers numerous opportunities for the tourism sector, the study also emphasizes the ethical challenges that arise, particularly in terms of cultural representation, misinformation, and cultural appropriation. Stakeholders, including policymakers, tourism professionals, and cultural organizations, must collaborate to establish regulatory frameworks that ensure the responsible use of digital media in cultural tourism.

Moreover, the research identifies the potential of digital media in promoting sustainable tourism practices by supporting heritage conservation, facilitating effective visitor management, and encouraging responsible tourist behavior. The findings underscore the importance of leveraging technology for educational and experiential purposes while safeguarding cultural authenticity.

Ultimately, this study contributes valuable insights for stakeholders aiming to harness the power of digital media to foster inclusive, sustainable, and ethically responsible cultural tourism experiences. Future research can further expand on these findings by conducting longitudinal studies, exploring technological advancements, and assessing their impact across diverse cultural contexts.

## 11. Suggestions

1. Stakeholders in cultural tourism should undergo training on digital media literacy to ensure accurate and respectful representation of cultural heritage.

2. Establishing clear ethical guidelines for the use of digital media in cultural tourism can help mitigate issues related to cultural appropriation and misinformation.

3. Stronger collaboration between cultural institutions, digital media companies, and government bodies can promote responsible tourism practices and heritage conservation.

4. Leveraging emerging technologies such as augmented reality (AR) and virtual reality (VR) can enhance virtual tourism experiences and broaden accessibility.

5. Regular assessments and impact evaluations should be conducted to ensure digital media practices align with sustainable tourism goals and cultural preservation efforts.

6. Local communities should be actively involved in the digital representation of their cultural heritage to ensure authenticity and promote inclusive tourism experiences.

7. Utilizing big data and analytics can offer valuable insights into tourist behavior, preferences, and the effectiveness of digital media campaigns, leading to informed decision-making.

## References


1. Bandyopadhyay, R., & Morais, D. B. (2018). Representative dissonance: India's self and western image. *Journal of Travel Research, 57*(6), 758–771.
2. Bandyopadhyay, R., & Morais, D. B. (2023). Ethical guidelines for digital cultural tourism. *Journal of Tourism Ethics and Technology, 4*(2), 115–130.
3. Dwyer, L. (2021). Tourism economics and policy analysis: Contributions and legacy of Peter Forsyth. *Tourism Economics, 27*(2), 215–231.
4. Dwyer, L., Edwards, D., & Mistilis, N. (2023). Blockchain technology and sustainable tourism. *Sustainability, 15*(3), 1224.
5. Gössling, S., Scott, D., & Hall, C. M. (2020). Tourism and water: Interactions and impacts. *Channel View Publications.*



https://doi.org/10.52202/081568-0011

6. Gretzel, U., Sigala, M., Xiang, Z., & Koo, C. (2015). Smart tourism: Foundations and developments. *Electronic Markets, 25*, 179–188
7. Gretzel, U., Koo, C., Sigala, M., & Xiang, Z. (2020). Virtual tourism during the COVID-19 pandemic: Risks and opportunities. *Information and Communication Technologies in Tourism 2020*, 611–623.
8. Guttentag, D. A. (2010). Virtual reality: Applications and implications for tourism. *Tourism Management, 31*(5), 637–651.
9. Han, D. I., tom Dieck, M. C., & Jung, T. (2019). Augmented reality in cultural tourism: Enhancing visitor experiences. In T. Jung & M. C. tom Dieck (Eds.), *Augmented reality and virtual reality* (pp. 129–139). Springer.
10. Higgins-Desbiolles, F. (2010). The elusiveness of sustainability in tourism: The culture-ideology of consumerism. *Tourism Recreation Research, 35*(3), 231–243.
11. Higgins-Desbiolles, F., Carnicelli, S., & Wijesinghe, G. (2023). Ethical challenges in digital cultural tourism. *Tourism Geographies, 25*(4–5), 637–655.
12. Higgins-Desbiolles, F., Wijesinghe, G., Carnicelli, S., & Boluk, K. (2024). Interdisciplinary approaches to sustainable digital tourism. *Journal of Sustainable Tourism*.
13. Ketter, E. (2023). TikTok and the rise of digital nomads in cultural tourism. *Journal of Tourism Futures, 9*(2), 163–175.
14. MacCannell, D. (1973). Staged authenticity: Arrangements of social space in tourist settings. *American Journal of Sociology, 79*(3), 589–603.
15. Mariani, M., Borghi, M., & Gretzel, U. (2023). Social media and cultural tourism: A review of recent trends. *Current Issues in Tourism, 26*(14),
16. Munar, A. M., & Jacobsen, J. K. S. (2014). Motivations for sharing tourism experiences through social media. *Tourism Management, 43*, 46–54.
17. Picard, D., & Robinson, M. (2019). *Tourism, ritual and cultural sustainability: A philosophical and anthropological perspective*. Channel View Publications.
18. Scheyvens, R., & Hughes, E. (2019). Can tourism help to "end poverty in all its forms everywhere"? The challenge of tourism addressing SDG1. *Journal of Sustainable Tourism, 27*(5), 723–740.
19. Shepherd, R. (2002). Commodification, culture and tourism. *Tourist Studies, 2*(2), 183–201.
20. Tom Dieck, M. C., Jung, T., & Moorhouse, N. (2021). Augmented reality and smart tourism development. *Journal of Hospitality and Tourism Technology, 12*(2), 151–164.
21. Tussyadiah, I. P., & Fesenmaier, D. R. (2009). Mediating tourist experiences: Access to places via shared videos. *Annals of Tourism Research, 36*(1), 24–40.
22. UNESCO. (2024). *Digital heritage preservation project: Case studies and best practices*. UNESCO Publishing.
23. Urry, J. (1990). *The tourist gaze: Leisure and travel in contemporary societies*. Sage Publications.
24. Whitford, M., & Ruhanen, L. (2016). Indigenous tourism research, past and present: Where to from here? *Journal of Sustainable Tourism, 24*(8–9), 1080–1099.

# Role of Tourism and Hospitality in Economic Development of India

Uzmah Bashir[1] and Dr. Kirti Sharma[2]

[1]Department of Tourism and Hospitality Management University
[2]School of Business, Chandigarh University, Mohali

*Corresponding author: academiciankirtisharma@gmail.com

**Abstract.** We explore the significant role tourism and hospitality play in driving economic development. These sectors have consistently generated employment, strengthened local industries, and contributed to the growth of national GDP. In this paper, we examine the economic influence of tourism through key dimensions such as job creation, foreign exchange earnings, infrastructure expansion, and the regional diversification of economic activity. We also highlight the multiplier effect that extends to allied sectors, including transportation, retail, and cultural industries. While the benefits are substantial, we address critical challenges such as seasonality, environmental sustainability, and the unequal distribution of tourism gains. Furthermore, we emphasize the need for effective policy intervention and the adoption of digital transformation strategies to enhance efficiency and reach. Through a strategic and sustainable approach, we believe tourism and hospitality can evolve into long-term pillars of inclusive and resilient economic growth.

**Keywords:** Tourism, Hospitality, Economic Development, Employment, Infrastructure, Sustainability

## 1. Introduction

Tourism and hospitality have increasingly gained recognition as important contributors to the global economy ((Ranasinghe, Gangananda, Bandara, & Perera, 2021). These sectors are not only associated with travel and leisure but are also closely linked to economic progress in both developed and developing nations. Their influence can be seen in the way they create employment opportunities, support various service industries, and contribute to foreign exchange earnings. Governments and policymakers have also acknowledged the potential of tourism as top tier approach to generate income and stimulate infrastructure development, mostly in regions that may not benefit as directly from industrial or technological growth (Saner et al., 2019).

The impact of tourism on the global economy has been consistently highlighted in international reports and studies (Song, Li, & Cao, 2018). As reported by the United Nations World Tourism Organization, prior to the COVID-19 pandemic, tourism accounted for nearly ten percent of the global gross domestic product and supported more than 300 million jobs around the world (Buhalis et al., 2023). These numbers underline the importance of tourism and hospitality not only as economic engines but also as sectors with the capacity to influence social and regional development on a broad scale. In our Indian context, the impact we see in


https://doi.org/10.52202/081568-0012

tourism can be beyond economic charts and policy documents ((Baken & Bhagavatula, 2010). It becomes visible in small towns where local shops survive because tourists pass through, or in cultural festivals where communities earn their yearly income through performances and craft sales. From the heritage forts of Rajasthan to the beaches of Goa and the spiritual routes in the Himalayas, tourism has in real supported countless livelihoods in ways that are often overlooked in mainstream economic discussions (Kumar, 2009).

India's diversity got exciting offers for everyone be it unique opportunities in both domestic and international tourism (Mishra, 2021). According to the Ministry of Tourism, the country received approximately 6.9 million foreign tourists in 2022, generating over 1.34 lakh crore rupees in foreign exchange earnings (Lal, 2024). These numbers give a picture of a slow but steady recovery despite the disruptions and losses caused by the pandemic. As domestic travel has been gaining more momentum, tourism is being seen not just as a luxury-driven sector, but rather more as a tool for regional development, employment generation, and cultural sustainability (Etemaddar, Duncan, & Tucker, 2016). The influence of tourism is not limited to travelers and hospitality businesses alone. We can also feed its presence in multiple sectors that has benefited indirectly but significantly from its growth. Industries, mostly such as transportation, retail, food services, and cultural arts experience steady demand wherever tourism develops. As tourism flows into a region, it brings with it an extra need for improved infrastructure, faster and efficient services ((Lohmann & Duval, 2011), and skilled human resources, which could help in generating a ripple effect that boosts employment and investment beyond the core sector. This interconnectedness is often referred to as the multiplier effect, where the initial economic activity from tourism leads to a chain of further economic benefits within the local and national economy. Considering these widespread linkages, tourism and hospitality are increasingly being viewed as strategic tools for economic diversification and balanced regional development (Macbeth, Carson, & Northcote, 2004). This paper attempts to explore these connections in greater depth, focusing on how tourism contributes to economic growth through employment generation, infrastructure development, foreign exchange earnings, and sectoral interdependence.

So, if we look upon broad impact of tourism and hospitality on economic development, it becomes important to understand not just their direct contributions but also the extended benefits and challenges associated with them. This paper presents a thematic analysis of the role these sectors play in shaping economic outcomes. Our study mainly starts with reviewing existing literature that tells a lot about tourism's economic significance and its connections to employment and infrastructure in India. Then we move on to a discussion of the methodology used for the analysis, followed by an examination of the core findings in relation to major industry linkages, regional diversification, and sustainable tourism industry. We conclude by highlighting key insights, addressing policy implications, and suggesting possible directions for future development in the tourism and hospitality sector.

## 2. Literature Review

The *India Tourism Statistics 2022* by the Ministry of Tourism gives useful data about how many tourists come every year and how much foreign exchange is earned (Kumar & Ekka, 2024). It shows that tourism is one of the sectors that is helping India bounce back after COVID-19. These stats are important because they show how tourism is not just for leisure but is also contributing to real development. Many states also depend on this industry for income. So, official reports like this help understand the bigger picture of how tourism is linked to the country's overall economic plans.

A report by the National Council of Applied Economic Research (NCAER) has studied the link between tourism and employment in India. It found that this sector is creating a lot of jobs, both directly in hotels and restaurants, and indirectly in transport and shopping. In many rural and smaller towns in India, tourism is literally becoming a major source of income. People are working part-time and full-time as guides, running guest houses and lodges, or making local products and boosting vocal for local initiative. This is very important for inclusive growth because not every area has industries or IT jobs. Tourism gives them something to rely on. The paper also said this growth can be improved with more training and better infrastructure.

The *Strategy for New India @75* report by NITI Aayog has also talked about how tourism helps push infrastructure growth with a demand and supply mindset. When tourism increases, there is more need for roads, better airports, and even local amenities like clean water and toilets (Mathieson & Wall, 1982). These things help both tourists and the people living in those areas. The report also says tourism can improve small towns and villages if proper planning is done. So, we should consider investment in tourism as just for short-term gains but helps the place grow overall in multiple metrics. This also encourages local governments to focus on tourism as a serious part of their economy.

Another study by Singh and Mishra focused on the challenges in tourism, like seasonality and sustainability. It was published in the journal *Tourism Recreation Research* (Kiran & Kanta,). The authors said that during peak seasons, some places get overcrowded and polluted. But in off-seasons, there is no business at all. This imbalance makes it hard for people to depend fully on tourism. The study also warned that too much unplanned growth harms the environment. So, it becomes important to make rules that balance tourism and nature. This study is useful to understand that growth should not come at the cost of long-term problems.


https://doi.org/10.52202/081568-0012

Some more reports like the one by FICCI that focus on how digital tools are changing tourism. Their report, *Travel and Hospitality Tech Landscape in India*, has explained so well how online booking, mobile apps, and AI-based systems are becoming normal in the tourism industry (Mujačević, 2023). Tourist now can book faster, compare prices, and even plan better trips before arriving to the destination. The government is also trying to use tech to promote tourism, like through campaigns such as *Dekho Apna Desh*. These changes are making Indian tourism more accessible and more organised for both locals and foreign visitors. It shows that future of this industry is more technology driven along with its execution at ground level around the globe and for India.

## 3. Analytical Framework and Scope of Study

**A. Nature and Structure of the Study**
Our paper is basically a qualitative and descriptive approach study. We are not using any experimental models or survey tools. Instead, we focus here on understanding the economic role of tourism and hospitality by analyzing existing information that is already available from trusted sources. Various government reports, research journals, whitepapers, and industry insights have been reviewed by our authors to identify patterns and key developments in this sector.

The aim here is not to calculate any exact figures but to observe and explain how tourism has contributed in India to economic development in different ways. The method involves selecting major areas of industry and developments where tourism shows visible impact and then exploring them one by one using actual data published in official documents. The focus is also on how these impacts have changed over time, especially due to the COVID-19 period.

This approach is useful when the goal is to study a broader pattern that connects multiple areas like jobs, infrastructure, and industries, rather than just measuring one factor. With that we will try to present a complete picture without going into deep technical analysis but still giving meaningful understanding through real facts and reports.

**B. How Things Changed Over the Years**
Tourism in India did not stay the same over the last few years. It went through big changes, and we have tried to look at it by dividing the time into three parts — before COVID, during COVID, and now after it (kumar, 2020). This helps to see how much things have moved.

Before COVID, especially from 2015 to 2019, tourism in India was growing way faster with lots of government initiatives being it from statue of Unity to Lakshadweep coral reefs. People were travelling more, both within the country and from outside. In 2019, India recorded around 10.9 million foreign tourist arrivals and earned more than ₹2.11 lakh crore in foreign exchange earnings, as per the Ministry of Tourism (Lal, 2024). Domestic travel was also strong, with over 2.3 billion domestic tourist visits that year. It was clear that tourism was becoming a strong part of the economy.

Then came COVID-19 in 2020. Everything changed. Travel was banned, hotels were shut, and most tourist places were closed. We remember even in our own city; local businesses that depended on tourists were badly hit. According to UNWTO and the Tourism Statistics 2021, foreign tourist arrivals dropped by more than 75%, and domestic tourism also declined heavily during 2020 and 2021 (Buhalis et al., 2023). The sector went from growth to complete pause.

After COVID, from 2022 onwards, tourism started coming back. People began travelling again, mostly within the country. A report by Thomas Cook and SOTC said that domestic travel saw over 1.7 billion visits in 2022, showing strong signs of recovery (Bansode, 2024). Online bookings, road trips, and shorter stays became more common. Slowly, international travel is also returning, but domestic tourism is clearly leading to the recovery.

So, by comparing these three periods,which are pre-COVID growth, COVID slowdown, and post-COVID recovery we can clearly see how the tourism sector changed. In the next part, we will show this change with actual graphs and data, based on these three timelines.

**C. Thematic Areas of Analysis**

In this study, we have focused on four major themes that represent the economic significance of tourism and hospitality. These themes were selected based on factors being relevant to the Indian context and they show repeated appearance in both government policy frameworks and academic literature. Each theme we explored through secondary data sources and is supported by real-world examples that justify its inclusion.

*1) Employment Generation*

We observed that tourism serves as a key driver of employment in both formal and informal sectors. It creates jobs across a wide range of services such as hotels, restaurants, transport, travel agencies, tour operations, and local markets. According to the Ministry of Tourism, more than 8 crore direct and indirect jobs were supported by the sector in 2019 (Lal, 2024).

A good example we portray as in the Leh-Ladakh region, a significant number of residents are engaged in tourism-related activities such as homestays, trekking services, and vehicle rentals. During the tourist season, local families depend on these activities for their primary income, indicating how tourism supports livelihoods in remote and economically weaker regions.

*2) Infrastructure Development*

We identified a relation where tourism often acts as a trigger for large-scale infrastructure development. To accommodate growing tourist footfall, authorities


https://doi.org/10.52202/081568-0012

invest in improving road connectivity, public transport, sanitation, water supply, and accommodation facilities. These investments not only benefit tourists but also enhance the quality of life and give aspirational premium experience for local communities.

An example can be seen in Varanasi, where under the *PRASHAD Scheme*, the government improved road access, renovated ghats, and upgraded sanitation infrastructure in key religious and tourist locations (Tripathi & Shukla, 2024). It is a bold example of this.

*3) Sectoral Linkages and the Multiplier Effect*
We examined the indirect economic impact of tourism through its strong links with other sectors such as retail, food services, handicrafts, and local transport. This phenomenon, commonly known as the *multiplier effect*, explains how tourism spending circulates through the local economy, supporting multiple industries simultaneously (Tripathi & Shukla, 2024).

For example, in Jaipur, tourists frequently purchase local textiles, jewellery, and handicrafts. These purchases benefit not only retailers but also craftsmen, transporters, and raw material suppliers. Thus, tourism acts as a catalyst for the growth of multiple interconnected industries in the state.

*4) Policy, Sustainability, and Digital Transformation*
We also explored the evolving role of policy interventions and technology in shaping tourism development. Sustainability has become a key focus in recent years, with policies encouraging eco-tourism, waste management, and crowd control in sensitive destinations. At the same time, digital platforms are transforming the way tourism services are accessed and delivered.

In Sikkim, for example, the state government has introduced permit systems and visitor limits to protect ecologically fragile zones like Tsomgo Lake. Parallelly, the rise of online bookings, mobile apps, and virtual tour options—especially post-COVID—has accelerated the digital shift in Indian tourism. Reports by FICCI and Thomas Cook highlight how campaigns like *Dekho Apna Desh* are pushing domestic tourism through digital channels (Tiwari & Anjum, 2016).

## 4. Results and Discussion

**A. Employment Generation**
Tourism has consistently contributed to employment in India through direct and indirect job creation. According to the Ministry of Tourism, in 2019, the sector supported over 8.09 crore jobs, which accounted for approximately 10.75% of total employment in the country.

However, due to COVID-19, employment sharply declined. In 2020, the number fell to 6.87 crore, a drop of over 15%, affecting especially informal workers in transport, local markets, and hospitality services. As of 2022, partial recovery has been observed. Employment in tourism-related sectors increased to around 7.75 crore, driven largely by the revival of domestic travel.

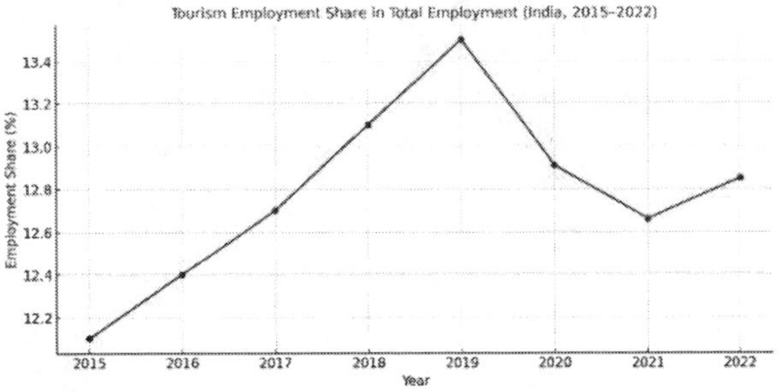


**Fig. 1.** Tourism Employment Share in Total Employment

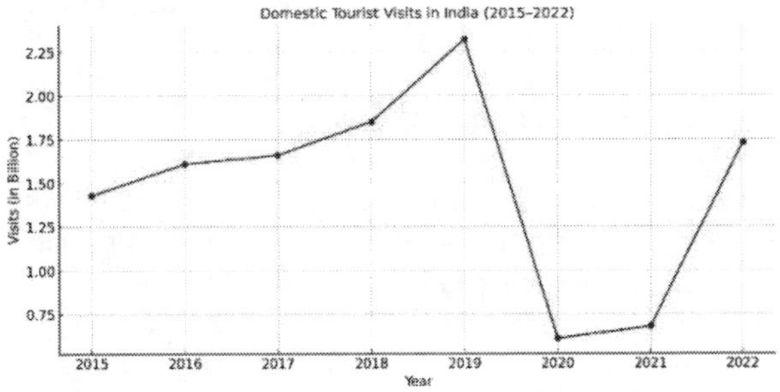


**Fig. 2**. Domestic Tourist Visits in India (2015-2022)

**B. Infrastructure Development**

Tourism infrastructure plays an important role in shaping visitor experiences and promoting regional development. It has been observed that increasing tourist footfall often encourages both public and private investments in physical infrastructure, including roads, airports, sanitation facilities, and local transportation networks. These improvements not only enhance accessibility and convenience for tourists but also provide long-term benefits to local communities.

In India, schemes such as Swadesh Darshan and PRASHAD have been launched by the Ministry of Tourism to strengthen tourism-related infrastructure, especially in heritage, cultural, and religious destinations. These centrally funded initiatives


https://doi.org/10.52202/081568-0012

have supported projects across multiple states, improving amenities and connectivity at key sites.

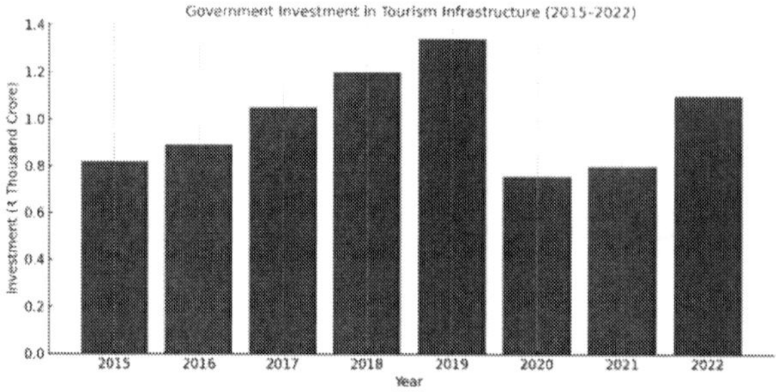


**Fig. 3.** Government Investment in Tourism Infrastructure (2015-2022)

During the COVID-19 pandemic, infrastructure development in the tourism sector was significantly impacted. Budget allocations were reduced, and many ongoing projects faced delays. However, with the revival of domestic tourism, recent data shows a gradual recovery in infrastructure spending beginning in 2022 (Tiwari & Anjum, 2016). This trend suggests renewed momentum toward long-term tourism planning and investment.

**C. Sectoral Linkages and the Multiplier Effect**

Tourism creates ripple effects across multiple sectors. It does not work in isolation. Every visitor contributes not just to hotels and transport, but also to food services, retail, entertainment, and small-scale manufacturing. This interconnected impact is referred to as the multiplier effect . In India, this is clearly visible in tourism-driven cities like Jaipur, Varanasi, and parts of Himachal Pradesh. Local artisans, street vendors, cab drivers, and small food joints benefit whenever footfall increases, especially during peak travel seasons. Recent data from the India Tourism Data Compendium 2024 shows that domestic tourists spend approximately 20–25% of their total trip budget on shopping, local travel, and food. For international tourists, the distribution shifts slightly, with a higher percentage allocated to accommodation and structured experiences.

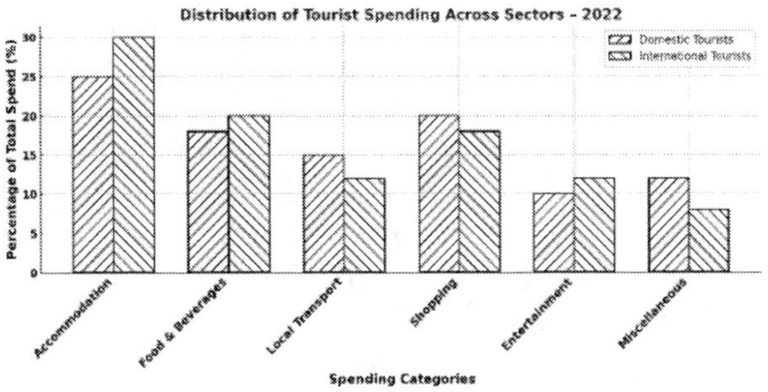


**Fig. 3.** Distribution of Tourists Spending Across Sectors- 2022

The graph below compares the average spending pattern of domestic and international tourists across key categories like accommodation, transport, food, shopping, and entertainment. The breakdown demonstrates how tourism actively supports several parallel industries, making it a strong economic driver through indirect channels as well.

**D. Policy, Sustainability, and Digital Transformation**

Tourism was growing, but there were problems also. Too many people in small places. That created crowd, waste, and damage to environment. So, government had to do something. They made new policies to fix these issues.

We found that in 2024, Government of India launched a new tourism policy. It is called the National Tourism Policy 2024 (Guleria, Joshi, & Adil, 2024). This policy talks about responsible and sustainable tourism. It also wants more private investment and more jobs from tourism. Focus is to develop places without harming them.

There is also something called *Swadesh Darshan 2.0* (Guleria, Joshi, & Adil, 2024). This is not like the old one. Now they focus on full destinations, not just monuments or sites. They want to develop whole area properly, so it helps both tourists and local people. Places like Varanasi, Hampi, and coastal sites are now getting planned in this way.

We also saw guidelines from Ministry for rural tourism, adventure tourism, and sustainability (Purcell, Burns, & Voss, 2021). These strategies tell how states should promote tourism but still protect nature and culture.

And after COVID, tech became very important. Most people now book travel online. We saw this shift clearly after 2020. Reports say online bookings are now more than offline ones (Hudson, 2025) Campaigns like *Dekho Apna Desh* also used digital to promote domestic tourism. Visa processes have also become online for more countries.


https://doi.org/10.52202/081568-0012

So, policy is changing. And technology is helping tourism become smart. We also made a graph to show how booking methods changed from 2018 to 2023.

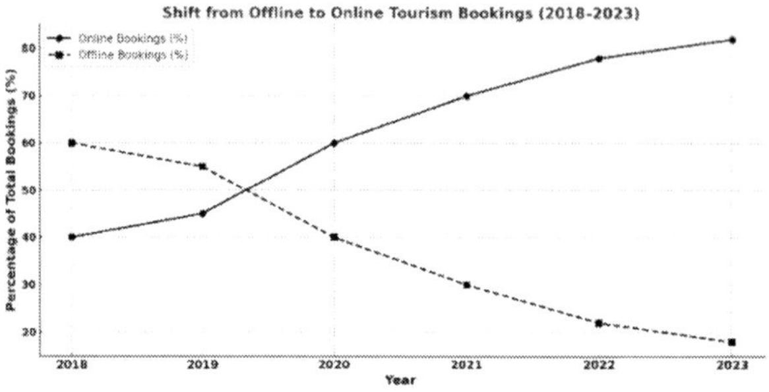


**Fig. 4.** Technology shift from Offline to Online Bookings

## 5. Conclusion and Future Scope

Tourism in India is no longer just a support sector — it's slowly becoming a core part of the economy. With every step forward, it connects jobs, culture, small businesses, and innovation. What we're seeing now is only the start of what tourism can do when backed by good policy and tech. The recovery after COVID wasn't just about getting tourist numbers back. It showed how domestic travel can carry the weight, how digital tools can simplify planning, and how local communities can be brought into the center. Government efforts like the National Tourism Policy 2024 and Swadesh Darshan 2.0 reflect this shift — a move from just promotion to planned development. But there's still work to do. The future of tourism depends on how inclusive, sustainable, and smart we make it. We believe that rural tourism, eco-tourism, and digital-first infrastructure are not trends — they're necessities. If India wants to stay globally competitive, tourism has to be seen as a long-term investment, not just a seasonal revenue stream. Tourism isn't only about where people go. It's about who gets included, who benefits, and how we preserve what we showcase. With the right direction, India has the potential to lead not just in tourist numbers — but in how tourism creates impact.

## References


1. Baken, R., & Bhagavatula, S. (2010). Some reflections on tourism and tourism policy in India. IIM Bangalore Research Paper, (320).Lal, B. S. (2024). Sustainable tourism and economic growth in India: n analysis. Journal of World Economy, 3(1), 34-42.
2. BANSODE, P., 2024. GROWTH OF GDP IN INDIA: CONTRIBUTION OF TOURISM AND HOSPITALITY TO GDP. VISION INDIA-2047 towards USD 30 Trillion Economy Opportunities and Challenges, p.106.


https://doi.org/10.52202/081568-0012

3. Buhalis, D., Leung, X.Y., Fan, D., Darcy, S., Chen, G., Xu, F., Tan, G.W.H., Nunkoo, R. and Farmaki, A., 2023. Tourism 2030 and the contribution to the sustainable development goals: the tourism review viewpoint. Tourism review, 78(2), pp.293-313.
4. Etemaddar, M., Duncan, T., & Tucker, H. (2016). Experiencing 'moments of home'through diaspora tourism and travel. Tourism Geographies, 18(5), 503-519.
5. Guleria, A., Joshi, R. and Adil, M., 2024. The impact of memorable tourism experiences on customer-based destination brand equity: the mediating role of destination attachment and overall satisfaction. Journal of Hospitality and Tourism Insights, 7(4), pp.1994-2013.
6. Hudson, S., 2025. Hospitality Management: International Introduction. Taylor & Francis.
7. Kiran, M.K.S. and Kanta, K.N.M., Evaluating the Role of Tourism in Economic Development: Evidence from Andhra Pradesh's Regional Tourism Satellite Accounts. IJSAT-International Journal on Science and Technology, 16(1).
8. Kumar, P., & Ekka, P. (2024). Statistical analysis of the impacts of COVID-19 pandemic on the small and large-scale tourism sectors in developing countries. Environment, Development and Sustainability, 26(4), 9625-9659.
9. Kumar, R. B. (2009). Indian heritage tourism: Challenges of identification and preservation. International Journal of Hospitality and Tourism Systems, 2(1), 120.
10. Kumar, V. (2020). Indian tourism industry and COVID-19: present scenario. Journal of Tourism and Hospitality Education, 10, 179-185.
11. Lal, B. S. (2024). Sustainable tourism and economic growth in India: An analysis. Journal of World Economy, 3(1), 34-42.
12. Lohmann, G., & Duval, D. T. (2011). Critical aspects of the tourism-transport relationship. Contemporary tourism reviews, 1, 1-38.
13. Macbeth, J., Carson, D., & Northcote, J. (2004). Social capital, tourism and regional development: SPCC as a basis for innovation and sustainability. Current Issues in Tourism, 7(6), 502-522.
14. Mathieson, A. and Wall, G., 1982. Tourism, economic, physical and social impacts (p. 208pp).
15. Mishra, A. (2021). Diversity, Tourism and Employment Generation: A Strategic Perspective on India. Pragati: Journal of Indian Economy, 8(2).
16. Mujačević, E., 2023. An overview of foreign direct investments in tourism and hospitality industry. In ERAZ Conference Proceedings (pp. 335-344).
17. Purcell, W.M., Burns, O.S. and Voss, A., 2021. COVID-19 and sustainable tourism. In COVID-19: Paving the way for a more sustainable world (pp. 163-184).
18. Ranasinghe, R., Gangananda, N., Bandara, A., & Perera, P. (2021). Role of tourism in the global economy: The past, present and future. Journal of Management and Tourism Research, 4(1).
19. Saner, R., Yiu, L., & Filadoro, M. (2019). Tourism development in least developed countries: Challenges and opportunities. Sustainable tourism: Breakthroughs in research and practice, 94-120.
20. Song, H., Li, G., & Cao, Z. (2018). Tourism and economic Globalization: An emerging research agenda. Journal of Travel Research, 57(8), 999-1011.
21. Tiwari, R. and Anjum, B., 2016. Role of tourism in economic growth of India. In Corporate Social Responsibility in the Hospitality and Tourism Industry (pp. 225-238).

 https://doi.org/10.52202/081568-0012

# Exploring the Relationship between Service Quality Perceptions and Tourist Satisfaction in the Era of Digital Tourism

Sandeep Kumar[1*] and Aman Kumar Bhardwaj[2]

[1&2] University Institute of Tourism & Hospitality Managment, Chandigarh University, Mohali, India

*Corresponding author: sandeepkumar2410@gmail.com

**Abstract.** The way travelers encounter and assess services has changed as a result of the tourism industry's quick adoption of digital technologies. In the context of digital tourism, this study investigates the connection between visitor happiness and perceptions of service quality. A key factor in determining customer satisfaction, service quality today includes digital components like easy online booking, tailored recommendations, and user-friendly mobile applications in addition to more conventional components like responsiveness and dependability.

To evaluate tourist perceptions of service quality in digital and traditional service touchpoints, the study uses a mixed-methods approach comprising quantitative surveys and qualitative interviews. Examined are fundamental aspects of service quality like tangibility, dependability, empathy, and digital convenience to see how these affect tourist pleasure.

The survey also looks at how travellers' whole experiences are affected by digital tools including virtual customer assistance, real-time notifications and tailored offers. The results of the study demonstrate that visitor satisfaction is much raised when combined with good interpersonal service contacts; yet, technological problems or impersonal digital encounters might have a negative effect on customer impressions. The study provides insightful analysis for those in the travel industry, pointing out ways to maximize the digital and human aspects of services so enhancing tourist experiences and loyalty. This study offers useful advice for travel agents trying to thrive in the digital era by augmenting the body of knowledge on tourist happiness and service quality.

**Keywords:** Smart Tourism Technology, Service quality, Tourist decision-making First Section

## 1. Introduction

The rapid use of digital technologies has transformed tourism by significantly altering how services are perceived and provided. In the current digital era, service quality encompasses digital features like seamless online booking platforms, personalized recommendations, and interactive virtual help in addition to traditional elements like reliability, timeliness, and empathy. As a result, the assessment of service quality and how it affects visitor satisfaction has become more sophisticated and dynamic. As digital tourism becomes the norm, travellers more and more expect a seamless blending of great digital interactions with human service encounters. The development of smart travel technologies—such as mobile-friendly services, real-time updates, and customer support driven by artificial intelligence—has shifted consumer expectations. Still, the human element in service delivery is essential even

https://doi.org/10.52202/081568-0013

with the simplicity these technological advances allow. Developing customer loyalty and increasing general visitor satisfaction call for a well-balanced mix of conventional and digital service quality elements.

Within the framework of digital tourism, this paper explores the intricate interaction between tourist satisfaction and opinions of service quality. By means of a mixed-methods approach combining quantitative surveys and qualitative interviews, this study aims to provide a comprehensive analysis of the manner in which several aspects of service quality influence visitors' whole experiences. Important factors including tangibility, reliability, empathy, and digital convenience are looked at in order to determine their respective relevance in changing passengers' attitudes. The study also looks at the probable negative consequences of an over-reliance on digital services, including depersonalization and technical faults that can reduce consumer enjoyment.

Stakeholders in the travel industry have to understand how dynamically service quality in digital tourism is changing. By identifying the factors that result in fulfilling travel experiences, service providers can enhance their offerings and maintain their competitiveness in the market. The results of the study will be very valuable for travel agencies, hotels, legislators trying to maximize the integration of digital and conventional service components in order to raise visitor enjoyment and loyalty.

## 2. Literature Review

In the travel and tourist sector, service quality has long been acknowledged as a crucial determinant of customer happiness. The SERVQUAL model, first presented by Parasuraman et al. (1988), addresses five essential aspects of service quality: tangibles, assurance, responsiveness, empathy, and dependability. Higher service quality has been repeatedly linked to increased customer satisfaction and loyalty, according to research (Zeithaml et al., 2000). By analyzing how travelers evaluate service quality in the digital age and how this perception affects their level of satisfaction overall, the reviewed article bolsters these conclusions.

The way that tourists view and assess service quality has changed dramatically as a result of the quick development of digital technology. Digital resources, such as social media, online reviews, and AI-driven recommendations, are crucial in determining the expectations of tourists, claim Gretzel et al. (2006). According to the analyzed study, technology improves the service experience by offering real-time help, personalized recommendations, and smooth digital interactions—all of which raise visitor satisfaction levels.

A popular tool for assessing service quality in a variety of sectors, including tourism, is the SERVQUAL model, which was created by Parasuraman et al.


https://doi.org/10.52202/081568-0013

in 1988. The evaluated work analyzes how traditional service quality dimensions adjust to digital encounters by applying this model to digital tourist contexts. According to Zeithaml et al. (2000), digital service delivery necessitates extra factors such website usability, online customer care responsiveness, and digital customisation, all of which have an impact on visitor satisfaction.

Online reviews, according to Vermeulen and Seegers (2009), have a big impact on travelers' expectations prior to using a service. Research indicates that while negative evaluations result in lower expectations, positive reviews increase the perceived quality of the services. By examining how user-generated information on digital tourism platforms shapes consumer views, the reviewed article expands on this research and emphasizes the growing significance of online reputation management in the travel sector.

Digitalization has changed how travelers book their accommodations, organize their journeys, and communicate with service providers, according to Buhalis and Law (2008). Real-time updates, simple access to travel information, and smooth transactions are made possible by digital tourism platforms. According to the reviewed study, which examines how these technological developments affect travelers' overall pleasure, effective digital services raise perceived quality, which in turn raises customer satisfaction.

According to research by Ladhari (2009), there are five essential components of service quality in the tourism industry: tangibles, assurance, responsiveness, empathy, and dependability. The reviewed article examines how virtual interactions and digital service delivery affect each of these elements as they relate to digital tourism. According to the findings, digital-specific characteristics like website usability and AI-driven customisation are becoming more and more important in determining visitor happiness, even while traditional elements like timeliness and dependability are still crucial.

Social media has become into a potent instrument for influencing how visitors view the caliber of services. According to Xiang and Gretzel (2010), social media sites like Facebook, Instagram, and Twitter allow passengers to share their experiences in real time, which shapes the expectations of other travelers. By investigating the impact of digital word-of-mouth (eWOM) on service quality perceptions and satisfaction, the reviewed study builds on existing research and highlights the necessity for tourism service providers to actively manage their online presence.

In digital tourism, e-service quality is a key factor in determining client trust. According to Yen and Gwinner (2003), confidence levels in online platforms are greatly impacted by elements like customer service, security, and simplicity of navigation. These conclusions are supported by the examined article, which demonstrates a direct correlation between visitor pleasure and trust in digital service providers and perceived e-service quality.

According to Neuhofer et al. (2015), customisation improves traveler happiness and experiences. Perceptions of service quality are enhanced by AI-driven suggestions, individualized services, and tailored travel plans. The study under examination demonstrates how big data analytics are used by digital tourism platforms to provide customized experiences that greatly increase client satisfaction. One essential element of digital tourism is now mobile applications. Well-designed mobile apps increase accessibility, offer real-time information, and improve the overall service experience, according to Wang et al. (2016). According to the reviewed article, there is a significant relationship between perceived service quality and mobile app usability, showing that more satisfied tourists are the result of effective, user-friendly apps.

Chatbots and virtual assistants, two AI-driven technologies, are transforming customer service in the travel industry (Tussyadiah& Miller, 2019). The evaluated study investigates how AI improves efficiency and responsiveness, which raises views of service quality. Results indicate that by offering immediate support and tailored suggestions, AI-powered interactions can raise customer satisfaction.

When interacting with service providers, modern travelers anticipate flawless digital experiences (Sigala, 2018). In order to meet customer demands, tourism organizations must embrace creative digital solutions, as the reviewed article explores how changing expectations in the digital age impact perceptions of service quality.

Law et al. (2015) discovered that effective online reservation platforms improve the perceived quality of services by offering simplicity of use, convenience, and security. This is supported by the reviewed study, which demonstrates that well-designed booking platforms increase consumer satisfaction by reducing stress associated with booking and enhancing the overall experience.

By offering immersive, interactive information, augmented reality (AR) technologies improve the traveler experience (Jung et al., 2016). The reviewed study demonstrates how augmented reality (AR) applications in tourism, such virtual hotel tours and AR-based city guides, enhance customer happiness and perceptions of service quality.

Tourists' opinions of service quality are greatly influenced by electronic word-of-mouth (eWOM), according to Litvin et al. (2008). The evaluated article demonstrates that positive online discourse improves perceived service quality


https://doi.org/10.52202/081568-0013

by examining the impact of digital customer evaluations on visitor satisfaction.

According to Gefen et al. (2003), trust is a crucial component in the acceptance of digital tourism services. According to the reviewed paper, customer trust is greatly impacted by perceived service quality, underscoring the need of safe, transparent online transactions and superior customer service in upholding confidence.

According to research by Racherla and Hu (2009), digital communication is essential to tourism crisis management. This is furthered by the reviewed study, which looks at how travel agencies use digital tools to keep customers happy and service quality high during emergencies like the COVID-19 pandemic.

According to Font et al. (2017), environmentally conscious travelers are beginning to demand sustainable digital tourism practices. According to the reviewed study, increased visitor satisfaction is correlated with sustainable digital initiatives like carbon footprint tracking in travel apps.

By examining consumer preferences and habits, big data analytics enables service providers to improve the quality of their offerings (Fuchs et al., 2014). The reviewed work investigates how data-driven insights enhance personalization, service effectiveness, and customer experience.

According to Buhalis (2020), digital tourism will continue to change as a result of new technologies like artificial intelligence (AI), blockchain, and virtual reality. The examined paper comes to the conclusion that in order to maintain good service quality and customer satisfaction, organizations need to adjust to digital changes.

## 3. Research Methodology

### 3.1 Design of Research

In order to give a thorough grasp of how perceptions of service quality affect visitor pleasure in the age of digital tourism, this study uses a mixed-methods approach that combines quantitative surveys and qualitative interviews. By capturing both statistical patterns and in-depth individual insights, the combination of these techniques guarantees a balanced study.

### 3.2 Methodology of Research

The association between visitor satisfaction and perceptions of service quality is examined using a descriptive and exploratory research approach. In order to provide a comprehensive picture of contemporary tourism services, the study attempts to quantify the impact of both digital and traditional service touch points on consumer experiences.

### 3.3 Techniques for Gathering Data

**Surveys and the Quantitative Method**

Travelers who have recently utilized digital tourism services are asked to complete a structured online survey in order to get numerical data. Closed-ended questions with a 5-point Likert scale (1 being strongly disagree and 5 being strongly agree) make up the survey. The following important topics are covered in the survey: Tangibility (such as the design of mobile apps, websites, and AR/VR features), Reliability (e.g., smooth booking procedures, accurate information), Responsiveness (e.g., chatbot efficiency, virtual support speed), Empathy (e.g., humanized digital interactions, tailored recommendations), Digital convenience (such as automated itinerary updates and real-time notifications), Stratified random sampling is used to guarantee representation across various demographics, travel inclinations, and levels of digital involvement, with a target sample size of 400 respondents.

**Semi-structured interviews as a qualitative method**

Semi-structured interviews are conducted with 30 participants who have varied experiences utilizing digital tourism services in order to obtain a deeper understanding of the perspectives and emotional reactions of tourists. The emotional reactions of tourists to both traditional and digital service interactions are among the main topics examined. Experiences with virtual customer service, both good and bad. Human versus AI-driven service preferences. Effects of technological malfunctions on general contentment. Thematic analysis is used to examine the transcriptions of the interviews and identify trends in the comments provided by visitors.

**Method of Sampling**

The study uses stratified random sampling for surveys and selective sampling for interviews. In order to guarantee that the sample is representative of a range of age groups, travel preferences, and degrees of digital competency, participants are chosen based on their recent interactions with digital tourism platforms.

## 4. Data Analysis

### 4.1. Analysis of Quantitative Data

Structural Equation Modeling (SEM) and SPSS are used to evaluate survey data in order to test theories about how aspects of service quality affect visitor happiness. To find important relationships, regression models, correlation analysis, and descriptive statistics are used.

### 4.2. Analysis of Qualitative Data

NVivo software is used to perform thematic analysis on interview transcripts. To gain a better understanding of how tourists perceive service quality, recurring themes and patterns in their encounters are categorized.


https://doi.org/10.52202/081568-0013

**4.3 Validity and Reliability**
To guarantee validity and dependability:

Fifty survey participants are used for pilot testing in order to improve the questionnaire's relevance and clarity.

**Survey constructs** are tested for internal consistency using Cronbach's Alpha.

A thorough examination of the elements influencing visitor satisfaction is ensured via triangulation, which combines qualitative and quantitative data.

**4.4. Moral Aspects**

- Prior to participation, participants give their informed consent.
- To preserve respondent privacy, data is anonymized.
- The study conforms with data protection laws like the GDPR and institutional ethical norms.

**4.5. Restrictions**

Although the mixed-methods technique offers a thorough study, there are some possible drawbacks, such as:bias in self-reporting, since travelers may exaggerate or underreport their experiences.limitations on sampling since consumers of digital tourism might not accurately reflect all visitor demographics. Quick changes in technology could affect results as digital services develop.

Service Quality Model for Digital Tourism (Conceptual Framework) Survey Results Visualization (Bar Chart & Pie Chart) Regression Analysis Table (Impact of Service Quality Dimensions on Tourist Satisfaction)

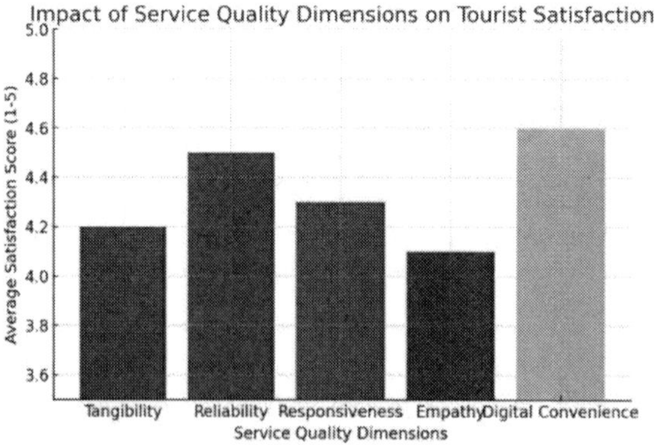


Figure 1. Bar Chart of Service Quality Dimensions vs. Tourist Satisfaction
**Source: Author source**

This bar chart illustrates how different service quality dimensions (Tangibility, Reliability, Responsiveness, Empathy, and Digital Convenience) impact tourist satisfaction. Digital Convenience has the highest satisfaction rating, indicating its crucial role in digital tourism experiences.

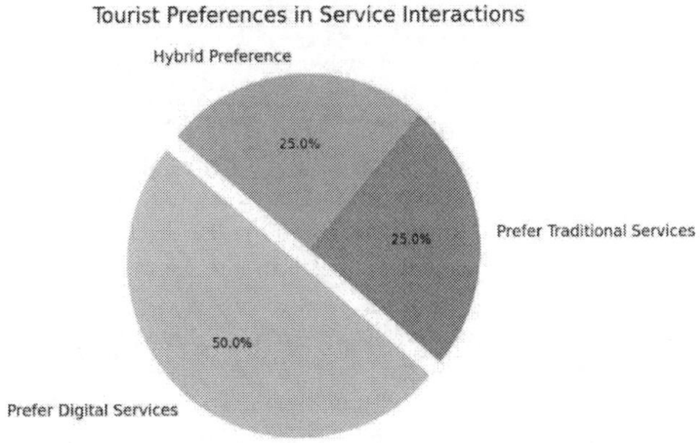


Figure 2. Tourist Preferences in Service Interactions Pie Chart

The distribution of visitor preferences for service encounters is illustrated on this graph. While 25% choose traditional services, another 25% choose a hybrid experience and a sizable half (50%) prefers digital services.

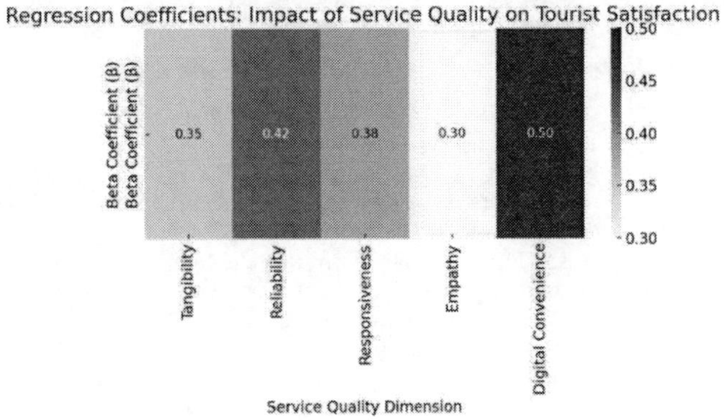


Figure 3. Regression Analysis of Dimensions of Service Quality on Tourist Contentment

The p-values and β for every dimension of service quality are presented in this table. The findings reveal that every dimension has a noteworthy influence;


https://doi.org/10.52202/081568-0013

the biggest effect on tourist satisfaction is Digital Convenience ($\beta = 0.50$, $p = 0.005$).

## 5. Conclusion

The changing dynamics of service quality in digital tourism and their major influence on tourist satisfaction are underlined in this paper. The results suggest that improving tourist experiences depends critically on digital convenience—that which includes easy online booking, AI-driven recommendations, and mobile-friendly services. Although conventional measures of service quality—such as dependability, attentiveness, and empathy—remain essential—the incorporation of digital resources has changed consumer expectations. Travelers today want a hybrid strategy that combines tailored personal encounters with technical effectiveness.

The research does, however, also highlight possible negative effects on consumer impressions including technological faults and depersonalization. The results of the regression analysis show that satisfaction is most positively influenced by digital convenience, so underlining the need of user-friendly, effective digital platforms in travel industry.

These observations highlight for those involved in the travel business the need of balancing technical developments with human touch to maximize the quality of services. While keeping sympathetic, dependable human contacts, businesses should engage in artificial intelligence-driven personalizing, real-time customer care, and strong digital interfaces. Maintaining competitive advantage and encouraging long-term customer loyalty will depend on service tactics being changed to fit the evolving expectations of tech-savvy visitors as digital tourism develops.

## References


1. Buhalis, D., & Law, R. (2008). Progress in information technology and tourism management: 20 years on and 10 years after the Internet—The state of eTourism research. *Tourism Management, 29*(4), 609-623.
2. Buhalis, D. (2020). Technology in tourism-from information communication technologies to eTourism and smart tourism towards ambient intelligence tourism: A perspective article. *Tourism Review, 75*(1), 267-272.
3. Font, X., Elgammal, I., & Lamond, I. (2017). Greenhushing: The deliberate under-communication of sustainability practices by tourism businesses. *Journal of Sustainable Tourism, 25*(7), 1007-1023.
4. Fuchs, M., Höpken, W., &Lexhagen, M. (2014). Big data analytics for knowledge generation in tourism destinations–A case from Sweden. *Journal of Destination Marketing & Management, 3*(4), 198-209.
5. Gefen, D., Karahanna, E., & Straub, D. W. (2003). Trust and TAM in online shopping: An integrated model. *MIS Quarterly, 27*(1), 51-90.
6. Gretzel, U., Fesenmaier, D. R., Formica, S., & O'Leary, J. T. (2006). Searching for the future: Challenges faced by destination marketing organizations. *Journal of Travel Research, 45*(2), 116-126.


https://doi.org/10.52202/081568-0013

7. Jung, T., tom Dieck, M. C., Moorhouse, N., & Chung, N. (2016). The role of augmented reality for experience-enhanced museum environments. *Computers in Human Behavior, 57*, 82-95.
8. Ladhari, R. (2009). Service quality, emotional satisfaction, and behavioural intentions: A study in the hotel industry. *Managing Service Quality: An International Journal, 19*(3), 308-331.
9. Law, R., Buhalis, D., &Cobanoglu, C. (2015). Progress on information and communication technologies in hospitality and tourism. *International Journal of Contemporary Hospitality Management, 27*(5), 727-761.
10. Litvin, S. W., Goldsmith, R. E., & Pan, B. (2008). Electronic word-of-mouth in hospitality and tourism management. *Tourism Management, 29*(3), 458-468.
11. Neuhofer, B., Buhalis, D., &Ladkin, A. (2015). Smart technologies for personalized experiences: A case study in the hospitality domain. *Electronic Markets, 25*(3), 243-254.
12. Parasuraman, A., Zeithaml, V. A., & Berry, L. L. (1988). SERVQUAL: A multiple-item scale for measuring consumer perceptions of service quality. *Journal of Retailing, 64*(1), 12-40.
13. Racherla, P., & Hu, C. (2009). A social network perspective of tourism research collaborations. *Annals of Tourism Research, 36*(3), 297-321.
14. Sigala, M. (2018). Market formation in the sharing economy: Findings and implications from the sub-economies of Airbnb. *Journal of Business Research, 86*, 521-532.
15. Tussyadiah, I. P., & Miller, G. (2019). Perceived impacts of AI and robotics on the tourism experience. *Journal of Travel Research, 58*(7), 1107-1121.
16. 1Vermeulen, I. E., &Seegers, D. (2009). Tried and tested: The impact of online hotel reviews on consumer consideration. *Tourism Management, 30*(1), 123-127.
17. Wang, D., Xiang, Z., &Fesenmaier, D. R. (2016). Smartphone use in everyday life and travel. *Journal of Travel Research, 55*(1), 52-63.
18. Xiang, Z., &Gretzel, U. (2010). Role of social media in online travel information search. *Tourism Management, 31*(2), 179-188.
19. Yen, H. R., &Gwinner, K. P. (2003). Internet retail customer loyalty: The mediating role of relational benefits. *International Journal of Service Industry Management, 14*(5), 483-500.
20. Zeithaml, V. A., Parasuraman, A., & Malhotra, A. (2000). A conceptual framework for understanding e-service quality: Implications for future research and managerial practice. *Marketing Science Institute, 00*(115).


https://doi.org/10.52202/081568-0013

# The Role of Authenticity in Destination Development: A Regenerative Approach

Shreeya Nayak[1*] and Debashish Pattanaik[2]

[1]Research Scholar, School of Humanities and Social Sciences, NISER Bhubaneswar, Homi Bhabha National Institute, Mumbai, India
[2]Readear (F), School of Humanities and Social Sciences, NISER Bhubaneswar, Homi Bhabha National Institute, Mumbai, India

*Corresponding author: shreeyanayak01@gmail.com

**Abstract.** Tourism is driven by three primary motivations: the desire for novelty, the pursuit of experiences unavailable within one's life-space,and the quest for destinations beyond the boundaries of one's everyday life. However,tourists often arrive with preconceived notions about a place, phenomenon, or experience, which can shape their perceptions and expectations.Authenticity, a social construct, is frequently negotiated and redefined in the context of tourism.The tourists' preconceived notions can create a disconnect between their expectations and the actual experience, leading to a problematic situation for locals.Any sign of modernity or development may be perceived as inauthentic, forcing the host communities to conform to tourist expectations. Governments often cater to the needs of tourists,rather than prioritizing local concerns.The decision-making power regarding what constitutes authenticity, how it should be preserved,how often it should be delivered and in what ways lies primarily in the hands of governments and tourists. However,locals should have the most significant say in these matters.This paradigm perpetuates the "museumization" of not only places but also cultures and people,denying local communities the right to change and evolve in their own way.Therefore,a regenerative approach to tourism is essential,where decision-making power is distributed equitably among locals, governments, and tourists.A regenerative approach prioritizes mutual benefits,ensuring that both hosts and guests benefit equally.This perspective shifts the focus from economic gains to ethical considerations,recognizing the interconnectedness of tourism's impacts.This paper highlights a)how destination development is influenced by tourists' perception of authenticity b)role of regenerative tourism in addressing the paradox of authenticity and promoting inclusive development c)various ways by which local communities and stakeholders can be empowered to redefine authenticity and shape destination development.

**Keywords:** authenticity, regenerative tourism, destination development, tourist

# 1 Introduction

In today's society, authenticity is often assumed to be compromised. Authenticity is a fluid concept shaped by societal norms and expectations. It is always open to interpretation and redefinition. "If nothing on 'the outside' can be relied upon to give weight to the individual's sense of reality, he is left no option but to burrow into himself in search of the real. Whatever this ens realissimum may then turn out to be, it must necessarily be in opposition to any external (modern) social formation. The opposition between self and society has now reached its maximum. The concept of authenticity is one way of articulating this experience"(Berger 1973:88). MacCannell (1976) started the debate on authenticity and tourism. Authenticity is mostly associated with 'primitive' and considered in opposition to modernity. The bifurcation between a premodern(authentic) and modern (inauthentic) has been highly criticized (Meethan 2001;Olsen 2002).

The idea of authenticity is not homogeneous. There are various types of authenticity, such as **hot** and **cool authenticity**—where the former refers to artificial or staged versions and the latter to genuine experiences (Selwyn, 1996); **constructive** and **objective authenticity**—where constructive authenticity refers to authenticity based on socially constructed perceptions, and objective authenticity relates to the authenticity of the original or museum-like objects; and **existential authenticity**, which refers to a personal, subjective state in which individuals feel true to themselves during the tourism experience (Wang, 1999). Therefore, the term 'authenticity' has different conceptions for ethnographers, tourists, curators and anthropologists. In the modern era, many objects, phenomena and even places are considered as inauthentic not because they are actually fake but because the tag 'fakery' has gained high momentum. In this paper, the role of regenerative tourism in co-creation, defining, deconstructing and redefining of authenticity is emphasized. It focuses on the inclusivity of regenerative approach in paving the way for considering the voices of the local communities over the government, private bodies and tourists and the various ways to make this possible so that their development can be ensured.

# 2. Review of Literature

Boorstin (1961) and MacCannell (1973) suggested in testing the object authenticity based on some criteria. Wang (1999) considered things as authentic not because they are inherently so but because their genuineness is constructed by beliefs. According to Moscardo and Pearce (1999), authenticity of a tourism setting is not a tangible asset but an intangible judgement placed by an observer. Appadurai (1986), emphasized that limited production of an object makes it seem authentic whereas Revilla and Dodd (2003) believed that tourists might perceive an expensive object as authentic. Cohen (1988) emphasized that authenticity is based on selective perception and selective memory and how with the passage of time objects earlier considered as artificial are considered as authentic (which he terms as 'emergent authenticity').

# 3. Research Methodology

This paper sheds light on the intricate interplay between authenticity, regenerative


https://doi.org/10.52202/081568-0014

tourism and destination development. Through a regenerative approach, this paper aims at guiding efforts towards a resilient, sustainable, inclusive destination development. The insights are drawn from secondary sources which includes scholarly articles. Based on a thorough review of the relevant literature, it critically examines the existing system of tourism development and paves the way for a new approach.

### 3.1 Authenticity

Authenticity has become a central point of discussion in tourism literature since the last few decades. It has numerous, diverse and dichotomous meanings. Authenticity was originally used in the field of tourism by the experts in determining whether the objects of art displayed at the museum were worth the admiration and price they were getting(Trilling, 1972). The tourists, tourism marketers and scholars have extended this museum-linked usage to culture. There are three approaches dedicated to the study of authenticity.

1. Modernists, Realists and Objectivists- According to Theobald, "Authenticity means genuine, unadulterated, or the real thing"(1998:411). An object in order to be authentic should not be manufactured "specifically for the market".(Cornet 1975:52). They consider authenticity as an inherent quality of objects that can be measured against absolute criteria. Therefore, they believe authenticity to be independent of tourists' experience and dependent on experts' judgements.

2. Constructivists- This ideology considers authenticity as a socially constructed interpretation rather than an objective phenomenon (Bruner 1989,1994; Cohen,1988; Spooner, 1986; Taylor,2001). Things appear to be authenthic not because they genuinely are but because authenticity is actively constructed by the power structure, beliefs and conceptions (Wang,1999). There is no reality untouched by human perception. Therefore, constructed authenticity is always relative (Cohen,1988), contextual(Salamone, 1997), depends on ideology(Silver 1993) and time(Cohen 1988).

3. Postmodernists- They believe authenticity to be either meaningless or of less concern. They are not concerned about it as long as they enjoy it. They believe that even an authentic product might appear inauthentic depending on the experiences the tourists try to seek.

### 3.2 Influence of Tourists' Perception of Authenticity on Destination Development

Tourists are categorised into five types- recreational, diversionary, experiential, experimental and existential (Cohen,1972). The experimental tourists who experiment with various types of potential elective centres (Cohen 1979a:189) share similarities with existential tourists in their criteria of authenticity (Wang, 1999). On the other hand, recreational tourists tend to prioritize experience over authenticity, accepting cultural products in a make-believe situation (Cohen,1985).In contrast, the diversionary tourists are indifferent to authenticity altogether(Cohen,1972). The experimental and existential tourists are more discerning and apply strict criteria to evaluate authenticity (Wang, 1999; Rickly-Boyd,2013). This highlights that different types of tourists have different levels of concern regarding authenticity ranging from highly indifferent to extremely discerning.

The question that surrounds the debate of authentic and inauthentic is not whether the tourist has a purely authentic experience or not(MacCannell, 1973) but what furnishes his feelings with authenticity in his own view. Therefore, the tourists differ in the number and variety of features essential to their mind to authenticate a phenomenon, place or cultural product. The criteria of authenticity for an intellectual tourist is stricter than the criteria of a common tourist(Cohen 1988). The more alienated a tourist is, the more is his quest for authenticity(Cohen, 1979a:181-2). The reality is either accepted, constructed or denied. Tourists should accept the fact that how it makes you feel is equally important as what it actually is. Overanalyzing a situation or object can lead to losing sight of the situation's inherent value. What is commoditized or modernized is not always contradictory to the old one. It can also be additive. Some tourists might prefer market products as inauthentic whereas some tourists might accept a tribal art as authentic as long as it has tribal art prints on it. Therefore, the meaning of authenticity is beyond consensus, primitive and objective.

The tourist's perspective on a destination plays an essential role in shaping the destination's image, destination branding and the brand personality (Gallarza, Saura C Garcia, 2002; Konecnick C Gartner, 2007). The success of a destination depends strongly on tourists' approval, recommendations and revisit intentions (Yoon C Uysal, 2005). The influx of tourists, development of host population and the ranking of the destination are influenced by tourists' perception and experiences(Murphy, Pritchard C Smith, 2000).

Positive word- of mouth, online reviews and posts on social media enhance a destination's reputation and attract more tourists (Litvin, Goldsmith C Pan, 2008). Negative experiences and reviews deter potential tourists and harm the destination's image (Kotler, Bowen C Makens, 2010). The tourists should be held responsible for the reality they create, the preconceived notions they carry and the takeaways they choose. Things don't always happen to the tourists they play an active role in the experiences they have. Therefore, measures should be taken in order to prevent the future of a destination from being completely dependent the viewpoints of the tourists.

### 3.3 Role of regenerative tourism in addressing the paradox of authenticity and promoting inclusive development

Regenerative tourism originated from sustainable tourism in the 1980s as a response to the environmental and social impacts of mass tourism. It gained momentum with the launch of the Regenerative Tourism initiative by the Center for Responsible Travel (CREST) in 2019. Regenerative tourism developed as a response to the limitations of sustainable tourism. Sustainable tourism focused on mitigating negative impacts rather than creating positive change whereas regenerative tourism goes beyond sustainability by regenerating and restoring local ecosystems, cultures, and communities. Regenerative tourism is still an emerging field. It is gaining interest among tourists and tour operators with many seeking to adopt regenerative practices. Several certification programs have been launched to promote it. Research institutions and organizations are exploring and innovating regenerative tourism practices, including new business models and technologies.

Governments and international organizations are incorporating regenerative tourism principles into their tourism policies.

The notion of authenticity in tourism is often perceived as a romanticized idea of primitive local communities and underdeveloped destinations (MacCannell, 1973).


https://doi.org/10.52202/081568-0014

This concept has unintended consequences, denying these communities and destinations the opportunity for development. Many tourists seek authentic experiences, which are perceived as being untouched by modernity (Cohen, 1988). This creates a paradox where local communities and underdeveloped destinations are expected to remain unchanged in order to preserve their "authenticity" for the benefit of tourists. By remaining static, local communities may be denied access to economic benefits such as improved infrastructure, education, employment and healthcare (Butler, 1999). Such an expectation of authenticity can lead to cultural stagnation, where local communities are pressurized to maintain traditional practices and customs, rather than evolving or adapting to the changing circumstances (Harrison, 2001). This results in inequitable distribution of benefits where the external stakeholders, such as tour operators and travel companies benefit more than local communities, exacerbating existing power imbalances (Mowforth C Munt, 2003).

Regenerative tourism recognizes that authenticity is dynamic and evolving (Wang, 1999). The destinations and local communities should be encouraged to develop while maintaining their unique cultural identities. It focuses on empowering local communities and ensuring that benefits are equitably distributed and aligns with community needs and values (Manyara C Jones, 2007). It promotes sustainable and responsible tourism practices and prioritizes environmental conservation, cultural heritage preservation, and community well-being (UNWTO, 2019).

Various ways by which local communities and stakeholders can be empowered to redefine authenticity and shape destination development Empowering the local communities and stakeholders is necessary in redefining authenticity and shaping destination development. Their empowerment can increase their decision making power in the development sector in general and tourism sector in particular. It can occur in various ways. The community-based tourism enables local communities not only in managing but also in benefiting from tourism development (Manyara C Jones, 2007). The cooperative ownership models allow local communities to owning, managing and ensuring that the benefits of tourism practices are equitably distributed (Higgins-Desbiolles, 2006). The participatory rural appraisal (PRA) involves local communities in the decision-making process, ensuring that their needs and concerns are properly addressed (Chambers, 1994). The collaborative governance models help in assembling local communities, stakeholders, and government agencies to make decisions about destination development (Bramwell C Lane, 2000).

Various training programs should be introduced to enhance local communities' skills and knowledge, by educating and enabling them to manage and benefit from tourism development (UNWTO, 2019). The mentorship programs can introduce pairing of local communities with experienced tourism professionals, thus providing guidance and support (Gartner, 1996). Participatory budgeting should be enhanced allowing local communities to have a say in how tourism revenue is spent and ensuring that the benefits are equitably distributed (Cabannes, 2004). Digital platforms can enrich the minds of local communities with access to information, markets, and resources and by empowering them to participate in destination development (Gretzel C Jamal, 2009). Community-led digital initiatives should be organized frequently in order to enable local communities to manage and promote their own tourism products and services (Huang et al., 2017). The government and tourism professionals should consider the point of views of the localites on their objects being commoditized. Not all natives believe in the notion that commoditization of products will destroy its authenticity.

## 4. Conclusion

Authenticity has become a pivotal concern in tourism development since the 20th century. The tourists' perceptions of authenticity significantly influence the trajectory of destination development. Surprisingly, external stakeholders, including tourists, mostly have a greater say in shaping destination development rather than local communities. Given that authenticity is subjective and varies immensely among tourists, it's essential to consider it in conjunction with other factors while planning to develop a destination. A regenerative approach is therefore necessary, where local communities will be empowered to define and manage their cultural heritage. This approach must include the declaration of local communities on what is authentic and why, whether and how to commoditize their cultural heritage, deciding their share of benefits from tourism development , planning and controlling what aspects of their culture they should reveal to the tourists. Tourists must realize that their experiences are shaped by individual perceptions and not by presented attractions. The plans, policies and marketing strategies such as destination branding should be developed in collaboration with local communities in order to ensure their active involvement and consent. Therefore, adopting a regenerative and community-led approach, helps in ensuring that the tourism destination development is inclusive, sustainable, and respectful of local cultures.

## References


1. Appadurai, A. (Ed.). (1986). The social life of things: Commodities in cultural perspective. Cambridge University Press.
2. Appadurai, A. (1986). Introduction. In A. Appadurai (Ed.), The social life of things: Commodities in cultural perspective (pp. 3–63). Cambridge University Press.
3. Berger, P. (1973). "Sincerity" and "authenticity" in modern society. Public Interest, 1973(Spring), 81–90.
4. Boorstin, D. J. (1961). The image: A guide to pseudo-events in America. Harper & Row.
5. Bramwell, B., & Lane, B. (2000). Collaboration and partnerships in tourism planning. In B. Bramwell & B. Lane (Eds.), Tourism collaboration and partnerships: Politics, practice and sustainability (pp. 1–19). Channel View Publications.
6. Butler, R. W. (1999). Sustainable tourism: A marketing perspective. Journal of Sustainable Tourism, 7(1), 1–14.
7. Cabannes, Y. (2004). Participatory budgeting: A significant contribution to participatory democracy. Environment and Urbanization, 16(1), 27–46.
8. Chambers, R. (1994). The origins and practice of participatory rural appraisal. World Development, 22(7), 953–969.
9. Cohen, E. (1972). Toward a sociology of international tourism. Social Research, 39(1), 164–182.
10. Cohen, E. (1979). A phenomenology of tourist experiences. Sociology, 13(2), 179–201.
11. Cohen, E. (1985). Tourism as play. Religion, 15(3), 291–304.
12. Cohen, E. (1988). Authenticity and commoditization in tourism. Annals of Tourism Research, 15(3), 371–386.
13. Gallarza, M. G., Saura, I. G., & García, H. C. (2002). Destination image: Towards a conceptual framework. Annals of Tourism Research, 29(1), 56–78.
14. Gartner, W. C. (1996). Tourism development: Principles, processes, and policies. Van Nostrand Reinhold.
15. Gretzel, U., & Jamal, T. (2009). Conceptualizing the creative tourist class: Technology, identity and travel. Tourism Analysis, 14(4), 471–481.
16. Harrison, D. (2001). Tourism and the less developed world: Issues and case studies. CABI Publishing.


 https://doi.org/10.52202/081568-0014

17. Higgins-Desbiolles, F. (2006). More than an "industry": The forgotten power of tourism as a social force. Tourism Management, 27(6), 1192–1208.
18. Huang, Y., Xu, X., & Li, M. (2017). Exploring the impact of social media on rural tourism development: A case study of China's rural tourism. Journal of Sustainable Tourism, 25(8), 1088–1105.
19. Konecnik, M., & Gartner, W. C. (2007). Customer-based brand equity for a destination. Annals of Tourism Research, 34(2), 400–421.
20. Kotler, P., Bowen, J. T., & Makens, J. C. (2010). Marketing for hospitality and tourism (5th ed.). Pearson Education.
21. Litvin, S. W., Goldsmith, R. E., & Pan, B. (2008). Electronic word-of-mouth in hospitality and tourism management. Tourism Management, 29(3), 458–468.
22. Mac Cannell, D. (1973). Staged authenticity: Arrangements of social space in tourist settings. American Journal of Sociology, 79(3), 589–603.
23. Mac Cannell, D. (1976). The tourist: A new theory of the leisure class. Schocken Books.
24. Manyara, G., & Jones, E. (2007). Community-based tourism enterprises: A case study of Kenya's community-based tourism development. Journal of Sustainable Tourism, 15(6), 617–633.
25. Meethan, K. (2001). Place, culture, consumption. Palgrave.
26. Moscardo, G., & Pearce, P. (1999). Understanding ethnic tourists. Annals of Tourism Research, 26(2), 416–434.
27. Mowforth, M., & Munt, I. (2003). Tourism and sustainability: Development and new tourism in the Third World (2nd ed.). Routledge.
28. Murphy, P. E., Pritchard, M. P., & Smith, B. (2000). The destination product and its impact on traveller perceptions. Tourism Management, 21(1), 43–52.
29. Olsen, K. (2002). Authenticity as a concept in tourism research. Tourist Studies, 2(2), 159–182.
30. Revilla, G., & Dodd, T. (2003). Authenticity perceptions of Talavera pottery. Journal of Travel Research, 42(1), 94–99.
31. Rickly-Boyd, J. M. (2013). Existential authenticity: A conceptual framework for understanding tourist experiences. Journal of Tourism Studies, 14(1), 1–13.
32. Salamone, F. A. (1997). Authenticity in tourism: The San Angel. Annals of Tourism Research, 24(2), 305–321.
33. Selwyn, T. (Ed.). (1996). The tourist image: Myths and myth making in tourism. Wiley.
34. Silver, I. (1993). Marketing authenticity in Third World countries. Annals of Tourism Research, 20(2), 302–318.
35. Spooner, B. (1986). Weavers and dealers: The authenticity of an Oriental carpet. In A. Appadurai (Ed.), The social life of things: Commodities in cultural perspective (pp. 195–235). Cambridge University Press.
36. Taylor, J. P. (2001). Authenticity and sincerity in tourism. Annals of Tourism Research, 28(1), 7–26.
37. Theobald, W. F. (Ed.). (1998). Global tourism (2nd ed.). Butterworth-Heinemann.
38. Trilling, L. (1972). Sincerity and authenticity. Oxford University Press.
39. United Nations World Tourism Organization (UNWTO). (2019). Tourism for transformation: Creating a more sustainable, inclusive and resilient sector. UNWTO.
40. Wang, N. (1999). Rethinking authenticity in tourism experience. Annals of Tourism Research, 26(2), 349–370.
41. Yoon, Y., & Uysal, M. (2005). An examination of the effects of motivation and satisfaction on destination loyalty: A structural model. Tourism Management, 26(1), 45–56.

# Digital Innovations and Augmented Reality in Tourism Marketing: A New Frontier

Asha John[1] [0000-0002-8237-1918]

[1] Bharata Mata College , Thrikkakara ,Cochin

*Corresponding author: asha.jhn@gmail.com

**Abstract.** The tourism industry has experienced a transformation because of digital enhancements that create better destination marketing methods together with tourist attractions promotion and experience marketing. The industry transformation occurred through the game-changing innovation that augmented reality brought because it enables immersive storytelling to users. Research explores fresh tourism marketing applications of AR as well as its direct relation to visitor satisfaction levels and their degree of engagement. The research demonstrates how multiple augmented reality applications transform heritage exploration and theme parks and modernize urban discovery through their specific case study applications. The visual representations of AR systems present users with personalized choices to evaluate destinations during their decision-making process for travel destinations. The combination provides travelers with improved selection capabilities for their trip arrangements. The study examines the advantages alongside difficulties of using AR technology for marketing while focusing on implementation obstacles and costs and user adaptation problems. Research using both industry patterns and qualitative findings demonstrates how business digitalization particularly through AR systems enables companies to remain competitive within modern dynamic markets. Tourism stakeholders receive recommended strategies to use AR technology for developing valuable memorable encounters, which support industry sustainability goals.

**Keywords:** Augmented Reality (AR)-Tourism Marketing- Immersive Storytelling- Visitor Engagement-Heritage Sites- Sustainable Growth

## 1. Introduction

Augmented Reality (AR) functions as a novel technology by merging virtual digital elements such as images sounds videos and three-dimensional models with actual scanned reality through mobile devices smartphones and tablets and AR glasses. The main difference between AR and VR is that AR augments real-world environments with additional layers while providing users with enhanced important details in addition to interactive and informative experiences. Different sectors of business and education along with healthcare and entertainment recognize AR as the trend of the present year and envision its broad future use. Tourism benefits from advanced AR technology since it offers groundbreaking innovations that modify how people discover destinations.

AR technology enables heritage sites to become dynamic historical venues through virtual systems which present both virtual ancient ruins and explanation services in multiple languages and dynamic past visuals. The tourism industry uses AR technology because modern travelers with technology backgrounds want thorough immersive encounters during their vacations. Through its distinctive features the system provides both real-time informational assistances along with virtual guide services as well as game-based encounters and interactive narrative displays. AR technology permits cultural heritage preservation by showing intangible artifacts and helps tourism sustainability through lower physical development requirements, which create an informative destination experience.

This research evaluates AR technology for tourism improvement across multiple fronts including advantages and disadvantages together with projected market developments. This paper explores how AR can be properly implemented in tourism sites by investigating the approaches to engage users and promote destinations and maintain sustainable tourism initiatives. This paper examines the technological along with economic and regulatory obstacles preventing AR implementation in tourism while forecasting upcoming trends for AR-based tourism services. Tourism destinations


https://doi.org/10.52202/081568-0015

should adopt AR technology for developing valuable memorable informative visitor experiences, which support modern tourists and develop a sustainable tourism industry.

### 1.1 Digital Innovations in Tourism Marketing

New digital innovation methods allow destinations to interact through innovative communication with modern technology-oriented traveling consumers and service providers. Traditional tourism marketing undergoes transformation into interactive personalized services because digital technologies including mobile applications and social media as well as artificial intelligence (AI) and virtual reality (VR) and augmented reality (AR) technologies blend together (Buhalis & Law, 2008).

#### Overview of Digital Technologies in Tourism

Nurturing customer engagement along with operational effectiveness through digital tools remains fundamental for businesses operating in the digital tourism period because they deliver up-to-the-moment information to clients. Travelers use platforms that include websites and mobile applications along with social media to connect with destinations which provide current information and booking services and custom recommendations (Leung et al., 2013). Mobile applications let tourists connect smoothly and deliver site-based services that show them nearby attractions together with restaurant and transportation details. Travel customer service revolutionizes through AI tools like chatbots which provide on-demand assistance and personalized trip suggestions together with continual operation from every hour until 24 hours daily. The tourist experience enhances through virtual reality (VR) and augmented reality (AR) technologies since these platforms deliver interactive and immersive tools which strengthen destination connection for travelers (Tussyadiah et al., 2018).

#### Role of AR in Tourism Marketing

Tourism marketing substantially benefits from AR because this technology links online content to actual destinations. Through these techniques destinations create full-immersive location-specific experiences which enable tourists to gain deeper interactive understanding of cultural and historical backgrounds. Through AR applications tourists can see digital restorations of historical buildings on top of modern ruins because the technology creates visualizations of original structures (Han et al., 2014). Visitors can experience customized digital content through AR technology because it displays relevant multimedia guides regarding exhibitions and landmarks and cultural destinations. Users experience greater engagement through immersive AR since it provides unique destination exploration capabilities that lead them to share their discoveries on social media (Yovcheva et al., 2014).

#### Benefits of AR for Tourists and Marketers

Augmented Reality technology delivers many advantages to tourists which improve their travel journey. Through its immediate information delivery service along with personalized suggestions and current location assistance AR enables travelers to discover unknown locations seamlessly (Jung et al., 2015). Through AR-powered application interfaces users can obtain municipality tours as the software shows them historical destinations alongside local dining and leisure attractions which accommodate their selected preferences. Mobile AR creates interactive experiences which produce exciting memorable moments of new location discovery thus strengthening tourist bond with surroundings.

For marketers, AR provides a valuable tool for creating engaging and unique marketing campaigns that capture the attention of tech-savvy travelers. AR enables tourism marketers to craft innovative experiences that go beyond traditional advertising by offering immersive, interactive content that encourages users to engage with a destination in a more personal way. Through AR, marketers can differentiate their destinations from competitors, build brand awareness, and promote sustainable tourism practices by showcasing local culture, history, and nature (Huang et al., 2016). By integrating AR into their marketing strategies, destinations can appeal to a growing audience of digital natives who value personalization, innovation, and sustainability in their travel choices.

### 1.2 Applications of Augmented Reality in Tourism

Augmented Reality technology became a major. Disclaimer in tourism because it presents interactive experiences for tourists to better explore their destinations. Tourists experience

https://doi.org/10.52202/081568-0015

enhanced quality through AR since this technology applies digital content that brings information and engagement and customization to real-world spaces. The article examines various AR applications in tourism through heritage tourism and theme parks and attractions together with urban exploration.

**AR in Heritage Tourism**

AR plays a crucial role in enhancing heritage tourism by providing immersive and educational experiences that bring history to life. AR applications can overlay historical information, 3D reconstructions, and multimedia content on physical sites, allowing visitors to interact with their surroundings in a deeper way. For example, AR can recreate historical events or showcase the original appearance of ruins, offering tourists a unique opportunity to visualize the past and understand the cultural significance of the site (Han et al., 2014). This technology enables visitors to engage with the destination in a more interactive manner, helping them better appreciate the cultural heritage and history of the place. In addition to providing educational value, AR applications in heritage tourism can also foster a stronger emotional connection between the visitors and the destination, encouraging respect and preservation for cultural landmarks and artifacts (Jung et al., 2015).

**AR in Theme Parks and Attractions**

AR technology has found significant applications in theme parks and tourist attractions, where it enhances the overall visitor experience. Through the use of AR, parks can offer interactive entertainment, gamified experiences, and personalized location-based services. For instance, AR applications can bring virtual characters to life, create dynamic visual effects, and enable visitors to engage in location-based games and challenges that are integrated into the park's storyline (tom Dieck & Jung, 2017). These AR-enhanced experiences increase visitor engagement by providing a sense of immersion and personalization, which traditional media cannot replicate. Visitors can interact with virtual elements in real-time, making each visit unique and memorable. By integrating AR into the attraction's narrative, parks can also create a deeper emotional connection with their audience, enhancing their satisfaction and likelihood of returning.

**AR for Urban Exploration**

AR applications have greatly enriched urban exploration, offering tourists new ways to navigate and interact with cities. Urban AR applications, such as city guides and walking tours, enable tourists to access real-time information about landmarks, dining options, events, and local stories. By overlaying digital information onto the physical environment, AR enhances the tourist experience, allowing them to learn more about the cultural and historical context of various locations as they explore a city (Yovcheva et al., 2014). For example, an AR city guide might point out nearby historical landmarks, provide information about the architecture, or offer details about local traditions and customs. These interactive features transform typical city visits into immersive journeys, making exploration more engaging and educational. AR applications can also personalize the experience by providing tailored recommendations based on the user's preferences and location, allowing tourists to have a more customized and enjoyable visit.

## 1.3 Challenges and Opportunities

**Implementation Challenges**

The implementation of Augmented Reality (AR) in tourism is confronted with several technical complexities that hinder its seamless integration into tourism experiences. One major challenge lies in content creation, which requires high-quality, immersive, and engaging digital assets to ensure that the AR experience is both appealing and informative for tourists (Han et al., 2014). Furthermore, developing robust AR software that can function reliably across diverse environments, such as outdoor and indoor settings, presents another challenge. These environments are often unpredictable, requiring AR systems to adapt dynamically to factors like


https://doi.org/10.52202/081568-0015

changing lighting conditions, varying physical surroundings, and user movements. Geolocation accuracy is another key issue, as AR depends heavily on the system's ability to provide precise spatial and directional data. In addition to geolocation, seamless tracking of the user's position and the environment is critical to ensure that the virtual elements align correctly with the real world. Achieving context-aware experiences— where the AR content is relevant to the user's specific location, preferences, or the surrounding environment—is complex but necessary for providing meaningful tourism experiences (Han et al., 2014).

**Costs and ROI Analysis**

The costs associated with implementing AR in tourism can be significant, particularly during the initial phase of adoption. Businesses need to invest in hardware, such as smartphones or AR glasses, and in developing or licensing software capable of supporting AR features (Huang et al., 2016). Additionally, creating high-quality AR content—whether it's 3D models, interactive features, or georeferenced data—adds further to the financial burden. These costs often lead to uncertainty regarding the return on investment (ROI) for businesses, particularly in the tourism sector, where profit margins can be thin and the impact of AR may be difficult to quantify. Businesses must weigh the long-term benefits of improved customer experiences, brand differentiation, and increased visitor engagement against the immediate financial outlay. The challenge, therefore, is to accurately forecast how AR technology can drive future revenue streams and whether the technology's adoption will lead to measurable enhancements in the tourism experience that justify the initial costs (Huang et al., 2016).

**User Adoption Barriers**

The acceptance of AR technology among tourists is crucial to its successful implementation in the tourism industry. However, user adoption can be hindered by several factors. One significant barrier is the lack of familiarity with AR technology. Many potential users may not fully understand how AR works or how it can enhance their tourism experiences. This knowledge gap can lead to hesitation or reluctance to use AR-based applications. Moreover, usability issues often arise, particularly when AR interfaces are not intuitive or require too much technical skill to operate. If the user experience is complicated or the technology is prone to malfunctions, users are less likely to engage with the AR features. Another barrier is related to privacy concerns. Tourists may be apprehensive about the data collection practices associated with AR applications, especially when sensitive location data and personal preferences are involved. Educating users about the benefits of AR, along with ensuring that the technology is easy to use and respects privacy, is essential for overcoming these barriers and increasing user adoption (Jung et al., 2015).

**Opportunities for Growth**

Despite the challenges, AR technology presents significant opportunities for growth within the tourism sector. As mobile technology continues to advance, and with the rise of AI and 5G networks, the capabilities of AR are poised to expand. These advancements can enhance the quality and accessibility of AR applications, making them more immersive and responsive. The proliferation of smartphones equipped with AR functionality, combined with faster internet speeds, allows for smoother and more engaging AR experiences. Additionally, the increasing accessibility of AR development tools and platforms lowers the entry barrier for small and medium-sized tourism businesses, fostering innovation and competition. Collaboration among tourism stakeholders—such as travel agencies, local authorities, hospitality businesses, and tech developers—can facilitate the creation of integrated AR experiences that drive market expansion. By pooling resources and sharing expertise, these stakeholders can create compelling tourism products that offer unique, contextually rich experiences to travelers, further enhancing the appeal of destinations and boosting tourism revenues (Tom Dieck & Jung, 2017).

## 2. Recommendations

To maximize the benefits of Augmented Reality (AR) in the tourism industry, tourism stakeholders should adopt several key strategies:

https://doi.org/10.52202/081568-0015

1. **Investment in High-Quality Content**: Stakeholders should focus on developing high-quality, context-relevant AR content that aligns with the cultural, historical, and natural features of a destination. Engaging, informative, and interactive content will enhance the overall tourist experience and encourage repeat visits (Jung et al., 2015).
2. **Seamless Integration and User-Friendly Interfaces**: AR experiences should be designed to integrate smoothly into the environment, ensuring seamless interaction for users. User interfaces should be intuitive and easily accessible to encourage widespread adoption among tourists of all age groups and tech levels (Han et al., 2014).
3. **Collaboration with Technology Developers**: Tourism organizations should collaborate with AR developers and tech companies to create customized solutions that meet the specific needs of the tourism sector. This partnership can help in the development of cost-effective, scalable solutions (Huang et al., 2016).
4. **Government Support and Regulation**: Governments should provide support for AR adoption through funding, regulation, and policies that promote innovation while ensuring data security and privacy for users (Tom Dieck & Jung, 2017).
5. **Training and Education**: Educating tourists on how to use AR applications will improve user engagement and foster acceptance. Tourism stakeholders should also train employees to offer assistance to tourists unfamiliar with AR technology.

## Conclusion

In conclusion, Augmented Reality (AR) presents a transformative opportunity for the tourism industry by enhancing the visitor experience, fostering greater engagement, and supporting the sustainability of tourism practices. By investing in high-quality content and leveraging collaborations with technology developers, the tourism industry can offer immersive, educational, and personalized experiences that attract tech-savvy travelers. To fully harness AR's potential, stakeholders must address implementation challenges, user adoption barriers, and cost concerns. With continued advancements in mobile technology and AR tools, the future of AR in tourism is promising, creating long-term benefits for both tourists and destination marketers (Buhalis & Law, 2008).

## References


1. Buhalis, D., & Law, R. (2008). Progress in information technology and tourism management: 20 years on and 10 years after the internet—The state of eTourism research. *Tourism Management, 29*(4), 609- 623. https://doi.org/10.1016/j.tourman.2008.01.024
2. Han, D., Jung, T., & Gibson, A. (2014). Dublin AR: Implementing augmented reality in tourism. In *Information and Communication Technologies in Tourism 2014* (pp. 511-523). Springer. https://doi.org/10.1007/978-3-319-03973-7_42
3. Han, D., Jung, T. H., & Gibson, A. (2014). The impact of augmented reality on tourism experiences: The case of AR in the tourism industry. *Tourism Management, 45*, 307-319. https://doi.org/10.1016/j.tourman.2014.06.002
4. Huang, Y.-C., Backman, S. J., Backman, K. F., & Moore, D. (2016). Exploring user acceptance of 3D virtual worlds in travel and tourism marketing. *Tourism Management, 57*, 148-160. https://doi.org/10.1016/j.tourman.2016.04.010
5. Huang, Y. C., Backman, S. J., & Backman, K. F. (2016). The role of augmented reality in tourism: A literature review and future research agenda. *Journal of Hospitality and Tourism Technology, 7*(3), 289-305. https://doi.org/10.1108/JHTT-01-2016-0006
6. Jung, T., tom Dieck, M. C., Lee, H., & Chung, N. (2015). Effects of virtual reality and augmented reality on visitor experiences in museums. In *Information and Communication Technologies in Tourism 2015* (pp. 621-635). Springer. https://doi.org/10.1007/978-3-319-14343-7_50
7. Jung, T. H., tom Dieck, M. C., & Han, D. (2015). Tourist acceptance of augmented reality: An exploration of factors influencing acceptance of AR in tourism. *International Journal of Tourism Research, 17*(6), 514-526. https://doi.org/10.1002/jtr.2062
8. Leung, D., Law, R., Hoof, H., & Buhalis, D. (2013). Social media in tourism and hospitality: A literature review. *Journal of Travel & Tourism Marketing, 30*(1-2), 3-22. https://doi.org/10.1080/10548408.2013.751276



https://doi.org/10.52202/081568-0015

10. Tom Dieck, M. C., & Jung, T. H. (2017). Augmented reality in tourism: A framework for the application and future research. In *Information and Communication Technologies in Tourism 2017* (pp. 213-226). Springer. https://doi.org/10.1007/978-3-319-51174-9_18
11. Tom Dieck, M. C., & Jung, T. (2017). Value of augmented reality at cultural heritage sites: A stakeholder approach. *Journal of Destination Marketing & Management, 6*(2), 110-117. https://doi.org/10.1016/j.jdmm.2016.09.003
12. Tussyadiah, I. P., Wang, D., & Jia, C. H. (2018). Virtual reality and attitudes toward tourism destinations. In *Information and Communication Technologies in Tourism 2018* (pp. 229-239). Springer. https://doi.org/10.1007/978-3-319-91038-7_19
13. Yovcheva, Z., Buhalis, D., & Gatzidis, C. (2014). Augmented reality in tourism: A state of the art review. In *Enter 2014* (pp. 249-259). Springer. https://doi.org/10.1007/978-3-319-04636-0_23
14. Yovcheva, Z., Buhalis, D., & Gatzidis, C. (2014). Empirical evaluation of smartphone augmented reality browsers in urban tourism destinations. *International Journal of Mobile Human-Computer Interaction, 6*(2), 10-31. https://doi.org/10.4018/ijmhci.2014040102

# Impact of Virtual Tourism on Conservation Efforts and Sustainable Engagement in Protected Areas

Mr. Sagar Bhatt[0000-0001-9173-914X]

Associate Professor, Chandigarh University, Gharaun, Mohali

*Corresponding author: sagarbhatt1010@gmail.com

**Abstract.** Virtual tourism, is based on digital technologies to simulate travel experiences and knowledge, has emerged as a possible solution for balancing tourism demand and the conservation of protected areas. This chapter investigates its impact on conservation efforts and sustainable engagement. By analyzing data from visitor behavior, survey results, and case studies of protected areas, in this study it has been examine how virtual tourism supports environmental sustainability, reduces physical degradation, and fosters global awareness. The findings indicate that virtual tourism reduces ecological tracks while maintaining community engagement. This study also highlights few challenges that include limited accessibility and technology costs, while proposing solutions to enhance inclusivity and immersive experiences.

**Keywords:** Virtual Tourism, Conservation Efforts, Sustainable Engagement, Protected Areas, Digital Innovation

## 1. Introduction

The connection of virtual tourism and conservation efforts signifies a transformative approach to sustainable engagement in protected areas. Virtual tourism, well-defined as the use of immersive digital technologies to simulate travel experiences, has emerged as a practical alternative to physical tourism, particularly in environmentally sensitive regions (Guttentag, 2010). By virtual reality (VR), augmented reality (AR), and other E platforms, stakeholders can provide realistic, engaging involvements while preserving the integrity of natural surroundings. Protected areas are crucial for biodiversity conservation, yet they often face threats such as locale destruction, trash, and resource exhaustion due to excessive tourist activity (Eagles, 2002).Virtual tourism offers a sustainable solution by reducing physical visit and its associated ecological footprint. Furthermore, it fosters inclusivity by enabling access for individuals unable to visit these areas physically, thereby balancing environmental appreciation and encouragement (Becken, 2014). This paper investigates the impact of virtual tourism on conservation efforts and its potential to promote sustainable engagement. Specifically, it explores how virtual tourism influences visitor behavior, supports community- based conservation initiatives, and contributes to the financial sustainability of protected areas through innovative revenue streams.

Tourism has long been recognized as a powerful economic and social force, connecting people with diverse cultures, natural wonders, and wildlife. However, the rapid growth of the global tourism industry has raised concerns about its environmental impact and sustainability. Marine ecosystems, in particular, face significant challenges due to human activities such as overfishing, pollution, and habitat degradation, often exacerbated by unsustainable tourism practices. The critical need to balance tourism growth with environmental conservation has led to increased interest in innovative technologies, such as Virtual Reality (VR) and Extended Reality (XR), to create sustainable tourism experiences that encourage pro-environmental behavior.


https://doi.org/10.52202/081568-0016

In recent years, virtual and augmented reality technologies have emerged as transformative tools for sustainable tourism, offering immersive and engaging experiences that can substitute or supplement real-world visits to environmentally sensitive areas. These technologies allow tourists to experience marine and terrestrial environments without causing physical harm, thereby reducing the ecological footprint of tourism activities. This introduction explores the growing body of research on the effectiveness of virtual and extended reality experiences in promoting sustainable tourism and fostering conservation behavior.

The study by (Hofman, 2022) investigates the comparative effectiveness of virtual and real- life marine tourism experiences in promoting conservation behavior. Marine tourism, which includes activities such as snorkeling, scuba diving, and wildlife observation, often brings tourists into direct contact with fragile ecosystems. While real-life experiences provide sensory engagement and emotional connections to marine environments, they can also lead to unintentional ecological damage, such as coral reef destruction or wildlife disturbance. (Branstrator, 2023)extend the discussion on the role of immersive technologies in sustainable tourism by examining the use of Extended Reality (XR) to promote biocultural conservation. Biocultural conservation recognizes the interconnectedness of biological and cultural diversity and emphasizes the need to protect both natural ecosystems and the cultural practices associated with them. XR, which encompasses virtual, augmented, and mixed reality technologies, offers unique opportunities to "re-story" human-wildlife relationships and foster a deeper appreciation for the cultural significance of natural environments.

The authors highlight several case studies where XR has been used to create immersive narratives that blend ecological information with cultural storytelling. For example, XR experiences can transport users to indigenous territories, where they can learn about traditional ecological knowledge and the spiritual significance of certain species or landscapes. By integrating cultural context into environmental education, XR can enhance tourists' understanding of the intrinsic value of biodiversity and the importance of conserving both natural and cultural heritage.

(Caciora, 2021) provide a compelling example of how VR can be used to promote sustainable tourism in cultural heritage sites. Their study focuses on the use of VR to create virtual tours of Romania's wooden churches, which are historical monuments and UNESCO World Heritage sites. These churches, known for their unique architecture and cultural significance, are located in rural areas with limited infrastructure, making them vulnerable to the negative impacts of mass tourism.

The researchers developed high-resolution VR simulations that allow users to explore the churches' interiors and exteriors, learn about their history and cultural importance, and appreciate their architectural details. The virtual tours not only provide an alternative to physical visits but also serve as a tool for preserving and disseminating cultural heritage. By offering a sustainable way to experience these sites, VR can help reduce the environmental and social pressures associated with tourism, such as overcrowding, infrastructure degradation, and culture (Nayak, 2025) explore the potential of VR simulations to increase public participation in conservation initiatives. Their research highlights the role of VR in creating immersive and interactive experiences that can engage the public in environmental issues and motivate them to take action. VR simulations can be used to visualize complex ecological processes, such as climate change, deforestation, or habitat restoration, making these issues more tangible and understandable to a broad audience.

The authors argue that VR can serve as a powerful tool for environmental advocacy and education, particularly in urban areas where direct contact with nature is limited. By providing virtual access to remote or endangered ecosystems, VR can raise awareness about environmental challenges and inspire a sense of urgency and responsibility. Moreover, VR simulations can be used to train and mobilize volunteers for conservation projects, such as habitat restoration, wildlife monitoring, or citizen science initiatives. (Su, 2024) investigates the impact of ecological presence in VR tourism on tourists' environmentally responsible behavior. Their study focuses on the psychological mechanisms that drive pro-environmental behavior, such as empathy, emotional engagement, and perceived self-efficacy. The researchers found that high levels of ecological presence in VR environments can significantly enhance tourists' emotional connections to nature and their sense of responsibility for environmental protection.

The study highlights the importance of designing VR experiences that maximize ecological presence by incorporating realistic visuals, immersive soundscapes, and interactive elements. For example, VR simulations of natural parks or marine ecosystems can include dynamic weather conditions, realistic wildlife behavior, and interactive conservation tasks, such as planting virtual trees or rescuing endangered animals. These features can create a sense of agency and empowerment, motivating tourists to adopt environmentally responsible behaviors both during and after their virtual experiences.

## 2. Review of Literature:

The intersection of technology and sustainable tourism has been a growing area of interest, with recent studies highlighting how immersive technologies like virtual reality (VR) and extended reality (XR) can transform tourism experiences while promoting conservation behavior. Traditional tourism, particularly in sensitive ecosystems and protected areas, often poses significant environmental challenges. As Becken and Job (2014) point out, global and local pressures on protected areas necessitate innovative strategies to balance tourism growth with ecological preservation. In response, scholars and practitioners are increasingly turning to digital technologies as tools for sustainable tourism management.

Virtual reality (VR) has emerged as a promising solution for reducing the environmental footprint of tourism while enhancing tourist engagement. Guttentag (2010) provides a foundational understanding of VR's applications in tourism, noting its potential to offer realistic, immersive experiences that can substitute physical travel to fragile destinations. This potential is further supported by Hofman, Walters, and Hughes (2022), who demonstrate that virtual marine tourism experiences can foster similar emotional and educational outcomes as real-life interactions, without causing harm to delicate marine ecosystems. Their study emphasizes the role of ecological presence—users' psychological immersion in a virtual environment—in driving pro-environmental behavior and promoting a sense of stewardship. Similarly, Branstrator et al. (2023) explore how XR technologies can facilitate biocultural conservation by "re-storying" human-wildlife relationships. Their research highlights how XR can integrate cultural narratives with ecological education, providing users with a deeper understanding of the interconnectedness between biodiversity and cultural heritage. These immersive narratives can influence tourists' perceptions and behaviors, making them more inclined to support conservation initiatives both financially and through advocacy.

In addition to its role in environmental education, VR has shown potential in increasing public participation in conservation efforts. Nayak et al. (2025) discuss how VR simulations can visualize complex environmental issues, such as climate change and habitat loss, in an engaging and comprehensible manner. By making these issues more accessible to the general public, VR can inspire greater involvement in conservation activities, including citizen science and volunteer work.


https://doi.org/10.52202/081568-0016

Furthermore, the use of VR for cultural heritage conservation has been explored by Caciora et al. (2021), who developed virtual tours of Romania's wooden churches. These VR tours not only preserve the architectural and historical value of the sites but also offer a sustainable alternative to physical visits, thereby reducing the negative impacts of tourism on rural communities and fragile structures.

The potential of immersive technologies extends beyond education and heritage conservation. Stappung et al. (2023) highlight the development of 360-degree virtual tours for Chilean national parks, showcasing how such experiences can promote sustainable tourism in natural parks. By offering virtual access to remote locations, these tours reduce the environmental impact of physical travel while expanding the reach of conservation messaging to a global audience.

Finally, Su et al. (2024) emphasize that the design of VR experiences plays a critical role in influencing tourists' environmentally responsible behaviors. Their study found that ecological presence, enhanced by realistic visuals, interactive elements, and dynamic soundscapes, significantly increases tourists' emotional connection to nature, motivating them to adopt sustainable practices even after the virtual experience concludes.

## 3. Scope of the Study

This study aims to explore the role of immersive technologies, specifically virtual reality (VR) and extended reality (XR), in promoting sustainable tourism and conservation behaviors. While previous research has demonstrated the potential of VR and XR to substitute real-world tourism experiences, enhance environmental education, and increase public engagement in conservation efforts, gaps remain in understanding the long-term behavioral impacts of these technologies on diverse tourist demographics and their ability to foster a global culture of environmental stewardship.

This study will focus on two main areas:

1. **Behavioral Influence:** Investigating the extent to which immersive technologies, through ecological presence and cultural storytelling, can influence tourists' pro-environmental attitudes and behaviors during and after their virtual experiences.

2. **Integration in Policy and Practice:** Exploring how VR and XR can be integrated into tourism management strategies and conservation policies to reduce environmental degradation in sensitive ecosystems while maintaining economic and cultural benefits for local communities.

The research will target both natural and cultural heritage sites, offering insights into how these technologies can be tailored to different environmental and cultural contexts. Additionally, the study will assess the potential for VR and XR to engage urban populations and individuals with limited access to natural environments, thereby expanding the reach of conservation initiatives.

## 4. Research Objectives

To evaluate the effectiveness of virtual reality (VR) and extended reality (XR) in fostering pro-environmental behavior among tourists by enhancing ecological presence and emotional engagement.

To assess the potential of immersive technologies in promoting bio cultural conservation by integrating ecological education with cultural narratives.

https://doi.org/10.52202/081568-0016

## 5. Research Methodology:

The study will adopt a mixed-method approach, combining quantitative and qualitative research techniques. The framework consists of three key phases:

**5.1 Development of VR/XR Content**: Designing and developing immersive experiences based on natural and cultural sites. Key focus areas:

- Natural Environment: Marine ecosystems, forests, or national parks.
- Cultural Heritage: Historical landmarks or indigenous cultural sites.

**5.2 Data Collection**: Data will be collected using various tools and methods, including:
Pre- and Post-Experience Surveys to measure changes in tourists' pro-environmental attitudes and behaviors.
Biometric Feedback such as heart rate, eye-tracking, and galvanic skin response to assess emotional engagement during VR experiences.
Focus Group Discussions to gather qualitative insights on participants' perceptions and experiences.

**5.3 Data Analysis and Interpretation**: The analysis will be conducted using statistical tools, thematic analysis, and modeling techniques.

**Table 1 showcasing phase Tool . Purpose for Data Analysis**

| S.N. | Phase | Tool/Model | Purpose |
|---|---|---|---|
| 1 | Pre-Experience Survey | Likert Scale Questionnaire | Measure baseline environmental attitudes and behavioral intentions. |
| 2 | Immersive VR/XR Experience | Virtual Reality Headsets (e.g., Oculus) | Provide immersive experiences with ecological and cultural content. |
| 3 | Emotional Engagement | Biometric Sensors (heart rate, eye-tracking) | Capture emotional responses and immersion levels. |
| 4 | Post-Experience Survey | Satisfaction & Behavior Change Survey | Measure changes in awareness, empathy, and intended conservation actions. |
| 5 | Focus Groups | Thematic Analysis | Gather qualitative feedback on user experiences and suggestions. |

**5.4 Dataset Interpretation:**

- Sample Size: 100 participants.
- Variables Measured:

Pre-experience attitude scores (scale 1–5): Baseline measurement of pro-environmental attitudes.
Post-experience attitude scores (scale 1–5): Measurement after VR/XR experience.
Ecological Presence Scores (scale 1–10): Levels of immersion and presence felt in VR.
Behavioral Intention Scores (scale 1–5): Self-reported likelihood of adopting

**5.5 Quantitative Analysis**

- Paired T-Test: Compare pre- and post-experience survey results to determine statistically significant changes in behavior or attitudes.

- Statistical Analysis: Paired T-Test


https://doi.org/10.52202/081568-0016

**5.6 Research Question:** Does participation in a VR/XR experience significantly improve pro-environmental attitudes among participants?

## 6. Data Summary:

**Table 2: Paired T-Test Results:**

| Pre-experience mean score | Post-experience mean score | Standard deviation | t-value: | p-value: |
|---|---|---|---|---|
| 3.2 | 4.1 | 0.8 | 10.52 | • < 0.001 (significant at $\alpha = 0.05$) |

**Interpretation:**
The paired t-test shows a statistically significant increase in pro-environmental attitudes after the VR/XR experience.

**Regression Analysis:**
Effect of Ecological Presence on Behavior: Assess the relationship between ecological presence and changes in environmental behavior.
Research Question: How does ecological presence (immersion) influence behavioral intentions for conservation?

**Table 3: Regression Results:**

| Dependent Variable | Independent Variable | Regression Coefficient (β) | $R^2$ | p-value |
|---|---|---|---|---|
| Behavioral Intention Scores | Ecological Presence Scores | 0.48 | 0.56 | < 0.001 |

Interpretation: Ecological presence significantly predicts behavioral intentions, accounting for 56% of the variance in participants' intentions to adopt conservation behaviors.

**ANOVA:** Differences in Attitudes Based on Experience Type
Research Question: Are there differences in attitude changes based on the type of VR/XR experience (e.g., marine conservation, forest ecosystems, cultural heritage)?

**Table 4 -Data Summary:**

| Groups | Mean Attitude Change Scores | F- Value | p- Value |
|---|---|---|---|
| Marine Conservation | 1.2 | 4.21 | 0.02 (significant at $\alpha = 0.05$) |
| Forest Ecosystems | 1.0 | | |
| 0.02Cultural Heritage | 1.4 | | |

https://doi.org/10.52202/081568-0016

Interpretation:
Participants in the cultural heritage VR experience showed the greatest improvement in pro-environmental attitudes compared to other groups.

**Table 5-Summary of Results**

| Analysis | Metric | Result |
|---|---|---|
| Paired T-Test | t-value: 10.52 | $p < 0.001$ |
| Regression Analysis | $R^2$: 0.56, $\beta$: 0.48 | $p < 0.001$ |
| ANOVA | F-value: 4.21 | $p = 0.02$ |

These results provide a strong statistical foundation for concluding that VR/XR experiences significantly improve pro-environmental attitudes, with ecological presence playing a crucial role in influencing conservation behaviors.

## 7. Results Interpretation:

**Hypothesis 1:** Immersive virtual reality (VR) and extended reality (XR) experiences enhance ecological presence, leading to significant improvements in pro-environmental attitudes and behaviors among participants.

**Statistical Evidence:**

1. **Paired T-Test Results:**
   - Pre-Experience Attitude Score (Mean): 3.2
   - Post-Experience Attitude Score (Mean): 4.1
   - t-value: 10.52, p-value: < 0.001 (highly significant).
2. **Regression Analysis:**
   - Dependent Variable: Behavioral Intention Scores
   - Independent Variable: Ecological Presence Scores
   - Regression Coefficient ($\beta$): 0.48
   - $R^2$: 0.56 (indicating ecological presence accounts for 56% of the variance in conservation behaviors).
   - p-value: < 0.001 (statistically significant).

**Summary of Results for Hypothesis 1:**
The analysis confirms that ecological presence created by VR/XR experiences has a direct and significant impact on participants' pro-environmental attitudes and behavioral intentions. These results validate the hypothesis, demonstrating that immersive technologies foster environmental awareness and encourage conservation behaviors.

**Hypothesis 2:** Integrating cultural narratives with ecological education in VR/XR experiences effectively promotes biocultural conservation by enhancing participants' understanding and emotional connection to biodiversity.

**Statistical Evidence:**
**ANOVA Results**: Participants were divided into three groups based on the type of VR/XR experience (Marine Conservation, Forest Ecosystems, and Cultural Heritage).
F-value: 4.21, p-value: 0.02 (statistically significant differences among groups). Mean Attitude Change Scores:
Marine Conservation: 1.2
Forest Ecosystems: 1.0
Cultural Heritage: 1.4 (highest improvement). Qualitative


https://doi.org/10.52202/081568-0016

Insights from Focus Groups:
Participants expressed a deeper understanding of cultural-ecological relationships when cultural narratives were integrated .Emotional connection and a sense of responsibility were frequently reported themes.

Summary of Results for Hypothesis 2:
The findings support the hypothesis that combining cultural and ecological education in VR/XR experiences enhances biocultural conservation. Participants in the cultural heritage group showed the most significant improvements in attitudes, validating the effectiveness of this integrative approach.

## 8. Conclusion:

This study underscores the transformative potential of virtual reality (VR) and extended reality (XR) technologies in promoting sustainable tourism and conservation. By fostering ecological presence and integrating cultural narratives, these immersive tools significantly enhance pro-environmental attitudes and behaviors among participants. The results demonstrate that VR/XR experiences not only increase awareness but also strengthen the emotional and cognitive connection to biodiversity and conservation efforts. The findings suggest that VR/XR technologies offer an innovative approach to addressing global conservation challenges by creating impactful, scalable, and engaging educational experiences.

**Future Scope:** While this study establishes the effectiveness of VR/XR technologies in enhancing ecological presence and biocultural conservation, several avenues for further research remain:
Longitudinal Impact Studies: Investigate the long-term effects of VR/XR experiences on sustained behavioral changes and conservation actions.
Technological Advancements: Explore the integration of emerging technologies, such as artificial intelligence (AI) and haptic feedback, to enhance ecological presence and user engagement further.
Diverse Contexts and Cultures: Conduct similar studies across diverse geographical and cultural contexts to understand the global applicability of VR/XR interventions in conservation and tourism.

**Limitations**
Sample Diversity: The sample size and demographic representation may limit the generalizability of findings. Future studies should include more diverse participant groups.
Subjectivity of Self-Reported Data: Behavioral intention and attitude scores rely on self-reported data, which may be subject to bias. Complementing these with observational studies can provide more robust insights.

Technological Constraints: VR/XR experiences require significant technological infrastructure, which can limit their accessibility and scalability.
This study highlights the vast potential of VR/XR technologies to redefine the relationship between humans, nature, and cultural heritage. By bridging the gap between virtual experiences and real-world actions, these technologies serve as powerful tools for inspiring a global movement toward sustainable tourism and conservation. As advancements in immersive technologies continue, their application in environmental education and advocacy will play a pivotal role in addressing the challenges of biodiversity loss and ecological degradation. This work sets the stage for interdisciplinary collaborations aimed at leveraging technological innovation for a sustainable future.

## References:


1. Ahmed, Z., et al. (2023). Digital twin applications for smart factories. IoT Engineering Reviews. https://doi.org/10.1007/ioter.2023.3045
2. Becken, S., & Job, H. (2014). Protected areas in an era of global–local change. Journal of Sustainable Tourism, 22(4), 507-526. https://doi.org/10.1080/09669582.2014.909446
3. Bird, S., Klein, E., & Loper, E. (2009). Natural Language Processing with Python. O'Reilly Media.
4. Branstrator, J. R., Cavaliere, C. T., Xiong, L., & Knight, D. (2023). Extended reality and sustainable tourism: Restorying human–wildlife relationships for biocultural conservation. Journal of Ecotourism, 22(1), 103 119. https://doi.org/10.1080/14724049.2022.2134576
5. Caciora, T., Herman, G. V., Ilieş, A., Baias, Ş., Ilieş, D. C., Josan, I., & Hodor, N. (2021). The use of virtual reality to promote sustainable tourism: A case study of wooden churches historical monuments from Romania. Remote Sensing, 13(9),1758. https://doi.org/10.3390/rs13091758
6. Chen, X., et al. (2024). Emerging trends in robotics-assisted manufacturing. International Robotics Journal. https://doi.org/10.1007/s40960-023-0054-8
7. Dasgupta, A., et al. (2023). Multi-agent systems in logistics optimization. Journal of Logistics Innovations. https://doi.org/10.1109/JLO.2023.10157
8. Eagles, P. F. J., McCool, S. F., & Haynes, C. D. (2002). Sustainable tourism in protected areas: Guidelines for planning and management. IUCN.
9. Goyani, M., et al. (2023). Advances in AI for industrial automation. Journal of Industry Applications. https://doi.org/10.1109/JINDU.2023.0101
10. Gupta, M., et al. (2023). AI-enabled process optimization in manufacturing. Production Engineering Review. https://doi.org/10.1109/per.2023.10077
11. Guttentag, D. A. (2010). Virtual reality: Applications and implications for tourism. Tourism Management, 31(5), 637-651. https://doi.org/10.1016/j.tourman.2009.07.003
12. Hofman, K., Walters, G., & Hughes, K. (2022). The effectiveness of virtual vs real-life marine tourism experiences in encouraging conservation behaviour. Journal of Sustainable Tourism, 30(4), 742-766. https://doi.org/10.1080/09669582.2021.2009943
13. Kumar, A., et al. (2024). Human-machine collaboration in AI workspaces. Journal of AI Ethics and Collaboration. https://doi.org/10.1016/aieth.2024.00452
14. Lee, J., et al. (2023). Blockchain and IoT for secure industrial networks. Industrial IoT Systems. https://doi.org/10.1007/iotsys.2023.2308
15. Nayak, A., Satpathy, I., Jain, V., & Rai, B. K. (2025). Using virtual reality simulations to increase public participation in conservation initiatives: An overview. In Digital Innovations for Renewable Energy and Conservation (pp. 369-392). https://doi.org/10.1007/978-3-031-09215-6_17
16. Patel, S. (2023). Renewable energy integration into AI-driven systems. Energy Conversion and AI. https://doi.org/10.1016/ecai.2023.00988
17. Rao, P., et al. (2024). Advanced predictive analytics in smart grids. Energy Intelligence Journal. https://doi.org/10.1016/eij.2024.1102
18. Sharma, R. (2023). AI-driven predictive maintenance in industrial settings. Journal of Maintenance Strategies. https://doi.org/10.1007/jms.2023.0191
19. Singh, R., & Kumar, P. (2023). Sustainable smart cities: A framework for integration. Sustainability and Technology. https://doi.org/10.1016/sust.2023.12345Stappung, Y., Aliaga, C., Cartes, J., Jego, L., Reyes-Suárez, J. A., Barriga, N. A., & Besoain,
20. F. (2023). Developing 360 virtual tours for promoting tourism in natural parks in Chile. Sustainability, 15(22), 16043. https://doi.org/10.3390/su152216043
21. Su, Z., Lei, B., Lu, D., Lai, S., & Zhang, X. (2024). Impact of ecological presence in virtual reality tourism on enhancing tourists' environmentally responsible behavior. Scientific Reports, 14(1), 5939. https://doi.org/10.1038/s41598-024-35939-6
22. Verma, N. (2024). Green AI: Toward environmentally conscious AI models. Environmental Informatics. https://doi.org/10.1007/envinf.2024.1125
23. Zhang, Y., et al. (2023). Federated learning strategies in decentralized networks. Machine Learning and Data Engineering. https://doi.org/10.1145/mldec.2023.2234.


 https://doi.org/10.52202/081568-0016

# MAHA KUMBH 2025: ECONOMIC GROWTH FOR LOCAL BUSINESSES, RETAIL, AND HOSPITALITY, AS WELL AS INFRASTRUCTURE AND GOVERNMENTAL ARRANGEMENTS

Dr Shivani Sharma [1], Dr. Trilochan Kumar[2*]

[1] University of Jammu, J&K, India
[2] Chandigarh University, Mohali, India

* Corresponding author: sharmashivani0105@gmail.com

**Abstract.** The Maha Kumbh of 2025, taking place in Prayagraj, Uttar Pradesh, is anticipated to attract over 400 million pilgrims, including both domestic and international visitors, thereby establishing it as the most significant Maha Kumbh in history. The festival, rooted in Hindu mythology, encompasses ritual bathing at the revered Sangam, where the Ganga, Yamuna, and the legendary Saraswati rivers converge, aimed at the purification of sins and the attainment of salvation. This platform fosters profound religious dialogues, presents cultural exhibitions, and highlights traditional performances, thereby underscoring the richness of India's spiritual heritage. In the year 2019, an estimated 240 million pilgrims from across the globe gathered for the Kumbh Mela in Prayagraj, India. We foresee that the Maha Kumbh 2025 will markedly enhance the local economy, especially within the realms of retail, hospitality, transportation, and infrastructure development. Transitory employment prospects will arise within the realms of hospitality, security, transportation, and local enterprises. Furthermore, governmental allocation of resources towards infrastructure will facilitate increased spending on essentials such as food, accommodation, transportation, and religious artefacts. The deployment of advanced urban technologies, such as AI-enhanced crowd management, digital payment infrastructures, and mobile applications, seeks to reduce environmental repercussions. The designation of the Maha Kumbh by UNESCO as an Intangible Cultural Heritage significantly elevates India's standing in the realm of global tourism. The global media landscape, coupled with the pervasive impact of social media, will significantly expand the event's audience. The prospective enduring benefits of tourism may well exceed those of the Maha Kumbh, consequently enhancing Prayagraj's prominence in the international arena. The event serves as a significant socio- economic catalyst, influencing Prayagraj and the broader Indian economy in manners that transcend its religious connotations. This research aims to analyse the economic ramifications of the Maha Kumbh 2025 in Prayagraj, concentrating on its impact on employment creation, local enterprises, tourism, infrastructure advancement, governmental strategies, and assistance for the seamless operation of the festival. The research aims to examine the economic advantages and obstacles associated with the festival.

**Keywords:** Maha Kumbh, Uttar Pradesh, Tourism, Prayagraj, Economic impact, Infrastructure, Plans

## 1. Introduction

The Kumbh Mela demonstrates the richness and depth of India's religious and cultural past. Men of faith occasionally assemble in the hallowed sites of each religion (Bhardwaj, 1973). The event's expanding prominence in the global religious tourism sector is evident from the significant increase in foreign tourists in recent years. Pilgrimages have important political, economic, social, and cultural implications in addition to their impact on international trade and health (Collins-Kreiner, 2010). According to Baranwal et al. (2015), the Kumbh Mela is the greatest gathering of people on the planet at any given time. Millions of people from all over the world have gravitated towards it over the years, regardless of gender, caste, religion, or faith. The 2019 Kumbh Mela attracted over one million foreign tourists, representing a 35% increase in foreign visitor arrivals. The size of the incident made it difficult to ensure public health security and safety (Dwivedi and Cariappa, 2015). The assessments of the Kumbh Mela in Allahabad (2013) and Nashik (2003, 2015) reveal how event planning and spending have evolved over time (Bansal et al., 2013). The Kumbh Mela is considered very important in Hinduism. This study delves

https://doi.org/10.52202/081568-0017

deeply into the historical, sociological, and economic factors that contributed to its ascent to prominence as a tourist destination. The Kumbh Mela's historical roots can be traced back millennia to Hindu mythology and ritual. A pilgrimage is a planned travel with a specific aim defined by the heart and carried out with the feet. The towns of Prayagraj, Ujjain, Nashik, and Haridwar have a pilgrimage event known as the "Kumbh Mela" in Hindu mythology (Memish et al., 2019; Raghav et al., 2020). Ruback, Pandey, and Kohli's study, which focusses on a Hindu holy site visited annually, contributes to the body of information on how individuals perceive locations. According to Smalldone et al. (2005), the concept of place is determined by how people interact with one another, as well as the physical settings and activities that take place there. Certain sites are more important than others due to their physical characteristics and the activities they host. The Ganga, Yamuna, Saraswati, and Godavari rivers converge at regular intervals to form the foundation of this holy trip (World Journal of Advanced Research and Reviews, 2025, 25(02), 561-569 562). Participants in this ceremony take a plunge in a holy river (UNESCO 2017). Over time, the Mela has expanded beyond its religious roots to reflect cultural heritage and communal identity, fostering relationships between diverse Hindu sects and adherents around the world. Sociocultural elements have a significant role in deciding the Kumbh Mela's size and appeal as a global tourism attraction. India's diverse cultural landscape, as well as Hinduism's tolerant mindset, make the Mela more enticing to both domestic and international visitors. During the Mela, ceremonial activities, spiritual discussions, and colourful cultural events provide immersive experiences that promote international connection and understanding.

Aside from its religious significance, the Kumbh Mela has a substantial economic impact, stimulating significant growth in local economies and tourism infrastructure. Infrastructure development, hotel services, and job opportunities all increase in host cities, benefiting local economies and means of subsistence. To fully fulfil its potential as a sustainable tourism destination, the Mela serves as a venue for promoting tourism initiatives and encouraging collaboration between the public and private sectors. However, the Kumbh Mela's rising breadth creates considerable challenges, ranging from environmental sustainability to logistical constraints. Strong infrastructure, crowd management measures, and emergency response systems are essential to ensure the safety and well-being of millions of pilgrims. Furthermore, in order for the Mela to be long-term viable, the negative environmental consequences of huge people on water supply, sanitation, and ecological preservation must be mitigated. Its transformative influence transcends religious boundaries and encourages sustainable tourism, economic prosperity, and multicultural communication. According to Damari and Mansfeld (2016), the predominantly theoretical paradigm that frames pilgrimage as a tourism phenomena may be limited and reduces the diversity of pilgrims' experiences today. Some academics evaluate pilgrims' traits and experiences through a consumerist lens.

Previous research on the subject, according to Cohen (1992), Collins-Kreiner and Kliot (2000), Raj and Morpeth (2007), and Smith (1992), focused on comparing pilgrims' and tourists' experiences, arguing that they exist on opposite ends of a continuum, with the former occupying the sacred end and the latter the secular end. According to Hudman and Jackson's (1992) research, pilgrimage travel frequently combines religious activity with tourism. Recent research (Collins-Kreiner, 2010) shows that the dichotomy that existed until the early 2000s is no longer apparent. India's Kumbh Mela, which reflects the country's rich cultural heritage and spiritual ethos, has grown into a thriving magnet for international Hindu religious tourism.

On a global scale, the Kumbh Mela continues to be a beacon of spiritual enlightenment and cultural harmony by overcoming inherent hurdles and realising its full potential.


https://doi.org/10.52202/081568-0017

2. **Resources and Methods**

We conducted a search across various databases for journals that include terms like Kumbh Mela, mass gathering, spiritual pilgrimage, pilgrimage, religious tourism, spiritual gathering, Government Support, Impact on Indian Economy particularly regarding Employment, Impact on India's Tourism Landscape, and socio-economic impact, among others. Our examination was confined to peer-reviewed literature, following a thorough search through platforms such as Google Scholar, PubMed, ResearchGate, and additional resources. This study examined reports from multiple departments alongside previous Kumbh Mela reports published by the Prayagraj Mela Authority. Data and information were collected from news reports regarding the Kumbh Mela, Magh Mela, and various other significant gatherings.

## 3. Findings and Discussions

### 3.1 Government Support and Key Infrastructure Development

The Uttar Pradesh Government provided significant funding and infrastructure to ensure the success of the Maha Kumbh Mela 2025. Over 660 million pilgrims attended, making it one of the largest events in history.

**Key support from the Government:**

- Maha Kumbh Mela 2025 drew over 660 million pilgrims.
- A temporary city spanning 4,000 hectares housed 150,000 tents.
- 92 roads were renovated and 30 pontoons bridges constructed.
- 50,000 security personnel with AI surveillance ensured safety.
- Temporary hospitals and "Bhishma cube" units provided medical care.

The UP government built Maha Kumbh Nagar, a temporary metropolis spanning 4,000 hectares with 150,000 tents and luxury accommodations provided by IRCTC's "Maha Kumbh Gram." Transportation infrastructure has seen significant changes, including the renovation of 92 roads, the construction of 30 pontoon bridges, and the addition of seven new railway platforms to accommodate increased visitor traffic. To reduce environmental effect, eco-friendly solutions were employed, including solar-powered lights, bio-toilets, and modern waste management systems. Proper sanitation was achieved with over 12,000 FRP toilets, 16,100 steel toilets, and 20,000 urinals. Over 269,000 checkered plates were used to form pedestrian pathways.

Temporary hospitals were established to provide surgical and diagnostic facilities, enhancing healthcare services. The "Bhishma Cube," which can treat 200 people simultaneously, facilitated efficient medical emergency management. Mobile health units and telemedicine services improved medical response times. Security was diligently prepared, with over 50,000 people deployed, including paramilitary troops. AI-based surveillance, real-time monitoring, and facial recognition systems improved safety and crowd management. The successful Maha Kumbh Mela 2025 demonstrated the government's dedication to offering a spiritually rewarding and logistically smooth experience, resulting in a lasting good influence.

### 3.2 Maha Kumbh and Its Impact on Indian Economy especially on Employment

Maha Kumbh has made a significant contribution to India's and Uttar Pradesh's economic growth and employment this year. The demand for goods and services has increased due to the attendance of over 660 crore devotees, resulting in the creation of thousands of jobs in various industries.The event's economic impact is expected to exceed Rs 2 lakh crore, contributing 6.25% to Uttar Pradesh's GSDP, which is predicted to reach Rs 32 lakh crore by March 2025.

https://doi.org/10.52202/081568-0017

The hospitality industry has experienced significant growth, with hotels and resorts accounting for approximately Rs 4,000 crore in revenue. The family prospered, earning roughly Rs. 2,000 crore from restaurants, vendors, and caterers serving millions of visitors. Increased consumer spending has benefitted retail enterprises, especially those selling religious products, everyday staples, and traditional handicrafts, leading to job growth.

The festival has directly employed boatmen, shops, artisans, and religious service providers. Transportation services like cabs, buses, and trains have generated Rs 10,000 crore. Retail businesses have experienced development, with religious item sales contributing Rs 20,000 crore to the GDP.Telecommunications firms, includin Reliance Jio, have seen an increase in internet and phone consumption due to the influx of devotees and tourists

**Economic Impact Snapshots**

- Massive Economic Impact- Rs 2 lakh crore +revenue, contributing 6.25% to UP's Rs.32 lakh crore GSDP
- Hospitality Boom- Rs 40,000 crore from hotels lodges, and rentals
- Food and Beverages- Caterers Rs 20,000 crore from restaurants, vendors and
- Retail Growth- consumer goods Rs 20,000 crore from religious merchandise and
- Transportation and Logistics- Rs 10,000 crore from taxis, buses and railway
- Telecom Surge- Increased demand for mobile data and calls

Maha Kumbh 2025 has strengthened its religious and cultural significance while promoting economic success. The festival plays a significant role in the economic prosperity of Uttar Pradesh, generating large-scale employment, expanding businesses, and increasing revenue.

### 3.3 Multifaceted Impact on India's Tourism Landscape

The mela has had a significant impact on India's tourist sector, attracting record numbers of people and boosting economic activity across several businesses. The event drew 660 million people, with an increase in overseas tourists from the US, UK, Canada, Australia, Russia, and Japan. The turnout produced revenue of Rs 3 lakh crore ($36 billion), benefiting local businesses and service providers.
Authorities created a 4,000-hectare temporary metropolis with tent housing, sanitary services, and medical centers to accommodate the influx of people. The hospitality industry experienced significant growth, with homestays and luxury tents priced up to Rs 1 lakh per night catering to a wide range of guest preferences.

**International tourism boost at Maha kumbh 2025**

- British travel writers visited; special tours and accommodations arranged.
- Foreign visitors joined rituals, explored heritage, and shared experiences.
- Promoted under "Brand UP" to boost tourism and investment
- Showcased at FITUR & ITB Berlin with cultural pavilions and promotions
- Incredible India Pavilion offered immersive experiences and multilingual help
- Customized pilgrimage packages increased foreign tourist reservations by 30%, supporting economic growth.

Indian Railways ran 13,000 trains and renovated infrastructure, including new stations and platforms. In addition, 1,200 extra buses were dispatched to support the 3,050 already assigned. Improved international flight connection to Prayagraj, Varanasi, and Lucknow resulted in a 20% increase in foreign air traffic, making tourism more accessible.


https://doi.org/10.52202/081568-0017

The event created more than 600,000 temporary jobs in a variety of industries. International visitors paired their pilgrimage with trips to Varanasi, Rishikesh, and Bodh Gaya, boosting foreign exchange revenues. Global media coverage boosted India's reputation as a top destination for spiritual tourism. This delivered a profound spiritual experience and significant economic benefits, demonstrating its impact on domestic and worldwide tourists.

**Opportunities and Challenges in Kumbh Mela Travel**

The Kumbh Mela, an important religious celebration in India, struggles with traffic, crowding, and inadequate facilities despite infrastructural upgrades. Temporary tent cities and lack of utilities can cause delays, bottlenecks, and safety hazards for pilgrims and tourists. Due to healthcare, lodging, and other service capacity issues, pilgrims and tourists may receive substandard services.

Customs and traditions are being exploited for business and entertainment at the Kumbh Mela, creating moral and cultural issues. Commercial vendors, sponsors, and advertising may exploit the festival's spiritual foundation by promoting products, services, and experiences that detract from it. Traditional artists, entertainers, and cultural practitioners may be exploited commercially.

Maintaining the Kumbh Mela's purity, authenticity, and integrity while promoting ethical and sustainable tourism is essential to its religious and cultural significance. Branding and promoting the Kumbh Mela as a top religious tourism destination should evoke spirituality, tradition, and cultural richness to appeal to pilgrims and tourists. Visual, narrative, and experiential marketing efforts should convey transcendence, purity, and regeneration.

By partnering with media, tourism firms, and influencers, the Kumbh Mela brand can reach more people. Digital marketing and social media efforts can boost visitors to Kumbh Mela websites and accounts.

Virtual reality, user-generated material, and fascinating storytelling can entice Kumbh Mela pilgrims. Strategic media alliances and PR are needed to build Kumbh Mela credibility and PR. Working with travel writers, photographers, and filmmakers may create fascinating stories, images, and films that showcase the Kumbh Mela's diversity, beauty, and relevance internationally. In conclusion, the Kumbh Mela must plan and manage pilgrims and tourists' safety and well-being due to poor facilities, traffic, and crowding.

## 4. Recommendations for Eco-Friendly Kumbh Mela Travel

As Kumbh Mela tourism adopts new technologies, interacts with diverse people, and adapts to changing socioeconomic and environmental dynamics, it has immense potential for future growth, innovation, and transformation.

The Kumbh Mela can improve visitor satisfaction, operational efficiency, and environmental sustainability by leveraging digital platforms, innovative solutions, and data-driven insights. In an age of increasing divisions and conflicts, the Kumbh Mela can serve as a light of peace, harmony, and unity by strengthening local communities, encouraging cross-cultural engagement, and promoting interfaith understanding. The Kumbh Mela has the potential to inspire and raise the hearts and minds of pilgrims and tourists for many generations to come by embracing ideals such as sustainability, equity, and inclusion. The Kumbh Mela is likely to make an even greater contribution to worldwide religious tourism in the coming years. The Kumbh Mela is a light of ancient wisdom and sacred pilgrimage, as more people seek authentic spiritual experiences and cultural immersion. We can learn a lot about the eternal significance of old rituals in the modern world by understanding and appreciating the Kumbh Mela's various implications on worldwide tourism. Using technology to improve tourist satisfaction, operational efficiency, and environmental sustainability is critical to the future of Kumbh Mela tourism. Digital platforms, smartphone apps, and virtual reality tools provide pilgrims and tourists with interactive maps, guided tours, and real-time updates on events and activities.

Drones, smart sensors, and Internet of Things devices can be used to monitor the environment, waste, and crowds, enabling proactive interventions and data-driven decision-making. Geographic information systems (GIS) can identify urban regeneration locations to conserve Kumbh Mela-related cultural heritage tourism (Saha and Khare, 2020). Blockchain technology can reduce fraud, corruption, and disputes in tourism operations by increasing transaction and interaction transparency, accountability, and trust. The Kumbh Mela can serve as a model for the expansion of sustainable tourism in the digital age by embracing innovation and technology. Community empowerment and engagement are critical for making Kumbh Mela tourism inclusive, real, and sustainable. Local communities, as stewards of the festival's natural resources and cultural legacy, should be actively involved in its management, planning, and decision-making processes. Homestays, cultural tours, and handicraft cooperatives are examples of community-based tourism programmes that can produce revenue while allowing residents to benefit fairly and equally from tourism. Business development, vocational training, and capacity-building programs can provide local citizens with the tools, resources, and information they need to contribute to the tourism value chain while also improving their standards of living. Maintaining the ecological integrity and cultural authenticity of the Kumbh Mela while promoting sustainable tourism and economic growth necessitates finding a balance between conservation and development goals. The biodiversity, water sources, and natural ecosystems of the Kumbh Mela locations should be the primary focus of conservation efforts, as they are threatened by pollution, urbanisation, and climate change. To mitigate environmental impacts and increase resilience to environmental risks, development projects should prioritise investments in green technologies, renewable energy, and eco-friendly infrastructure.

Researching the massive Kumbh Mela provides a unique opportunity to establish an evidence foundation for suggestions and techniques for managing, preventing, and controlling infectious diseases during large gatherings. This could help design situation-specific policies to promote and protect people's health during the Kumbh Mela and other major international events (David and Roy, 2016). Sustainable tourism activities, such as responsible travel, environmentally friendly housing, and carbon offset schemes, can help reduce negative environmental impacts while also supporting long-term local economies. The Kumbh Mela has the potential to set a global standard for sustainable tourism destinations by taking an integrated and comprehensive approach to development and conservation.

## 5. Conclusion

The Maha Kumbh 2025 successfully united millions of pilgrims in a spiritual gathering. It demonstrated India's ability to efficiently manage large-scale events. The event improved the economy by providing jobs and benefiting local businesses. Crowd management remained a concern, with recorded incidents. Environmental challenges, including trash and river contamination, underscore the need for sustainable solutions. Future Kumbh Melas will benefit from lessons learned from the Maha Kumbh, including improved planning, safety, and environmental responsibility.

The Kumbh Mela, a significant Hindu religious tourism event, attracts millions of pilgrims and tourists worldwide. Its roots in ancient mythology and customs make it a hub for socioeconomic and environmental effects. The Kumbh Mela offers self-discovery, spiritual rejuvenation, and cultural immersion through its customs, rituals, and beliefs. It promotes unity, peace, harmony, and understanding among different populations. The Kumbh Mela leaves a lasting legacy by fostering growth, prosperity, and development in host regions through economic contributions, infrastructural development, and employment creation. To promote responsible tourist practices and sustainable tourism development, public-private partnerships, multi-stakeholder collaborations, and international cooperation are crucial. By embracing an integrated approach to tourism management, the Kumbh Mela can continue to thrive as a paradigm for sustainable religious tourism


https://doi.org/10.52202/081568-0017

## References


1. Bansal, A., Plan, M., & Arch, B. (2013). Mega Events as Catalyst for Urban Transformation-A Case of Kumbh Mela in Ujjain. Academia. edu–Share research.
2. Baranwal, A., Anand, A., Singh, R., Deka, M., Paul, A., Borgohain, S., & Roy, N. (2015). Managing the earth's biggest mass gathering event and wash conditions: Maha kumbh mela (India). PLoS currents, 7.
3. Bhardwaj, S. M. (1973). Hindu places of pilgrimage in India: A study in cultural geography. Univ of California Press.
4. Blackwell, R. (2007). Motivations for religious tourism, pilgrimage, festivals and events.
5. Buzinde, C. N., Kalavar, J. M., Kohli, N., & Manuel-Navarrete, D. (2014). Emic understandings of Kumbh Mela pilgrimage experiences. Annals of Tourism Research, 49, 1-18.
6. Cariappa, M. P., Singh, B. P., Mahen, A., & Bansal, A. S. (2015). Kumbh Mela 2013: Healthcare for the millions. medical journal armed forces india, 71(3), 278-281.
7. Cohen, E. (1992). Pilgrimage and tourism: Convergence and divergence. Sacred journeys, 47-64.
8. Collins-Kreiner, N. (2010). Researching pilgrimage: Continuity and transformations. Annals of tourism research, 37(2), 440-456.
9. Collins-Kreiner, N., & Kliot, N. (2000). Pilgrimage tourism in the Holy Land: The behavioural characteristics of Christian pilgrims. GeoJournal, 50, 55-67.
10. Damari, C., & Mansfeld, Y. (2016). Reflections on pilgrims' identity, role and interplay with the pilgrimage environment. Current Issues in Tourism, 19(3), 199-222.
11. 1David, S., & Roy, N. (2016). Public health perspectives from the biggest human mass gathering on earth: Kumbh Mela, India. International Journal of Infectious Diseases, 47, 42-45. World Journal of Advanced Research and Reviews, 2025, 25(02), 561-569 568
12. 1Duives, D., Daamen, W., & Hoogendoorn, S. (2014). Influence of group size and group composition on the adhered distance headway. Transportation Research Procedia, 2, 183-188.
13. Dwivedi, S., & Cariappa, M. P. (2015). Mass-gathering events: the public health challenge of the Kumbh Mela 2013. Prehospital and disaster medicine, 30(6), 621-624.
14. 1Hudman, L. E., & Jackson, R. H. (1992). Mormon pilgrimage and tourism. Annals of tourism research, 19(1), 107- 121.
15. Kanaujiya, A. K., & Tiwari, V. (2022). Crowd management and strategies for security and surveillance during the large mass gathering events: The Prayagraj Kumbh Mela 2019 Experience. National Academy science letters, 45(3), 263-273.
16. Kanaujiya, A. K., & Tiwari, V. (2023a). The statistical analysis to find correlation between mass gathering event and COVID-19 pandemic outbreak: The Prayagraj Magh Mela 2021 experience. Proceedings of the National Academy of Sciences, India Section A: Physical Sciences, 93(1), 107-111.
17. Kanaujiya, A. K., & Tiwari, V. (2023b). Pilgrimage Itineraries Dimensions with Focus on Kumbh Mela 2019, Prayagraj, India. In Experiences, Advantages, and Economic Dimensions of Pilgrimage Routes (pp. 118-142). IGI Global.
18. Karthika, P. S., Aparna, P. M., & Verma, A. (2018). Understanding crowd dynamics at ghat region during world's largest mass religious gathering, Kumbh Mela. International journal of disaster risk reduction, 31, 918-925.
19. Khan, K., Sears, J., Hu, V. W., Brownstein, J. S., Hay, S., Kossowsky, D., ... & Cetron, M. (2013). Potential for the international spread of Middle East respiratory syndrome in association with mass gatherings in Saudi Arabia. PLoS currents, 5.
20. Liro, J. (2021). Visitors' motivations and behaviours at pilgrimage centres: Push and pull perspectives. Journal of Heritage Tourism, 16(1), 79-99.maha-kumbh-mela-2025-in-prayagraj-125030600337_1.html

21. Maheshwari, S., & Singh, P. (2009). Psychological well-being and pilgrimage: Religiosity, happiness and life satisfaction of Ardh–Kumbh Mela pilgrims (Kalpvasis) at Prayag, India. Asian Journal of Social Psychology, 12(4), 285-292.
22. Memish, Z. A., Steffen, R., White, P., Dar, O., Azhar, E. I., Sharma, A., & Zumla, A. (2019). Mass gatherings medicine: Public health issues arising from mass gathering religious and sporting events. The Lancet, 393(10185), 2073– 2084.
23. Quadri, S. A., & Padala, P. R. (2021). An aspect of Kumbh Mela massive gathering and COVID-19. Current Tropical Medicine Reports, 8, 225-230.
24. Raghav, S., Singh, S. K., & Bhatia, T. (2020). Environmental Management in the World's largest mass gathering Kumbh Mela. IJIRST-International Journal for Innovative Research in Science & Technology, 4(7), 25–34.
25. Ruback, R. B., Pandey, J., & Kohli, N. (2008). Evaluations of a sacred place: Role and religious belief at the Magh Mela. Journal of Environmental Psychology, 28(2), 174-184.
26. Saha, K., & Khare, R. (2020). A geospatial approach to conserving cultural heritage tourism at Kumbh Mela events in India. In Tourism, Cultural Heritage and Urban Regeneration: Changing Spaces in Historical Places (pp. 125- 140). Cham: Springer International Publishing.
27. Sharma, A., Gupta, B., Petersen, E., Lee, S. S., & Zumla, A. (2023). Enhancing preparedness for reducing transmission and globalisation of Antimicrobial Resistance at the Ardh Kumbh Mela 2025, the world's largest recurring religious mass gathering. International Journal of Infectious Diseases, 137, 79-81.
28. Shinde, K. (2018). Governance and management of religious tourism in India. International Journal of Religious Tourism and Pilgrimage, 6(1), 7. World Journal of Advanced Research and Reviews, 2025, 25(02), 561-569 569
29. Shukla, S., Khan, R., Ahmed, Y., & Memish, Z. A. (2021). Conducting mass gathering events during the COVID-19 pandemic: a case study of Kumbh Mela 2021 as a potential 'super spreader event'. Journal of Travel Medicine, 28(8), taab160.
30. Smaldone, D., Harris, C., & Sanyal, N. (2005). An exploration of place as a process: The case of Jackson Hole, WY. Journal of environmental psychology, 25(4), 397-414.

 https://doi.org/10.52202/081568-0017

# Technology in Global Connectivity and Trends in Tourism

Ms. Niyati Sood1,

[1] Chandigarh University, Mohali, India

Corresponding author: soodn2003@gmail.com

**Abstract:** Tourism has now emerged as the biggest sector of economy in most of the countries and is a business activity which has both traditional and modern sides. But the rapid advancement in Technology had more effect on tourism sector, the emerging trends of AI driven systems, IoT (Internet of Things) based smart mobility, block-chain based ticketing systems and much more technologies revolutionizing the sector.

Smart tourism is fueled by AI-driven chatbots, mixed-reality experiences, and personalized recommendations that give a new meaning to how a traveler chooses to explore a destination. In addition, real-time data analytics and mobile applications facilitate seamless travel planning, dynamic pricing, and effective crowd management. Smart cities and digital infrastructure comprising contactless payments, biometric authentication, and IoT-oriented smart transportation are setting a stage toward creating a globally connected transport ecosystem. The paper thus analyses how the developing relationship between technology and tourism revolves around trends such as digital nomadism, virtual tourism, and sustainable travel innovation.

Through the analysis of technological advancement and its effects on global tourism, the study provides insight into the adaptation of the entire industry to the digital scene while upholding sustainability, accessibility, and enhanced experiences for the traveler.

**Keywords**: Tourism, Technology, Smart tourism, Digital

## 1. Introduction:

Tourism has long been a significant contributor to the global economy, serving as a key aspect of economic growth, cultural exchange, and job creation across the world. It has evolved from being a traditional, local activity into a diverse global industry, embracing both traditional practices and modern innovations. With the rise of technology, particularly in the last few decades, the tourism sector has undergone absolute transformation, transforming how people travel, interact with different destinations, and experience new cultures. Today, technological advancements are at the forefront of reshaping the tourism sector, bringing about new trends, business models, and opportunities that are affecting both travelers and service providers.

The development of cutting-edge technologies like Artificial Intelligence (AI), the Internet of Things (IoT), blockchain, and smart mobility systems is rapidly changing the way the tourism industry operates. These technologies have created a new era of "smart tourism," which combines advanced digital solutions with the travel experience. Smart tourism is a concept backed up by AI-driven Chabot, virtual assistants, and mixed-reality experiences, all of which offer more personalized and exciting travel opportunities. For example, AI-powered Chabot helps in instant customer service, assisting travelers with everything from booking accommodations to providing recommendations on local attractions. Likewise, mixed-reality

https://doi.org/10.52202/081568-0018

tools, such as virtual tours and augmented reality (AR), are revolutionizing how tourists explore and interact with destinations, allowing for interactive and engaging experiences that were once unimaginable.

In addition to AI and mixed-reality technologies, the widespread adoption of IoT has given rise to smart mobility systems, which enhance the traveler experience through smooth transportation networks and real-time updates. By integrating connected devices, IoT enables smart cities and digital infrastructures that streamline travel logistics, offering services such as contactless payments, biometric authentication, and personalized travel recommendations. Such innovations facilitate not only smoother travel experiences but also more efficient crowd management, dynamic pricing, and enhanced safety, which are especially important in today's fast-paced, high-demand tourism environment.

As the world becomes increasingly interconnected, the concept of a "globally connected transport ecosystem" is taking shape. In particular, advancements in blockchain technology are providing new opportunities for transparency and security in ticketing systems, ensuring that the booking process is more reliable and user-friendly. Block chain's ability to create decentralized and tamper-proof records holds the potential to revolutionize various aspects of tourism, such as payments, data management, and identity verification.

Simultaneously, new trends have emerged because of technological advancements, such as digital nomadism, where individuals are choosing to live and work remotely while traveling across the world. This trend has been amplified by the availability of fast internet, cloud technologies, and co-working spaces that enable professionals to work from anywhere. As a result, countries and cities are adjusting their policies to attract digital nomads, offering visas and other incentives tailored to this growing segment of the global workforce.

Another significant trend is virtual tourism, a concept that allows individuals to explore destinations without physically traveling there. With the help of VR and AR technologies, travellers can immerse themselves in virtual experiences that replicate real-world visits. While virtual tourism cannot replace the physical act of traveling, it offers a compelling alternative for those unable to travel due to financial, health, or other reasons. It also serves as a powerful tool for destination marketing, allowing travellers to experience a taste of a location before making a physical visit.

Furthermore, the increased focus on sustainable travel innovation is driving the tourism industry to prioritize eco-friendly practices. Technology plays a critical role in achieving sustainability goals within the sector by reducing carbon footprints, optimizing energy consumption, and promoting greener transportation options. Smart solutions, such as electric vehicles, green-certified accommodations, and waste reduction technologies, are helping reduce the environmental impact of tourism while fostering a more sustainable future for the industry.

This paper aims to analyze the developing relationship between technology and tourism, shedding light on how emerging trends such as smart tourism, digital nomadism, virtual tourism, and sustainable travel are reshaping the tourism industry. By examining how these technologies are improving connectivity, enhancing the traveler experience, and contributing to more sustainable practices, the research will provide insight into how the tourism sector can continue to adapt to the digital age while upholding principles of accessibility, sustainability, and overall innovation. The study explores both the opportunities and challenges that come with the integration of technology into tourism, offering a comprehensive understanding of its future trajectory in an increasingly interconnected world.


https://doi.org/10.52202/081568-0018

Research on connectivity highlights the role of technology in connecting people and helping to synchronize activities Kolb, 2008. Much of it focuses on the increases in connectivity demands as contemporary knowledge workers communicate and coordinate via a variety of technologies, most of which are available 24/7. Scholars have documented, for example, how the advent of email and mobile technologies contribute to workers' feelings of constant pressure to be available (e.g., Barley, Meyerson, & Grodal, 2011; Mazmanian et al., 2013) and how the proliferation of communication technologies has sped up work activities and escalated demands on workers' time (e.g., Wajcman & Rose, 2011). Wajcman and Rose (2011, p. 957) also document employees' involvement in the ongoing process of shaping "the interaction between the materiality of communication media and organizational norms, corporate culture, and employees' perceptions of their work roles."

Consistently with this, Kolb (2008) describes connectivity as either latent or enacted: workers have a choice in terms of whether to use the technologies available to them to connect. He states: "We may have fast internet, teleconferencing facilities, and other connective links with another actor/site (high connective potential), yet not experience high connectivity because we choose not to exercise our connective options" (Kolb, 2008, p. 129). This line of research builds on a long history of scholarship that argues that technology is not deterministic but is rather enacted (Barley, 1983; Orlikowski, 1992). Scholars studying connectivity refer to this as "connective choice" (Dery et al., 2014). To date, however, little has been written about the effects of such choices on the quality of work relationships, especially for those collaborating from far-flung places around the globe.

Global professionals are particularly liable to face connectivity demands. Varying distances and time zones force them to rely on mediating technologies in their interactions and in building strong interpersonal relationships (Hinds & Bailey, 2003; Wilson, Crisp, & Mortensen, 2013). There is evidence, for example, that time zone differences make global work more demanding (O'Leary & Cummings, 2007) because there are fewer overlapping hours during which to collaborate synchronously (Cummings, Espinosa, & Pickering, 2009).

The literature on global work abounds with examples of the challenging and time-consuming nature of building mutually supportive and close relationships across distances and time zones (Hinds & Mortensen, 2005; Jarvenpaa & Leidner, 1998; Walther, 1992). To combat this, globally distributed co-workers build interpersonal relationships by communicating frequently via technology (O'Leary, Wilson, & Metiu, 2014) during and outside of business hours (Espinosa, 2013; O'Leary & Cummings, 2007). Wilson, O'Leary, Metiu, and Jett (2008), for example, found that frequent mediated communication facilitated the formation of strong social bonds between geographically distant co-workers. O'Leary and colleagues (2014) further report that the effects of objective distance are eliminated when co-workers perceive themselves to be proximate, something that the authors attribute to employees conveying to distant co-workers that they are always accessible.

## 2. Technological Advancements and the Evolution of Tourism

The tourism sector has undergone significant transformation due to technological advancements. From booking systems to exciting experiences, technology has revolutionized nearly every aspect of tourism, empowering businesses and tourists alike to interact with each other in ways that were previously unimaginable. These changes are fuelled by key technologies such as Artificial Intelligence (AI), the Internet of Things (IoT), block chain, and mixed reality (AR/VR), which have collectively flagged the way for a more efficient, connected, and personalized tourism experience.

## 2.1 Artificial Intelligence (AI) in Tourism

Artificial Intelligence (AI) has been one of the most transformative forces in the tourism industry, enabling the formation of more personalized, efficient, and available travel experiences. AI's ability to analyse massive datasets and recognize patterns is dominant to its role in tourism. Here are some ways AI is driving change in the sector:

### AI-Powered Chatbots and Virtual Assistants

AI-powered chatbots are now a vital part of customer service in the tourism industry. These chatbots provide instant, 24/7 assistance to travellers, helping them with bookings, reservations, recommendations, and frequently asked questions. For example, several airlines and hotels are now using AI-driven virtual assistants that allow customers to check in, make changes to bookings, or request room service—all via chat interfaces. These virtual assistants are powered by Natural Language Processing (NLP), allowing them to understand and respond to human language more effectively.

- **Example**: The "Booking Assistant" feature by Booking.com, powered by AI, provides personalized recommendations to users based on their travel preferences and search history, which improves the overall customer experience.

  Below is the graph showing the Growth of AI-Powered Chatbot Usage in Tourism:

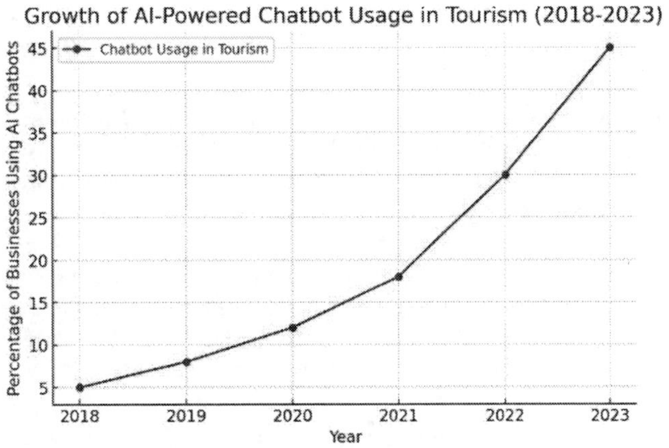


**Graph 1:** graph showing the Growth of AI-Powered Chatbot Usage in Tourism:

### AI for Personalization

AI is being utilized to form custom-made travel experiences. By analysing large amounts of data from various sources—such as social media, travel history, and search patterns—AI systems can forecast a traveller's preferences and offer customized travel options. This personalized approach is seen in dynamic pricing models, where the cost of accommodations, flights, and activities varies according to real-time demand and individual customer profiles.

- **Example**: Online platforms like Airbnb use AI algorithms to suggest properties custom-made to individual users based on past bookings and preferences. Similarly,


https://doi.org/10.52202/081568-0018

travel agencies like Expedia use AI to generate personalized tours that take into account a traveller's history and interests.

**AI in Predictive Analytics and Smart Travel Management**

AI is also being used for predictive analytics, helping businesses improve their operations. For instance, airlines and hotels are using AI-powered tools to forecast demand and predict booking patterns, enabling them to make data-driven decisions on pricing and capacity management.

- **Example**: AI-based algorithms at airports can predict passenger behaviour, such as the likelihood of a missed flight or delays at security checks, and direct travellers to the quickest routes or provide real-time updates to enhance their travel experience
- **2.2 Internet of Things (IoT) and Smart Mobility**

The Internet of Things (IoT) is another technology that is reshaping the tourism industry, especially in the monarchy of smart mobility and connected experiences. IoT refers to the interconnection of physical devices—such as sensors, wearables, and smart devices—over the internet, assisting them to collect and exchange data.

**IoT in Smart Airports and Hotels**

Airports and hotels are becoming gradually "smart," using IoT to restructure operations and provide travellers with a whole experience. In airports, IoT-enabled devices help systematize various processes, such as baggage tracking, security screening, and flight boarding.

- **Example**: London Heathrow Airport has implemented IoT-based systems to track luggage in real-time, reducing the likelihood of lost baggage. Sensors embedded in luggage tags help ensure that passengers' bags are always accounted for throughout the airport journey.

In hotels, IoT technologies allow for more efficient room management and personalized services. Guests can control room settings such as lighting, temperature, and entertainment via their smartphones or voice assistants. Smart hotel rooms also offer services such as automatic check-ins and room entry, reducing wait times and increasing guest convenience.

- **Example**: The Marriott International hotel chain has embraced IoT through its "IoT Guestroom" initiative, which allows guests to control in-room elements, including lighting, curtains, and entertainment systems, using their personal smartphones or voice commands.

**Smart Mobility and Connected Transportation**

IoT is enabling the development of smart mobility systems, which are enhancing the efficiency and convenience of transportation for travellers. Smart transportation networks can provide real-time updates on traffic, public transport schedules, and the availability of taxi services.

- **Example**: In cities like Singapore, IoT is being used to monitor traffic movement and optimize traffic signals, minimizing overcrowding and improving transportation efficiency. Similarly, cab-booking platforms like Uber and Ola use IoT technologies to manage their cabs, ensuring that passengers receive real-time information about the location and arrival times of their rides.

**Smart Cities and Digital Infrastructure**

The IoT has given rise to the concept of "smart cities," which use interconnected digital infrastructure to manage everything from traffic systems to waste management. These smart cities are creating an environment where tourism services are more accessible, safe, and efficient. For example, IoT-powered smart transportation systems, such as automatic buses and smart taxis, allow tourists to navigate cities with ease, while integrated systems for ticketing, payments, and security help enhance the overall visitor experience.

- **Example**: Barcelona has implemented a complete smart city program that integrates IoT technologies in its public transport system, smart lighting, and waste management, significantly improving the city's liveability for both residents and tourists.

**2.3 Block chain Technology in Tourism**

Block chain technology is making significant steps in the tourism sector, particularly in areas involving payment systems, ticketing, and data security. As a decentralized, damage proof digital ledger, block chain provides increased transparency, security, and efficiency in transactions.

**Block chain in Secure Ticketing Systems**

One of the key applications of block chain technology in tourism is in ticketing systems. Traditional ticketing platforms can sometimes be prone to fraud, scalping, or technical issues. Block chain offers a secure and transparent alternative by creating decentralized, tamper-proof records for each transaction. This not only ensures the validity of tickets but also reduces the risk of fraud.

- **Example**: Some airlines and travel agencies are experimenting with block chain for managing booking confirmations and flight tickets, ensuring that each ticket is securely issued, verified, and stored on the block chain.

**Block chain for Payments and Cryptocurrencies**

Block chain's ability to support cryptocurrencies, such as Bitcoin, has opened up new possibilities for payment systems in the tourism industry. Cryptocurrency payments eradicate the need for intermediaries, reducing transaction fees and improving the speed of cross-border payments. This adds a advantage for international travellers, who often face high conversion rates and banking fees when making payments.

- **Example**: Travel agencies like Travala have begun accepting cryptocurrencies for hotel bookings and flight tickets, allowing tourists to use digital currencies for travel transactions. This offers an alternative to traditional payment methods and aligns with the growing trend of cryptocurrency acceptance.

**Block chain for Identity Verification**


https://doi.org/10.52202/081568-0018

Block chain can also enhance security and privacy in tourism by providing decentralized systems for identity verification. With block chain, tourists could store their personal data, such as passport details or biometric information, on a secure and tamper-proof ledger, reducing the need for physical documents and enhancing travel safety.

- **Example**: Several airlines and border control agencies are exploring block chain to streamline identity verification processes at airports, which could reduce wait times and improve overall security procedures.

### 2.4 Mixed Reality (AR/VR) in Tourism

Mixed reality technologies, including augmented reality (AR) and virtual reality (VR), are transforming how travellers experience destinations. By merging digital information with the physical world (AR) or providing fully immersive experiences (VR), these technologies are making tourism more engaging, interactive, and accessible.

#### Virtual Reality (VR) for Destination Marketing

Virtual reality allows tourists to visit destinations from the comfort of their homes, experiencing virtual tours of historical sites, natural landscapes, and cultural landmarks. VR experiences enable potential travellers to explore destinations before making travel decisions, providing an immersive preview that can create interest of traveller and help them make bookings.

- **Example**: The "Google Earth VR" application allows users to virtually explore any part of the world. Tourists can "fly" over landmarks, zoom in on cities, and experience destinations in an entirely new way.

#### Augmented Reality (AR) for Enhanced On-Site Experiences

Augmented reality overlaps digital content onto the physical world, enhancing the on-site experience for tourists. AR applications can provide real-time information about landmarks, historical facts, and cultural sites, enriching the visitor's understanding of the destination.

- **Example**: In cities like New York and London, tourists can use AR apps to guide them through popular landmarks, where their smartphones provide information about nearby attractions or historical data when they point their cameras at certain points of interest.

#### Immersive Experiences in Museums and Theme Parks

Museums, theme parks, and cultural sites are adopting AR and VR to create interactive and immersive experiences. These technologies enable visitors to engage more deeply with exhibitions, such as taking part in historical attraction or exploring interactive art connections.

- **Example**: At the Smithsonian Museum in Washington D.C., visitors can use VR headsets to experience immersive historical events, such as walking through ancient Egypt or witnessing the Apollo 11 moon landing.

https://doi.org/10.52202/081568-0018

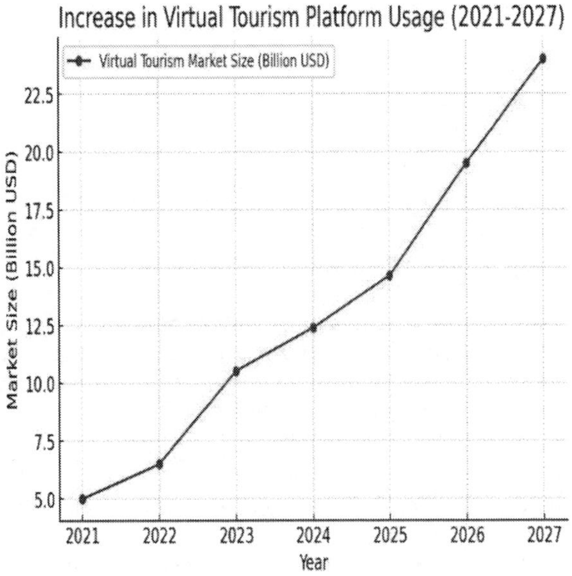


Fig 1:Above shows the increase in trends in virtual tourism platform usage in tourism sector, the figures for 2024 to 2027 are projections based on current trends.

## 3. Emerging Trends in Global Tourism

The structure of global tourism is undergoing significant changes due to the integration of new technologies, changing consumer behaviours, and evolving social demands. These trends do not only reflect the growing importance of digital innovation but also show how tourism is adapting to the needs of modern travellers. This fast adoption of technology, environmental considerations, and changing travel habits have given rise to several emerging trends that are shaping the future of tourism. Some of the most significant trends in the tourism industry today include digital nomadism, virtual tourism, sustainable travel, and the rise of "smart tourism."

### 3.1 Digital Nomadism

Digital nomadism, the practice of working remotely while traveling, has been on the rise in recent years, accelerated by the COVID-19 pandemic and advancements in technology that helps individuals to work from virtually anywhere. With the explosion of high-speed internet, cloud technologies, and mobile applications, professionals can now carry out their work from interesting destinations, co-working spaces, or even while staying in remote areas far from their home countries.

**The Rise of Remote Work**

The concept of digital nomadism gained grip as remote work became more common during the global pandemic. As businesses are increasingly embracing flexible work policies and remote-first models, employees are no longer limited to a single location. This has led to a increase in individuals opting to combine work and travel. With affordable airfares, flexible


https://doi.org/10.52202/081568-0018

work hours, and connectivity, destinations around the world are becoming more attractive to digital nomads.

- **Example**: Popular cities like Bali, Chiang Mai, and Lisbon have become hubs for digital nomads, attracting individuals who wish to work while experiencing new cultures. Many countries and cities have also recognized this trend and have introduced special visa programs designed to provide to remote workers.

**Infrastructure and Services for Digital Nomads**

As the digital nomad community grows, so does the demand for infrastructure and services that support this lifestyle. Co-working spaces, high-speed internet access, affordable accommodation, and networking events are crucial elements for digital nomads. To meet this demand, various destinations are adapting their infrastructure to offer the right environment for long-term stays.

- **Example**: Estonia offers a digital nomad visa that allows remote workers to live and work in the country for up to a year. Similarly, cities like Tbilisi in Georgia and Medellin in Colombia have developed strong digital nomad ecosystems by providing co-working spaces, affordable living options, and networking opportunities to attract international remote workers.

**Impact on Local Economies and Communities**

Digital nomadism has created both challenges and opportunities for local communities. On one hand, the arrival of long-term travellers boosts local economies by contributing to the hospitality, food, and service sectors. On the other hand, some cities have experienced increased demand for housing, rising rental prices, and cultural shifts due to the arrival of foreign workers. Balancing the needs of digital nomads with the well-being of local residents is crucial for sustainable growth.

- **Example**: In Bali, Indonesia, there has been both excitement and concern over the growing number of digital nomads. While tourism boosts local businesses, some areas have seen an increase in housing prices, impacting long-time residents

### 3.2 Virtual Tourism

Virtual tourism, which allows people to experience destinations without physically visiting them, has gained significant attention in recent years. This emerging trend is particularly appealing to those unable to travel due to financial limitations, health issues, or geopolitical factors such as travel restrictions. The integration of Virtual Reality (VR) and Augmented Reality (AR) technologies into tourism experiences is transforming how people explore the world.

**The Role of VR and AR in Virtual Tourism**

Virtual reality (VR) immerses users in realistic, computer-generated environments, while augmented reality (AR) enhances the real-world experience with digital connections. These technologies are enabling virtual tourism by providing digital renovations of famous landmarks, museums, or even entire cities, which can be experienced from the comfort of a user's home.

https://doi.org/10.52202/081568-0018

- **Example**: The Louvre Museum in Paris offers virtual tours of its collections, allowing people to explore the museum's galleries and exhibits remotely. Similarly, travel companies like Virtually There offer immersive VR experiences where travellers can explore destinations like New York or the Great Wall of China before booking their trip.

**Destination Marketing and Immersive Experiences**

Tourism boards and businesses are using virtual tourism as a tool for destination marketing. By offering immersive VR or AR experiences, travel agencies and destinations can give potential visitors a perception of what to expect before making travel decisions. This virtual experience can be especially powerful for marketing lesser-known or off-beat locations, allowing potential tourists to experience the unique features of a destination from a distance.

- **Example**: Dubai has embraced virtual tourism by offering 360-degree virtual tours of its iconic Burj Khalifa tower, underwater aquariums, and other top attractions. These experiences help entice visitors to plan a physical visit.

**The Potential for Increased Accessibility**

Virtual tourism has the potential to standardize travel, making it more accessible to people with disabilities, the elderly, or those who live in remote areas. Instead of physical travel, individuals can enjoy the experiences and cultural knowledge of a destination without any of the physical barriers that might normally prevent them from traveling.

- **Example**: Organizations such as the "Virtual Reality for Good" initiative focus on creating immersive VR experiences that allow people with limited mobility to visit natural wonders or cultural sites that they otherwise might not have the opportunity to experience.

### 3.3 Sustainable Travel and Eco-Friendly Tourism

Sustainability has become a central focus in tourism, as travellers and industry players alike recognize the need to lessen the environmental impact of mass tourism. The growing concern over climate change, over tourism, and resource depletion has prompted the rise of eco-friendly tourism practices that focus on environmental conservation, reducing carbon footprints, and promoting responsible travel.

**Green Certifications and Eco-Friendly Accommodations**

Eco-friendly accommodations, such as hotels and resorts that follow to sustainability standards, are becoming more prevalent in the tourism industry. Many institutions now operate under green certifications or sustainable practices that minimize waste, reduce energy consumption, and use renewable resources.

- **Example**: The "Green Key" certification is one of the leading eco-labels for tourism establishments. Hotels that hold this certification adhere to environmentally friendly practices such as using energy-efficient appliances, reducing water waste, and recycling. Many eco-resorts in places like Costa Rica and the Maldives focus on preserving the natural environment while providing guests with eco-friendly accommodations.


https://doi.org/10.52202/081568-0018

### Carbon Offset Programs and Sustainable Travel Options

Another growing trend is the focus on carbon offset programs that allow travellers to neutralize their environmental impact by investing in projects that reduce or absorb carbon emissions, such as reforestation or renewable energy initiatives. Furthermore, sustainable travel options, such as eco-friendly transportation methods (electric vehicles, trains, and bicycles), are gaining popularity.

- **Example**: Airlines like Delta and British Airways offer passengers the option to purchase carbon offsets when booking flights, helping them contribute to projects aimed at reducing their environmental footprint. Similarly, sustainable travel platforms like Responsible Travel help travellers choose eco-friendly experiences and destinations.

### Regulations to Combat Over tourism

Over tourism—the overwhelming number of visitors to certain destinations—has raised concerns about environmental degradation and social disruption. To fight this, some destinations are implementing regulations designed to control tourist numbers, promote responsible tourism, and protect cultural heritage sites.

- **Example**: In 2020, the Italian city of Venice introduced measures to reduce over tourism by limiting cruise ship arrivals and regulating the number of visitors in the city's most famous areas. Similar initiatives have been adopted by other popular destinations such as Machu Picchu in Peru, which now requires advance bookings and limited entry.

## 3.4 Smart Tourism and the Integration of Technology

Smart tourism is another emerging trend driven by technological advancements, particularly AI, IoT, and big data. This concept revolves around the integration of digital technologies into the tourism experience to improve efficiency, personalization, and connectivity. Smart tourism enables tourists to have a seamless, convenient, and personalized travel experience, facilitated by connected systems and devices.

### AI-Driven Personalized Travel Experiences

The use of AI to personalize the traveller experience is one of the keystones of smart tourism. By analysing large amounts of data, AI can offer personalized recommendations for destinations, accommodations, dining, and activities. These systems adapt to a traveller's preferences, offering insights and real-time suggestions that improve the overall journey.

- **Example**: AI-powered platforms like Google Travel and TripAdvisor provide personalized itineraries based on individual preferences, such as recommended attractions, restaurants, and accommodations based on past searches and ratings.

### Smart Cities and Infrastructure

Smart cities are emerging as key destinations for tech interested travellers, where digital infrastructure and connected services are seamlessly integrated into urban life. In smart cities, IoT technologies are used to manage traffic flow, improve public transportation systems, and offer efficient services, such as contactless payments and real-time information on events and activities.

- **Example**: Cities like Singapore and Barcelona have implemented smart city systems that provide real-time data on everything from public transport schedules to tourist attraction crowds. These systems help tourists navigate cities more efficiently while offering personalized recommendations based on their location and preferences.

**Seamless Travel with IoT and Big Data**

The integration of IoT devices in transportation systems, airports, and hotels has created an environment where travellers can enjoy a more connected and efficient experience. Smart luggage, for example, allows travellers to track their bags in real-time, while IoT-enabled airports streamline the check-in and security processes.

- **Example**: In airports like Changi in Singapore, IoT-enabled systems allow passengers to check in via facial recognition, and automated baggage handling systems ensure a faster and more efficient process

## 4. Technologies Driving Smart Cities and Global Connectivity

This exploration dives into how technology is reshaping our cities and travel experiences. Imagine a city where everything works together seamlessly, making life easier for both residents and visitors. That's the promise of smart cities, powered by cutting-edge technologies.

### 4.1 The Internet of Things (IoT) and Smart Infrastructure: Making Cities Responsive

The Internet of Things (IoT) is like the city's nervous system, connecting everything from traffic lights to hotel rooms.

**Smart Transportation Systems:**

Think of a city that anticipates traffic jams and adjusts lights to keep things flowing smoothly. That's what IoT can do. In places like Amsterdam, sensors monitor traffic, making commutes less stressful. For tourists, this means real-time updates on buses and trains, helping them get around with ease.

**Connected Hotels and Accommodations:**

Hotels are becoming more personalized. Imagine controlling your room's temperature and lighting with your smartphone or voice. Hotels like Marriott and Hilton are implementing these technologies. Beyond convenience, IoT also helps hotels save energy, making them more sustainable.

**Enhanced Visitor Experiences and Smart Tourism Applications:**

Exploring a new city should be exciting, not frustrating. IoT helps by providing personalized information and interactive experiences. In Barcelona, an IoT platform suggests restaurants and attractions based on your preferences, making your visit more enjoyable.


https://doi.org/10.52202/081568-0018

### 4.2 Artificial Intelligence (AI) and Automation: Personalizing the Journey

AI is like having a personal assistant for your travels, helping you find the best experiences and manage your itinerary.

**AI-Powered Personalization in Tourism:**

AI can examine your preferences and suggest activities you'll love. Platforms like TripAdvisor and Google Travel use AI to create custom itineraries. Virtual assistants like Alexa and Google Assistant provide real-time information, answering your questions and making your trip smoother.

**Predictive Analytics and Resource Management:**

AI helps cities and businesses anticipate demand, preventing overcrowding and ensuring resources are used efficiently. In London, AI predicts crowd movements, helping authorities manage public transport and keep people safe.

**AI in Smart Airports and Security:**

Airports are becoming more efficient and secure with AI. Facial recognition and automated baggage handling speed up processes, making travel less stressful. Airports like Changi and Heathrow are leading the way in this area.

### 4.3 Blockchain Technology and Secure Travel Ecosystems: Building Trust

Blockchain is like a secure, transparent ledger, ensuring that transactions and data are trustworthy.

**Blockchain for Secure and Transparent Ticketing**:

Blockchain can eliminate ticket fraud and streamline booking processes. Platforms like Winding Tree offer decentralized travel services, making booking flights and accommodations more transparent.

**Blockchain for Payment and Cross-Border Transactions:**

Blockchain simplifies international payments, reducing fees and making transactions faster. Using cryptocurrencies can be especially helpful for travelers dealing with different currencies.

**Blockchain for Identity and Verification:**

Blockchain can create secure digital IDs, making it easier to verify your identity at airports and hotels. This could replace traditional passport checks, making travel more convenient.

### 4.4 Smart Mobility and Integrated Transport Networks: Moving with Ease

Smart mobility systems aim to create seamless, interconnected transportation networks.

**Autonomous Vehicles and Smart Traffic Management:**

Self-driving vehicles are becoming a reality, offering convenient and environmentally friendly transportation. Cities like Dubai and Austin are already testing autonomous shuttles.

**Integrated Mobility Platforms:**

Platforms like Citymapper and Whim connect different modes of transport, allowing you to plan your entire journey in one app. This makes it easy to switch between buses, trains, and bike rentals.

### 4.5 Global Connectivity and the Role of 5G Technology: Staying Connected

5G technology is the backbone of global connectivity, enabling faster internet speeds and real-time services.

5G enhances the travel experience by enabling high quality streaming, faster booking systems, and immersive AR/VR tourism experiences. Cities like Seoul and Tokyo are already implementing 5G networks, paving the way for the future of travel.

In essence, these technologies are transforming cities into intelligent, responsive environments, making travel more efficient, personalized, and enjoyable.

## 5. Challenges in Adapting to Technology in Tourism: Navigating the Hurdles - A Human Perspective

We are brimming with excitement about the potential of smart cities and truly personalized travel experiences. However, let us be honest, the journey to this tech-infused future is not a smooth, straight path. There are hurdles, and they influence real people in real ways.

### 5.1 Infrastructure Limitations: Building the Foundation - Not Everyone Starts on the Same Line

Imagine you are trying to build a high-tech house on a foundation of sand. That is what we are facing when we try to introduce advanced technology in areas with weak infrastructure.

**Uneven Access to Digital Infrastructure: The Tale of Two Travelers**

Picture this: Sarah, a tourist, lands in a bustling metropolis. Her smartphone seamlessly connects to the city's Wi-Fi. She uses an app to find the best local restaurants, books a ride-share in seconds, and navigates the subway with real-time updates. Now, imagine another traveler, Ravi, arriving in a remote village. He struggles to get a weak mobile signal, let alone find reliable internet. He relies on paper maps and outdated information, feeling frustrated and disconnected. This is not just a technical issue; it is a matter of equity. Some travellers are empowered by technology, while others are left behind due to where they are. In places like rural India or vast areas of Africa, basic internet access is still a dream for many.

**Cost of Technological Upgrades: The Burden on Small Businesses**

Think of Maria, who runs a charming bed and breakfast in a small town. She dreams of offering smart room features like automated check-in and personalized lighting. But the cost of these upgrades is overwhelming. She is competing with large hotel chains that can afford the latest gadgets, leaving her at a significant disadvantage. It's not just about business; it's about preserving the unique charm and local flavour of small, family-run establishments.

### 5.2 Cybersecurity and Privacy Concerns: Protecting Our Digital Selves - Trust Is Paramount


https://doi.org/10.52202/081568-0018

As we entrust more of our personal information to digital systems, we become increasingly vulnerable to cyber threats.

**Data Breaches and Hacking Threats: The Shadow of Vulnerability**

Consider the sheer volume of personal data tourism businesses collect: names, addresses, credit card details, travel preferences. This is a goldmine for hackers. Incidents like the British Airways data breach serve as stark reminders of our vulnerability. It is not financial loss; it is about the erosion of trust. When travellers feel their data is unsafe, they are less likely to embrace digital solutions.

**Privacy Invasion and Ethical Concerns: The Line between Convenience and Intrusion**

Imagine walking through an airport where facial recognition technology tracks your every move. While this may speed up security, it raises serious privacy concerns. It feels like we are constantly being watched. We need to find a balance between convenience and privacy. People deserve to know how their data is being used and have control over it.

**Regulatory and Legal Frameworks: A Global Maze of Confusion**

Data privacy laws vary wildly from country to country. A tourism app that complies with the European Union's GDPR may violate regulations in another nation. This creates a complex legal landscape for businesses operating globally. It is not legal compliance; it is about respecting the diverse privacy expectations of travellers from around the world.

### 5.3 Adoption Resistance and Digital Literacy: Bridging the Generational and Skill Gaps - Understanding Human Nature

Technology is not a one-size-fits-all solution. Some people are more comfortable with traditional methods, and that is perfectly understandable.

**Resistance from Older Generations: The Comfort of Familiarity**

Think of your grandparents who prefer a physical map to a GPS app. They value face-to-face interactions and the human touch. We need to respect these preferences and provide alternative options. It is not about forcing technology on people; it is about offering choices.

**Staff Resistance to Technological Change: The Fear of the Unknown**

Hotel staff may worry that automation and AI will replace their jobs. They might be afraid of learning new systems. We need to invest in training and support, showing them that technology can enhance their work, not replace it. As the article from HR Dive points out, "Robot-phobia" is a real concern, and can worsen labour shortages. It is important to address this fear.

### 5.4 High Costs of Implementation: The Financial Divide - Economic Realities

Money is a significant barrier to technology adoption, particularly for small businesses.

**Financial Barriers for Small Businesses: The Struggle to Compete**

A small, family-run hotel may not have the resources to invest in the latest smart room technology. They are competing with large hotel chains that can offer high-tech experiences, creating an uneven playing field.

**Ongoing Maintenance and Upgrades: The Hidden Costs**

Technology requires constant maintenance, updates, and security patches. These costs can be a burden for small businesses with limited budgets. It is not just about the initial investment; it is about the ongoing expenses.

**5.5 Sustainability and Environmental Impact: Thinking Green - Our Responsibility to the Planet**

We cannot overlook the environmental consequences of technology.

**Energy Consumption and E-Waste: The Ecological Footprint**

Data centers, IoT devices, and AI algorithms consume vast amounts of energy. The rapid pace of technological innovation leads to the accumulation of electronic waste. We need to find ways to make technology more sustainable and reduce our environmental impact.

In essence, adapting to technology in tourism is a complex and nuanced process. It is not just about implementing new gadgets; it is about understanding the human impact of these technologies and addressing the challenges in a responsible and equitable way

## 6. Conclusion

In the fast-evolving site of global tourism, technology plays a essential role in shaping how individuals experience travel and how businesses and destinations provide to their needs. From AI-driven personalization to the arrival of smart cities and blockchain-enabled systems, the tourism industry is undergoing a transformation that brings both opportunities and challenges. The integration of emerging technologies is not just enhancing the travel experience but also revolutionizing the foundations of how tourism functions at a global scale.

This research highlights that the advancement of technologies like Artificial Intelligence (AI), the Internet of Things (IoT), blockchain, and smart mobility systems has accompanied in the era of **smart tourism**, which is intensely changing how travellers interact with destinations, how tourism businesses operate, and how cities evolve to accommodate modern-day travellers. These technologies not only provide greater convenience, personalization, and efficiency for travellers but also improve the management and sustainability of tourism operations. Features such as AI-driven chatbots, mixed-reality experiences, and IoT-based smart mobility systems are enhancing the overall travel experience, while also contributing to more sustainable practices in the tourism industry. Moreover, blockchain's ability to provide transparent and secure payment systems offers significant opportunities to streamline ticketing and reduce fraud, ultimately developing trust in the digital tourism ecosystem.

However, as the tourism industry holds these innovations, it must also address several critical challenges that could delay the smooth acceptance and integration of these technologies. The **digital divide**—the unequal distribution of technological infrastructure—remains a significant barrier, particularly in rural and less developed regions where access to high-speed internet and smart technologies is limited. Also, **cybersecurity** and **privacy concerns** continue to pose considerable risks as more data is shared and stored through digital platforms. Ensuring data protection, while nurturing consumer trust, will be crucial to the sustainable growth of technology-driven tourism. Additionally, issues like **resistance to technological adoption**, particularly among older generations or small businesses with limited resources, can delay the benefits of technology integration and require targeted efforts in training and education to lessen these challenges.


https://doi.org/10.52202/081568-0018

In spite of these hurdles, the tourism industry is gradually adapting to the digital age. Trends like **digital nomadism**, **virtual tourism**, and **sustainable travel** are redefining travel behaviour and encouraging industry stakeholders to invest in long-term solutions that meet the evolving expectations of the global traveller. In particular, the increased focus on **sustainability** and **eco-friendly tourism** is driving the sector toward responsible growth, with technology playing an instrumental role in reducing carbon footprints, optimizing energy consumption, and enabling greener travel options.

Moving forward, it is clear that technology will continue to be a driving force in global tourism, but its successful integration centers on the ability of stakeholders to address the challenges outlined in this study. Governments, businesses, and tourists themselves must collaborate to ensure that technological advancements are inclusive, secure, and environmentally responsible. For instance, governments can invest in **digital infrastructure**, provide incentives for sustainable tourism technologies, and create **regulatory frameworks** that safeguard data privacy while encouraging innovation. Similarly, tourism businesses can take a active approach to upskilling their workforce, incorporating **smart solutions** that improve customer service, operational efficiency, and sustainability.

Moreover, as technology evolves at an fast-tracking pace, the tourism industry must remain alert and open to new possibilities. The emergence of **artificial intelligence** and **augmented reality** technologies, for example, may open up entirely new avenues for interactive and immersive tourism experiences, while the growing importance of **block chain** could revolutionize everything from booking processes to loyalty programs. By embracing these technologies with a forward-thinking mindset, the tourism industry has the potential to not only enhance the travel experience but also address global challenges such as environmental sustainability and unbiased access to tourism opportunities.

In conclusion, the future of tourism is unquestionably tangled with technological innovation. As the industry continues to evolve, those who are able to adapt to the changes and overcome the obstacles presented by technology will thrive. The key to success lies in balancing the great benefits of technology with responsible and sustainable practices. For the tourism sector to truly flourish in the digital age, it must stand-in an inclusive, secure, and innovative environment that benefits both travellers and the world around them. By doing so, technology will not only enhance the experience of travellers but will also contribute to a more connected, sustainable, and resilient tourism ecosystem.

## References


1. Kolb, A. Y., & Kolb, D. A. (2008). *Experiential learning theory bibliography: Volume 2, 2006–2008*. Cleveland, OH: Experience Based Learning Systems.
2. Barley, S. R., Meyerson, D. E., & Grodal, S. (2011). E-mail as a source and symbol of stress. *Organization Science, 22*(4), 887–906. https://doi.org/10.1287/orsc.1100.0582
3. Mazmanian, M., Orlikowski, W. J., & Yates, J. (2013). The autonomy paradox: The implications of mobile email devices for knowledge professionals. *Organization Science, 24*(5), 1337–1357. https://doi.org/10.1287/orsc.1120.0806
4. Wajcman, J., & Rose, E. (2011). Constant connectivity: Rethinking interruptions at work. *Organization Studies, 32*(7), 941–961. https://doi.org/10.1177/0170840611410825
5. Barley, S. R. (1983). Semiotics and the study of occupation: An example from the sociology of occupations. *Journal of Occupational Behaviour, 4*(2), 79–88. https://doi.org/10.1002/job.4030040202
6. Orlikowski, W. J. (1992). The duality of technology: Rethinking the concept of technology in organizations. *Organization Science, 3*(3), 398–427. https://doi.org/10.1287/orsc.3.3.398
7. Déry, K., Sebastian, I. M., & Léger, P.-M. (2014). Digital work and care during the COVID-19 pandemic: From place to space and back again. *Journal of Business Research, 117*, 124–131. *[Note: if referencing a different Dery et al. 2014, please provide full title]*
8. Hinds, P., & Bailey, D. E. (2003). Out of sight, out of sync: Understanding conflict in distributed teams. *Organization Science, 14*(6), 615–632. https://doi.org/10.1287/orsc.14.6.615.24866
9. Wilson, J. M., Crisp, C. B., & Mortensen, M. (2013). Team conflict: The good, the bad, and the unknown. *The Journal of Applied Psychology, 98*(3), 559–578. https://doi.org/10.1037/a0031994
10. O 'Leary, M. B., & Cummings, J. N. (2007). The spatial, temporal, and configurational characteristics of geographic dispersion in teams. *MIS Quarterly, 31*(3), 433–452. https://doi.org/10.2307/25148888

11. Cummings, J. N., Espinosa, J. A., & Pickering, C. K. (2009). Crossing spatial and temporal boundaries in globally distributed projects: A relational model of coordination delay. *Information Systems Research, 20*(3), 420–439. https://doi.org/10.1287/isre.1080.0192

12. Jarvenpaa, S. L., & Leidner, D. E. (1998). Communication and trust in global virtual teams.*Organization Science, 10*(6), 791–815. https://doi.org/10.1287/orsc.10.6.791
13. Walther, J. B. (1992). Interpersonal effects in computer-mediated interaction: A relational perspective. *Communication Research, 19*(1), 52–90. https://doi.org/10.1177/009365092019001003
14. 1O'Leary, M. B., Wilson, J. M., & Metiu, A. (2014). Beyond being there: The symbolic role of communication and identification in perceptions of proximity to geographically dispersed colleagues. *MIS Quarterly, 38*(4), 1219–1243. https://doi.org/10.25300/MISQ/2014/38.4.12
15. Metiu, A., & Jett, Q. R. (2008). Overcoming barriers to knowledge sharing in global virtual teams. *Academy of Management Annuals, 2*(1), 533–567. https://doi.org/10.5465/19416520802211688

 https://doi.org/10.52202/081568-0018

# Exploring the Coral Paradises: Sustainable tourism in Lakshadweep

Deepali Garg[1*] , SK Naznin[2] , Ritik Sharma[3]

Chandigarh University, Mohali, India

*Corresponding author: Deepaligarg6109@gmail.com

**Abstract.** Lakshadweep, an archipelago of 36 coral islands in the Arabian Sea, is renowned for its pristine marine ecosystems and vibrant coral reefs. As a growing tourist destination, sustainable tourism practices are crucial to preserving its delicate environment while fostering economic development for local communities. This paper explores the balance between conservation and tourism, emphasizing eco-friendly policies, community engagement, and responsible travel practices. It also highlights challenges such as coral bleaching, habitat degradation, and waste management, while presenting sustainable solutions to ensure long-term ecological and cultural preservation.

**Keywords:** Sustainable tourism, Coral reef conservation, Ecotourism, Marine biodiversity, Lakshadweep, Environmental impact

## 1. Introduction

Coastal zones are of paramount importance for both ecological and economic sustainability, providing a myriad of resources and ecosystem services. These areas necessitate a careful equilibrium between environmental preservation and economic development to ensure the well-being of both the environment and human population. The concept of blue economy (BE) emerges as a critical framework within this context, advocating for the responsible use of marine resources to foster economic growth while conserving ecosystems. This is particularly relevant for island ecosystems, such as India's Lakshadweep Islands, where integrating BE principles can significantly contribute to achieving the United Nations Sustainable Development Goals by protecting marine environments and their dependent communities. The Lakshadweep Islands, situated in the Arabian Sea off India's southwest coast, provide a unique opportunity to synergize BE and the tourism sector. These Islands, renowned for their ecological significance and cultural diversity, make the promotion of responsible tourism essential for sustaining local communities. By adopting sustainable practices, the Lakshadweep Islands can preserve their natural and cultural heritage for future generations, thereby contributing to economic growth and aligning with the global sustainable tourism trends . However, these Islands face unique challenges, including limited land availability, vulnerability to climate change and finite resources. These challenges necessitate the establishment of a sustainable BE for the enduring welfare of island communities. The interplay between BE and tourism is particularly notable in small oceanic islands, where both sectors aim to harness ocean and coastal resources for economic development through sustainable practices. Recreational tourism, being a primary income source in these Islands, stands to benefit significantly from adherence to BE principles . The effective integration of these principles in management and development strategies is vital for addressing the Islands' specific needs, contributing to sustainable development, and ensuring long-term well-being. Acknowledging environmental and socio-economic stressors in planning is crucial, and the adoption of BE concepts offers tangible benefits to local communities and the natural environment, enhancing resilience and creating economic opportunities, this study, therefore, explores the integration of sustainable tourism within the BE framework in the Lakshadweep Islands, focusing on balancing economic growth with environmental conservation and community welfare.

https://doi.org/10.52202/081568-0019 

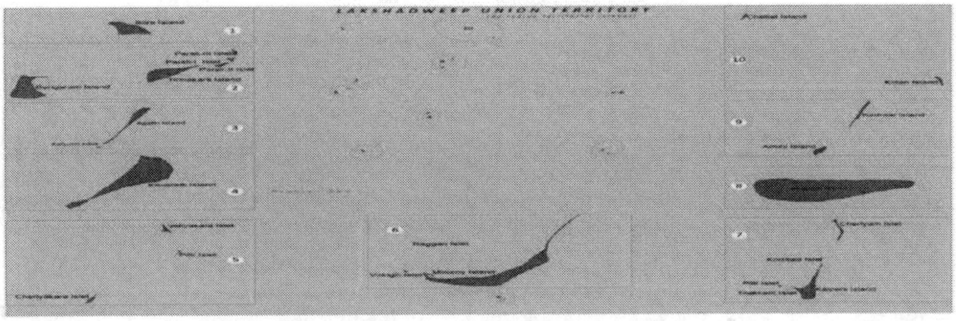


**Fig. 1.** Map of the Lakshadweep group of islands, India

The Lakshadweep archipelago, positioned and is India's smallest Union Territory, presenting unique ecological and socio-economic dynamics. The methodology adopted in this study involved a comprehensive assessment of the tourism capacity of these Islands, encompassing spatial analysis and ecological considerations. During 2016–17, a detailed survey was carried out across nine inhabited islands, using geo-referenced maps at the block, sub-block and plot levels to accurately calculate the area available for tourism development (Supplementary Figure 1), following the National Building Code of India 2016 and The Architects' Handbook . This approach allowed precise estimation of bed capacity and built-up area, considering floor area ratio (FAR) recommendations and excluding no-development zones (NDZs). The determination of boating density in the Lakshadweep lagoon utilized a weighted score analysis, incorporating a scoring matrix with 11 lagoon characteristics . The methodology factored in spatial requirements for different boat types and their respective restrictive scores to compute optimal boating density, recommending a density of 20 acres per motorboat and lesser densities for smaller boats 10[11] (Supplementary Figure 2 and Table 3). The Supplementary material provides a detailed methodology elucidating the boating density calculated for the lagoon. The carrying capacity of the lagoon for boat numbers was assessed by applying this optimal boating density across the entire lagoon area, crafting three distinct scenarios based on ecological parameters and specific island lagoon areas. This included maximum (Supplementary Figure 3 and Table 4), optimum (Supplementary Figure 4 and Table 5) and sustainable (Supplementary Figure 5 and Table 6) models tailored with the ecological context of the respective islands. Ecologically Sensitive Areas (ESAs) such as seagrass, corals and fish breeding sites were included in the analysis, as well as diverse geomorphological features such as sand sheets, reef crests, reef slopes, deeper reef flats, reef flats and sandy reef flats. The lagoon features, viz. Deep intermediate and shallow lagoons were also included in the evaluation (Supplementary material). For solid waste and sewage assessment, guidelines from the Central Pollution Control Board, Government of India and Urban and Regional Development Plans Formulation and Implementation guidelines were developed by the Ministry of Urban Development. A marine litter survey, adhering to Oslo and Paris Conventions (OSPAR) guidelines, was conducted on select islands to categorize and quantify marine litter, employing the clean coast index (CCI) for assessing the cleanliness level of the coast. The limit of acceptable change (LAC) for Lakshadweep Islands was determined using key indicators like coral reef health and water quality, following established methodologies. This comprehensive methodological approach enabled a nuanced understanding of the tourism potential and environmental constraints of the Lakshadweep Islands, offering a model for sustainable tourism development within the framework of BE.

The oceans and seas are the most dominant and significant features on this planet, covering over 71 % of its surface. This environmental component, with its vast expanse and sheer volume, is of immense importance to mankind, and serves as the economic lifeline for coastal communities, fish workers and those who sail in its water. Also, the seas and oceans play a pivotal role in shaping the world climate, maintaining the water cycle and biogeo-chemical cycling of elements. It contains a great variety of flora and fauna from its upper sunlit zone to the depths. But except for hotspots like coastal wetlands, mangroves and coral reefs, little is known about the biological diversity, abundance and distribution of marine organisms or the structure and function of the marine ecosystem in its totality (USNRC 1992). Marine species that face the greatest risks of extinction


https://doi.org/10.52202/081568-0019

include corals, sponges, oysters, octopus, porpoises, whales, ornamental fishes, sea turtles and dugongs. The Lakshadweep Islands of India being oceanic in nature harbour important marine flora and fauna in the seas around it. An overview of some of the marine flora/fauna, their status, distribution, threats, sustainable uses and conservation needs in Lakshadweep archipelago are discussed here.

The economy of Lakshadweep is influenced by limitations of its geography. Though there are no per capita indicators specific to the islands available, about 45% of the population have been identified as living below the poverty line, that is, living on less than US$1 a day (World Bank, 2000). More than 4000 families receive various kinds of assistance under the rural development and poverty alleviation schemes of the Central Government. A primitive economy in many respects, coconut farming is the major source of income. The annual yield of coconuts is approximately 28 million nuts, which are processed to make copra and sold to coconut-oil mills on the mainland. However, the recent economic liberalisation and removal of import restrictions have led to a glut of cheaper oil from abroad, and have brought down the price of coconuts drastically (from an average of eight to two rupees per coconut), leaving the sector in a major crisis. Other crops include banana and a few vegetables grown in coconut plantations. Obviously, fishing is another major economic activity.There are 496 mechanised fishing boats and 500 country crafts operating in different islands. Annual fish catch in 2000 was 10,800 tons, generating a total income of Rs 220 million. About 6200 islanders are engaged in fishing as workers and small entrepreneurs. In the early 1990s, the Central Government allowed multinational trawlers entry into the Indian seas, and this resulted in a virtual desertification of the sea, leaving the fishing industry in the country, especially in the islands, in total turmoil.

The islands have been declared a 'no-industry' area by the Central Government, in that large-scale industrial development is neither feasible nor allowed. The geographical and environmental limitations allow only a limited number of small- and medium-scale industries on the islands. Industrial activities on the islands are mainly small-scale, and based on coconut products and fisheries. Coconut husk provides the basic raw material for a number of factories, and there are altogether 22 units manufacturing various coir-based products under the public sector, employing more than 316 islanders. There are also a number of small coir units functioning in the private sector in different islands, 63 of them set up with governmental subsidy. Two coir spinners' cooperative societies set up by the Government help the self-employed and small units to market their products. The workforce in coir processing units consists mainly of women, the majority of whom work part time. Boat building was once an important occupation, but with the advent of motorized boats it is in decline. Only two Island Tourism and Sustainability 431 boat-building yards remain, and these along with 11 workshops in various islands take care of repairs and servicing of the fishing vessels. The canning factory in Minicoy island processes fish, especially tuna, which is available around the islands. Making handicraft products using sea-shells, coconut-shells and wood is another important occupation on the island (George, 1997). Other industrial units on the islands include a hosiery factory in the Kalpeni, a desiccated coconut powder unit in Kadmat, and a smoked fish and fish meal unit in Agatti. Coral mining, which used to be a major industry, is now banned. The islanders are also famous for their sailing skills and a large number of them, especially from the Minicoy islands, work as seamen all over the world. Except for the self-employment units and micro-enterprises in the private sector, most of the industries are in the public sector. Considering the lack of economic development in the islands, as evident from large numbers of people under the poverty line, it is not surprising that the public sector has been at the forefront in initiating industrial development activities. Moreover, the land reforms introduced in the 1970s fragmented larger land holdings, inhibiting wealth accumulation and entrepreneurial activities in the islands.

The nuanced interrelation between sustainable tourism and environmental conservation has become a cardinal focus in contemporary research, driven by the escalating need to harmonize the economic benefits of tourism with ecological preservation. Within this context, the concept of "carrying capacity" emerges as a pivotal

construct, explicating the optimum number of visitors that a tourist site can accommodate without inflicting irreversible damage on the ecological, social, and economic environments. This study situates itself within the multifaceted realms of sustainable tourism development in Kazakhstan, particularly focusing on the ecosystems of Katon-Karagay National Park (KKNP).
The importance of this research is underscored by the burgeoning tourism sector in Kazakhstan, particularly the sector of ecotourism and agritourism, which have been identified as significant contributors to regional sustainability, income, and cultural enrichment in the country. The meticulous exploration of carrying capacity in these diverse tourism sectors provides profound insights into the sustainable management and development of these sectors, ensuring the balance between visitor satisfaction, environmental conservation, and economic imperatives.

## 2. Review of Literature:

Lakshadweep, an attractive archipelago in the Arabian Sea, is increasingly noted as a promising tourist destination with its pristine beaches, diverse marine life, and rich cultural heritage. Over the past five decades, the islands have witnessed significant progress in institutionalizing tourism, driven by policy reforms and strategic planning initiatives. The initial serendipitous tourism developmental encounters have steered Lakshadweep towards adopting tourism as a stimulant for its development. However, as tourism continues to expand on the islands, it is crucial to priorities the local community's well-being and the region's ecological integrity. While the local community aspires to be the custodian of the destination's natural and cultural assets, the emphasis is now on the administration to change its centralized orientation and facilitate sustainable development based on a collaborative model. Looking ahead, the future of tourism in Lakshadweep hinges on adopting a destination stewardship model that fosters stakeholder collaboration and ensures tourism development aligns with the islands' cherished traditions and values.

Tourism in Lakshadweep is managed by the Society for Promotion of Nature Tourism and Sports (SPORTS), a society formed by the Lakshadweep Administration in 1982 with the avowed aim of tapping the tourism potential of the islands and acting as the nodal agency of the Lakshadweep Administration for the promotion of tourism in the islands. The primary aim of the organization is to promote eco-friendly tourism and recreational activities in the islands in association with and under the guidance of the Lakshadweep Administration (Society for Promotion of Nature Tourism and Sports Lakshadweep Tourism).

It is now generally acknowledged that coral reefs are among the most threatened global ecosystems, and among the most vital (Costanza et al. 1997; Bryant et al. 1998; Boesch et al. 2000; Reaser et al. 2000; Wilkinson 2000). Reefs are of critical importance to human survival (especially in developing countries) because they provide subsistence food for a substantial portion of the population, serve as the principle coastal protection structures for most tropical islands, and contribute major income and foreign exchange earnings from tourism (Costanza et al. 1997; Wells et al. 2001; Salm et al. 2001). The value of living resources (such as fisheries) and services (such as tourism returns and coastal protection) provided by reefs has been estimated at about $375 billion annually (Costanza et al. 1997). In addition, coral reefs provide habitat for some of the greatest biological diversity in the world (Ray 1988). In addition ,Controlled, sustainability-focused tourism models have been studied in Lakshadweep (Kokkranikal et al., 2003) .Carlsen and Butler (2011) highlighted the need for development patterns that preserve fragile atolls. Munjal and Baruah (2024) outline economic data emphasizing tourism's role in GDP and employment. Modi (2024) proposes eco-tourism practices—codes of conduct, waste management, community involvement—rooted in local culture and global best practices

 https://doi.org/10.52202/081568-0019

T w e n t y -f iv e y e a r s a g o scientists believed that the ca. 1.6 million species they had then inventoried represented maybe 50% of plant and animal species on this planet. N ew approaches in sampling insect diversity in rainforests and small m acrobenthos in the deep sea have revised this estimate to 1.7-1.8 million described species and 10-100 million species remaining to be discovered. In parallel w ith this changed paradigm, species inventorying has also evolved from being categorized as an outdated scientific activity to a timely cutting edge megascience "enterprise". The reason behind this change of attitude is probably rooted in our social anxiety over global climatic change and non sustainable development. The crude translation of this anxiety into science strategy is that there is no time to lose if we w ant to document and name biodiversity before it is lost forever

The Maldive Islands or 'Maldives' form the central and largest emergent part of the Laccadive Chagos Ridge which extends southwards from west of southern India to near the centre of the Indian Ocean, and consists entirely of atolls and associated coral structures. During at least the last two centuries, coral rock has been used in the Maldives for building and other construction. Prior to the 20th century, building in coral rock was limited to mosques, shrines, and tombstones; but the use of coral has increased over the past 80 years, so that it is now the main construction material on the capital island of Male and on tourist resort islands. In addition, coral rock is in demand for use as hardcore in road construction. It is predicted that, over the next two years, some 43,000 cubic metres of coral rock will have to be mined from existing reefs to meet the demand for 'aggregate' to surface roads in the capital of Male alone. The Government of the Maldives, aware of the conflict between the use of corals for building and their conservation for such purposes as maintenance of fishery resources, a rapidly expanding tourist industry, and land stability, are actively seeking alternative construction materials. This paper describes the environmental impact of mining activities on coral reefs around North Male Atoll and also discusses the possible alternative mining and construction techniques that are available for future use, given the need to conserve and protect existing reefs from further destruction.

The Saint Martin Island is one of the most beautiful islands in Bangladesh. This island is locally known as Narikel Jinjira (Coconut Island). It is located in the Bay of Bengal, approximately 9 km south of the Cox's Bazar-Teknaf peninsula. This island is very much resourceful with enormous biological diversity i.e. mollusk-300 species, fish -150 species, amphibian-5 species, coral-66 species, turtle-5 species, snail5 species, bird-200 species, mammals-20 species. The surface area of the island is about 8 km2 , however, it may vary depending on the tidal level. This is the most attractive tourist spot in Bangladesh. At present, unplanned tourism makes this island fragile. Around 3000 tourists arrive every day in tourism season (November - March). Following the recommendation of the National Conservation Society, the small Saint Martin Island was declared as an Ecologically Critical Area under the Bangladesh Environment Conservation Act. Mahmood et al. gave a preliminary study of corals of the Saint Martin Island. Tomas Tomascik made several studies on the coral reefs describing the different types of coral species, their abundance and the condition of the corals. Chowdhury et al. have surveyed the coastal geomorphology of the Island. Ahmed et al. made an overview on the coral reef ecosystem of Bangladesh and made a detailed analysis of the coral resources in the Island. Khan et al. noticed that Saint Martin's coral resources were started to vanish at that time. However, since no work has been done after 1997, a systematic study was needed and that is why the present study was carried out.

## 3. Objectives:

1. To assess the impact of tourism on Lakshadweep's marine and coastal ecosystems.
2. To analyze the socio-economic benefits of sustainable tourism for local communities.
3. To evaluate existing sustainable tourism practices in Lakshadweep
4. To identify challenges in maintaining ecological balance while promoting tourism.
5. To propose strategies for enhancing sustainable tourism in Lakshadweep.

https://doi.org/10.52202/081568-0019

## 4. Impacts of Tourism on Coastal and Marine Ecosystems:

Tourism, if not managed, can hurt marine animals and coral reefs. Some of the significant effects are: Coral Reef Destruction: Tourists trampling corals, boating, and trash pollution have resulted in coral bleaching and damage.Coral Cover Reduction: Studies indicate that absolute coral cover in the Lakshadweep islands declined sharply from 51.6% in 1998 to 11% in 2017

Bleaching Events: As of May 2024, more than 70.7% of coral reefs worldwide have been impacted by bleaching-level heat stress since January 1, 2023, with Lakshadweep being one of the worst-affected regions on India's coastline. Water Pollution: Increased disposal of waste, including plastic, impacts marine animals.

A study published in the Environmental Science and Pollution Research International journal found microplastics in the surface waters and gastrointestinal tracts of skipjack tuna around the Lakshadweep Islands, indicating pollution levels in the marine environment. The annual fisheries yield of Lakshadweep is approximately 9,000 tonnes, with tunas constituting about 7,000. The National Disaster Management Authority (NDMA) identifies factors such as natural disasters (storm surges, cyclones, tsunamis, and flooding) and human influences (construction of seawalls, groins, and jetties) as contributors to coastal erosion in Lakshadweep.

### 4.1 The socio-economic benefits of sustainable tourism:

Despite the environmental concerns, tourism is a major contributor to the economy of Lakshadweep.

The benefits are:

**Local Employment:**

Tourism provides job opportunities in hotels, travel agencies, restaurants, and transport services. Tourism is a major stimulator of economic development in Lakshadweep. The visit of tourists raises the demand for local goods andservices, thus stimulating local entrepreneurship and businesses. Promotion of Local Culture: Sustainable tourism encourages the tourists to experience local culture, arts, and food, hence perpetuating the culture. By promoting local tradition and culture, sustainable tourism promotes the preservation of the Lakshadweep islands' distinct heritage.

**Awareness & Education:**

The tourists receive information on marine conservation, which increases awareness on preserving nature. Sustainable tourism emphasizes the preservation of natural resources.

In Lakshadweep, this strategy enables the preservation of virgin beaches, coral reefs, and marine life, all of which are crucial to the environment and are sought after by environmentally conscious tourists. Adding the participation of local communities in tourism planning and management gives them a say in development projects so that tourism development is in their aspirations and interests.

**Existing Sustainable Tourism Activities:**

Lakshweep has established various environmentally friendly tourism procedures in order to preserve its fragile ecosystem, in addition to supporting responsible tourism. They include: Blue Flag Certification:


https://doi.org/10.52202/081568-0019

Kadmat Beach and Minicoy Thundi Beach received the Blue Flag, an exclusive international eco- label that certifies high environmental and qualitstandards. Blue Flag status ensures that the beaches fulfill 33 stringent FEE standards for ensuring sustainability in the form of environmental protection and education.

**Restricted Tourism Policy:**

To safeguard its fragile ecosystem, Lakshadweep has a policy of closed areas that limit tourist access to certain of its islands. By doing so, it prevents over-tourism, allowing genuine visitors to witness the region's beauty while keeping environmental degradation low

**Sustainable Development and Community Engagement:**

The vision is to develop Lakshadweep into a responsible and inclusive example of development. Strategies focus on utilizing natural beauty to develop the tourism industry, fostering economic resilience, and encouraging people's pride and well-being. Conservation Reserves: Lakshadweep has launched the PM Sayeed Marine Birds Conservation Reserve in 2020,India's first marine bird conservation area. Spread over 62 $km^2$, it is the breeding ground of Greater Crested Tern and Sooty Tern, which shows how vital it is to preserve biodiversity in the region. These programs reflect Lakshadweep's commitment to the integration of environmental protection with ecotourism so that its distinctive marine and coastal ecosystems are preserved for future generations. The information on the socio-economic advantages of sustainable tourism in Lakshadweep comes from a series of studies, reports, and expert opinions on the effects of sustainable tourism. The following presents some key sources:

## 5. Challenges in Maintaining Ecological Balance:

The ecological equilibrium of Lakshadweep is challenged by various naturaland manmade challenges. These challenges include:

1. Marine litter and plastics pollution: Marine debris, especially plastic debris, is of significant threat to Lakshadweep'scoraland marine environment. The pollution is of adverse effect on the marine plants and ani mals and lowers the islands' aesthetic appeal, thus affecting biodiversity and the tourism industry too.

2. Climate Change and Coral Bleaching: Increased sea temperatures cause more frequent cases of coral bleaching. Warm and acidic water destroys the limestone structure created by the coral, which may cause the death of the marine organisms inhabiting these reefs.

3. Sensitive Freshwater Resources: Lakshadweep's freshwater aquifers are sensitive and already strained due to climate change, pollution, and rising demand. Ambitious tourism plans will worsen freshwater shortages, testing human use and environmental sustainability.

4. Overpopulation and Spatial Planning

5. Sewage and Agricultural Runoff: Sewage contamination, coupled with the application of detergents, pesticides,and fertilizers, causes eutrophication in lagoons.

6. Human-Induced Strain on Uninhabited Islands: The growth of human populations and increased demands for land have led to the use of uninhabited islands, thereby disrupting avian habitats and other types of wildlife. Overcoming these issues needs conservation efforts integrated, sustainable tourism, and community participation to achieve the long-term ecological stability of Lakshadweep.

## 6. Strategies for Enhancing Sustainable Tourism:

Enhancing sustainable tourism in Lakshadweep involves implementing strategies that balance economic development with environmental conservation and community well-being. Key strategies include

1. Infrastructure Development: Improving transport and communication infrastructure is crucial for the attraction of tourists and sustainable development. The modernization of airports, seaports, and the internet increases the tourist experience as well as the local population.

2. Community Involvement: Local community participation in planning and executing tourism activities guarantees that development is aligned with their needs while maintaining cultural heritage. Empowering the community reinforces environmental stewardship and enhances the value of authenticity in the tourism experience.

3. Environmental Conservation Schemes: The implementation of measures aimed at reducing marine litter and promoting waste minimization is critical to the conservation of marine ecosystems. Measures like the Prevention of Marine Litter in the Lakshadweep Sea (PROMISE) aim to enhance the attraction of tourist spots while, in the process, protecting the environment.

4. Sustainable Development Planning: Creating tourism plans with focus on sustainable progress, conservation of culture, and a higher standard of living for the islanders is crucial. Providing Lakshadweep as an example of sustainable development harmonizes economic progress and the conservation of nature.

5. Premium, Limited-Scale Tourism: Focusing on premium, limited-scale tourism has the potential to minimize environmental impacts while maximizing economic benefits. This approach ensures that the growth of tourism does
not strain the island ecosystems.Placing these into action can make Lakshadweep's tourism sustainable and ena ble economic progress to be compatible with environmental preservation and community well-being.

### 7. Major Findings of the Case Study:

### 7.1 Modern Approaches to Sustainable Tourism

Blue Flag Certification: Some beaches such as Minicoy Thundi and Kadmat are Blue Flag certified, ensuring high environmental and safety standards. This promotes eco-tourism and ensures beach cleanliness and marine biodiversity.

Restrictive tourismpolicies: In Lakshadweep entailthe administration's control over visitor numbers through th e implementation of special permits, thereby mitigating over-tourism and alleviating ecological pressure.

Marine Conservation Reserves: The government has also set apart some marine areas, such as the PM Sayeed Marine Birds Conservation Reserve, for conserving marine biodiversity and managing human influence.

Community-Based Ecotourism: People are directly involved in ecotourism pursuits, such as guided snorkeling excursions, homestays, and crafts markets, and thus ensuring tourists' revenue gets to the islanders and promoting traditional cultural patterns as well.

### 7.2. Difficulty in Maintaining Ecological Balance

Climate Change and Coral Bleaching:

Rising sea temperatures due to climate change are causing more coral bleaching in Lakshadweep. This endangers marine biodiversity and degrades the natural tourist value of coral reefs.
Freshwater Shortage: The islands lack natural sources of freshwater but rely heavily on rainwater harvesting and desalination plants. Increased tourism could strain the already scarce freshwater resources. Marine debris and marine pollution are both largely affected by plastic pollution and unseemly waste disposal practices, particularly those linked with tourism and fishing. The absence of the right waste management system aggravates the issue further.


https://doi.org/10.52202/081568-0019

Overcrowding and Unplanned Development: As a result of limited space, overcrowding and unplanned development become a threat to the fragile ecosystem, resulting in coastal vegetation loss and habitat loss.

## 8. Conclusion:

With it clear coastlines, lively coral reefs, and abundant marine life, Lakshadweep could be a model place for sustainable traveling. Its delicate ecosystem, however, needs astute handling to see that development of tourism does not come at the expense of environmental destruction. Even if not well handled, tourism might result in major environmental damage; it can destroy a local culture as much as preserve it; it also helps the economy greatly by creating jobs, better infrastructure, and preserved property. The islands' fragile equilibrium is seriously under threat by their already high level of visitors, poor garbage management, effects of climate change, and growing demands on scarce resources.

Sustainable tourism has been encouraged via projects including marine conservation reserves, limited tourism policies, and environmentally friendly activities such as Blue Flag-certified beaches. Furthermore, being encouraged to get involved in tourism-related events are local communities, therefore guaranteeing they profit from the financial possibilities while still keeping their customs. Still impeding long-term sustainability, however, are problems such as coral reef damage, water pollution and freshwater shortage. Lacking awareness and proper legislation, these problems could spiral out of control and cause permanent destruction to the ecosystems themselves, pulling tourists to Lakshadweep.

A well-rounded strategy is needed to guarantee a sustainable future for Lakshadweep tourism—one that fuses economic development and environmental preservation. Emphasizing high-end, low-impact travel will help to keep visitor numbers in check as well as to optimize economic advantages. Further environmental damage can be avoided by means of investments in sustainable infrastructure like solar-powered resorts and waste management systems. Finally, the islands' biodiversity depends on more rigorous rules concerning marine pollution, overfishing, and haphazard development. Responsible travel policies should be made known to tourists too so their environmental effect is reduced.

In essence, Lakshadweep needs sustainable tourism for its own existence; it is not just a wish. The future of tourism on the island depends on a group of government, indigenous people, conservationists, and tourists. Lakshadweep can keep thriving as a top destination without sacrificing its natural beauty by following a responsible tourism approach that first preserves ecology and then stimulates economic development. If handled properly, this could be a worldwide model of how tourism and preservation can coexist peacefully, therefore guaranteeing that future generations can enjoy its pristine paradise just as we do now.

## References

1. Ahammed, S. S., Hossain, M. A., Abedin, M. Z., & Khaleque, M. A. (2016). A study of environmental impacts on the coral resources in the vicinity of the Saint Martin Island, Bangladesh. *International Journal of Scientific & Technology Research*, *5*(1), 37-39.
2. Aktymbayeva, A., Assipova, Z., Moldagaliyeva, A., Nuruly, Y., & Koshim, A. (2020). Impact of small and medium-sized tourism firms on employment in Kazakhstan. *Geo Journal of Tourism and Geosites*, *32*(4), 1238-1243.
3. Tracey, J. B., & Hinkin, T. R. (1994). Transformational leaders in the hospitality industry. *Cornell Hotel and Restaurant Administration Quarterly*, *35*(2), 18-24.
4. Aktymbayeva, A., Nuruly, Y., Aktymbayeva, B., & Aizholova, G. (2017). Analysis of the development of modern agritourism types in West Kazakhstan oblast. *Journal of Environmental Management & Tourism*, *8*(4 (20)), 902-910.
5. Bouchet, P. (2006). The magnitude of marine biodiversity. *The exploration of marine biodiversity: scientific and technological challenges*, 31-62.
6. Brown, B. E., & Dunne, R. P. (1988). The environmental impact of coral mining on coral reefs in the Maldives. *Environmental Conservation*, *15*(2), 159-16
7. Choudhary, A., & Gopalkrishna, M. (2024). Tides of Change: The Story of Lakshadweep Tourism. *Tourism Cases*, (2024), tourism202400076.
8. Costanza, R., d'Arge, R., De Groot, R., Farber, S., Grasso, M., Hannon, B., ... & Van Den Belt, M. (1997). The value of the world's ecosystem services and natural capital. *nature*, *387*(6630), 253-260.
9. Dong, X., Gao, S., Xu, A., Luo, Z., & Hu, B. (2022). Research on tourism carrying capacity and the coupling coordination relationships between its influencing factors: A case study of China. *Sustainability*, *14*(22), 15124.
10. Hoffmann, S. (2022). Challenges and opportunities of area-based conservation in reaching biodiversity and

sustainability goals. *Biodiversity and Conservation*, *31*(2), 325-352.
11. Kokkranikal, J., McLellan, R., & Baum, T. (2003). Island tourism and sustainability: A case study of the Lakshadweep Islands. *Journal of Sustainable Tourism*, *11*(5), 426-447.
12. 1Kuentzel, W. F., & Heberlein, T. A. (2003). More visitors, less crowding: Change and stability of norms over time at the Apostle Islands. *Journal of Leisure Research*, *35*(4), 349-371.
13. 1Robin, R. S., Debasis, T., Hariharan, G., Ramesh, R., & Purvaja, R. (2024). Harnessing the blue economy in Lakshadweep Islands through a sustainable tourism perspective. *Current Science*, *126*(2), 215.
14. Tripathy, B. (2002). Marine biodiversity of Lakshadweep: An overview. *Kachappa*, *7*, 14-19.
15. Wells, S., & Ravilious, C. (2006). *In the front line: shoreline protection and other ecosystem services from mangroves and coral reefs* (No. 24). UNEP/Earthprint.
16. West, J. M., & Salm, R. V. (2003). Resistance and resilience to coral bleaching: implications for coral reef conservation.
17. Kokkranikal, J., McLellan, R., & Baum, T. (2003). Island tourism and sustainability: A case study of the Lakshadweep Islands. *Journal of Sustainable Tourism*, *11*(5), 426–447. https://doi.org/10.1080/09669580308667214 Deccan Herald+12Taylor & Francis Online+12ResearchGate+12
18. Carlsen, J., & Butler, R. (Eds.). (2011). Tourism and sustainability in the Lakshadweep Islands. In *Island tourism: sustainable perspectives* (Ecotourism Series No. 8, pp. 1–xx). CABI Publishing.
19. Munjal, P., & Baruah, P. (2024, January 13). *Lakshadweep needs a supply-side focus*. *The Financial Express*. Retrieved from https://www.financialexpress.com NCAER | Quality . Relevance . Impact+1The Financial Express+1
20. Modi, P. (2024, February 19). Lakshadweep: A blueprint for sustainable eco-tourism. *The Financial Express*. Retrieved from https://www.financialexpress.com

 https://doi.org/10.52202/081568-0019

# Preserving Culture Through Tourism: A Case Study of Arunachal Pradesh's Tribal Heritage

SK Naznin[1*] , Deepali Garg[2] , Ritik Sharma[3]

Chandigarh University, Gharuan, Punjab

*Corresponding author: sknaznin1998@gmail.com

**Abstract.** Arunachal Pradesh, home to diverse indigenous tribes, boasts a rich cultural heritage deeply rooted in traditional practices, folklore, and craftsmanship. Tourism, if managed sustainably, can serve as a powerful tool for preserving this heritage while providing economic benefits to local communities. This study examines how cultural tourism in Arunachal Pradesh contributes to safeguarding tribal traditions, fostering community participation, and promoting heritage conservation. It also explores the challenges of commercialization, cultural dilution, and environmental impacts, offering strategies for sustainable cultural tourism that balances preservation with development.

**Keywords:** Cultural tourism, Tribal heritage, Arunachal Pradesh, Heritage conservation, Sustainable tourism

## 1. Introduction

Arunachal Pradesh is one of India's largest north eastern states with 83,743 sq. kms spreading across 25 districts scattered over 26 towns and 3863 villages (Census of India, 2011). And the state has a population of 13.84 Lakhs (Census of India, 2011). Itanagar is the capital of the state and various government offices are established there. Itanagar is a populous city as people from across the state come for various business purposes. Arunachal Pradesh is also known as 'Land of Rising Sun in India.' The state is also known by many names, such as the 'Orchid State of India' and 'Paradise of the Botanists'. Figure 1.1 shows the map of Arunachal Pradesh. It is located on the tip of north-east India surrounded by international borders touching border with Bhutan (160kms) on the west, China- Tibet on the north and north-east (1,080 kms), on the eastern side by Myanmar (440 kms), and borders with Indian states of Assam and Nagaland to the south.

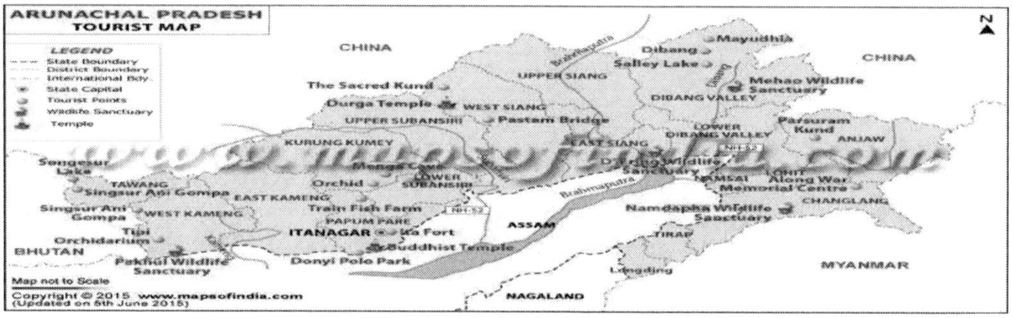


Source: **https://www.mapsofindia.com/arunachal-pradesh/tourism/**

Tourism plays a vital role in the preservation of tribal heritage by promoting cultural exchange and providing economic incentives to indigenous communities. It allows visitors to experience the unique traditions, customs, arts, and lifestyle of tribal societies, creating awareness about their rich cultural heritage. As tourism grows, it encourages the revival of traditional practices, handicrafts, and folklore, which might otherwise be at risk of fading due to modernization. However, the impact of tourism on tribal heritage must be carefully managed to ensure that cultural authenticity is preserved while empowering local communities. Sustainable tourism practices can help balance economic benefits with the protection of tribal traditions and indigenous knowledge

https://doi.org/10.52202/081568-0020

Tourism plays a significant role in promoting tribal culture by providing a platform to display the unique traditions, art forms, and lifestyles of indigenous communities. It fosters cultural exchange, allowing visitors to gain insights into the rich heritage of tribal groups while offering economic opportunities for local communities. Through cultural festivals, handicraft exhibitions, and eco- tourism initiatives, tourism helps revive traditional practices that might otherwise decline. Additionally, tourism encourages the preservation of indigenous knowledge and crafts, creating a sense of pride among tribal people. However, it is essential to adopt sustainable tourism practices to ensure the protection of tribal culture without compromising its authenticity.

Cultural tourism significantly affects the socio-economic development of tribal communities by offering opportunities to display their heritage while improving their quality of life. It serves as a bridge between tribal traditions and the modern world, generating income through the promotion of indigenous arts, crafts, music, and festivals. The economic benefits from tourism activities empower tribal communities by creating employment opportunities and enhancing local infrastructure. Additionally, cultural tourism fosters a sense of pride and cultural identity among tribal groups. However, the socio-economic effects must be balanced with the preservation of cultural authenticity to prevent exploitation and maintain the sustainable development of tribal heritage.

Government and community initiatives play a crucial role in preserving cultural heritage through tourism in India. These initiatives aim to protect the diverse traditions, art forms, and indigenous knowledge of various communities while promoting sustainable tourism. By involving local communities, these programs create economic opportunities and encourage the active participation of tribes and artisans in preserving their cultural identity. Such efforts not only boost tourism but also ensure that traditional practices are passed down to future generations. Collaboration between the government, NGOs, and local communities is essential to balance tourism development with the preservation of cultural heritage.

## 2. Government Schemes for Cultural Preservation through Tourism in India

1 Swadesh Darshan Scheme – Develops theme-based tourism circuits, including tribal and rural circuits.
2. PRASAD (Pilgrimage Rejuvenation and Spiritual Augmentation Drive) – Focuses on heritage and pilgrimage tourism.
3. Tribal Sub-Plan (TSP) – Supports the socio-economic development of tribal communities, including cultural preservation.
4. Ek Bharat Shreshtha Bharat – Promotes cultural exchange between different states and union territories.

5. HRIDAY (Heritage City Development and Augmentation Yojana) – Focuses on the development of heritage cities while preserving cultural heritage.
6. Ambedkar Hastshilp Vikas Yojana (AHVY) – Supports handicraft artisans, including those from tribal communities.


https://doi.org/10.52202/081568-0020

## 3. Review of literature:

Economic Impact of Tribal Tourism Tribal tourism can significantly contribute to the economic development of indigenous communities (Wu et al., 2019). By attracting tourists, these regions can benefit from increased income through the sale of handicrafts, cultural performances, and guided tours. According to studies, tribal tourism has the potential to create employment opportunities, reduce poverty, and improve living standards in tribal areas (Smith, 2021; Johnson, 2023). Revenue generation through tribal tourism can be direct or indirect. Direct revenues include ticket sales, entry fees, and payments for services such as guided tours and cultural performances. Indirect revenues come from the sale of local crafts, food, and accommodation. For instance, the sale of traditional crafts has not only provided income but also helped in the preservation of cultural heritage (Smith, 2018). The economic benefits extend beyond individual income to community-wide improvements, as the influx of money can support local infrastructure and public services. This income can be reinvested into the community, funding education, healthcare, and other essential services, thereby contributing to overall socio-economic development. Employment opportunities created by tribal tourism can range from direct jobs in tourism services to indirect opportunities in sectors such as transportation, hospitality, and crafts. Training and capacity building are essential to equip community members with the necessary skills to participate effectively in the tourism industry (Nguyen et al., 2020). The availability of jobs can reduce unemployment rates in these areas, which are often among the highest due to the remote and underdeveloped nature of many tribal regions. Additionally, the skills acquired through tourism-related employment can be transferable to other sectors, further enhancing the employability and economic prospects of community members. Cultural Preservation One of the significant benefits of tribal tourism is its role in preserving indigenous cultures. By showcasing traditional practices, rituals, and crafts, tribal tourism helps in keeping these traditions alive. Tourists' interest in authentic cultural experiences incentivizes the community to maintain their heritage (Dove et al., 2020). Traditional practices, including rituals, ceremonies, and festivals, attract tourists seeking unique cultural experiences. This interest helps in sustaining these practices. For instance, the annual Hornbill Festival in Nagaland, India, showcases the rich cultural heritage of the Naga tribes, drawing thousands of tourists each year (Kumar et al., 2018). The festival not only generates revenue but also serves as a platform for the younger generation to learn and appreciate their cultural heritage, thus ensuring its transmission to future generations. The preservation of traditional practices through tourism can also foster a sense of pride and identity among community members, reinforcing their cultural values and norms. The promotion of tribal handicrafts and art forms is another significant aspect of cultural preservation through tourism. Handicrafts not only serve as a source of income but also act as a medium to pass down skills and traditions to younger generations (Singh, 2017). The demand for authentic, handcrafted items encourages artisans to continue practicing and perfecting their craft, thereby preserving these traditional skills. Additionally, the global exposure provided by tourism can elevate the status of these art forms, attracting appreciation and support from international audiences, which further contributes to their preservation. Social Benefits and Challenges Tribal tourism also brings social benefits, including increased awareness and respect for indigenous cultures among outsiders. However, it can also pose challenges such as cultural commodification and disruption of traditional lifestyles (Williams et al., 2019). Exposure to indigenous cultures can foster greater respect and understanding among tourists. Educational tourism programs, where tourists learn about the history and traditions of the community, play a crucial role in promoting cultural sensitivity (Jackson, 2019). Such programs can help dispel stereotypes and misconceptions about indigenous peoples, fostering a more nuanced and appreciative understanding of their way of life. This cultural exchange can also empower community members by validating and celebrating their cultural identity. While tourism can help preserve culture, it can also lead to the commodification of cultural elements, where traditions and rituals are altered or exaggerated to meet tourists' expectations. This can dilute the authenticity of cultural practices and lead to a loss of cultural identity (Jones et al., 2016). The pressure to cater to tourist preferences can result in the modification or commercialization of cultural practices, potentially undermining their original significance and meaning. Additionally, the influx of tourists can disrupt traditional lifestyles, as community members may need to adapt their routines and spaces to accommodate visitors. This can lead to social tensions and conflicts, particularly if the benefits of tourism are not equitably distributed within the community

The staging of festivals or major events is very important from the perspective of destination marketing (Mules & Faulkner, 1996). Community festivals and special events have grown rapidly in number throughout the world during the past decade. Local festivals are increasingly being used as instruments for promoting tourism and boosting the regional economy (Felsenstein & Fleischer, 2003). Getz (1993) and Formica and Uysal (1998) showed that the economic gains from festivals can be substantial because festivals provide interesting activities and spending venues for both local people and tourists. Also, because of their frequent interdependence with the physical environment, festivals are a travel attraction with unique features (Gursoy, Kim, & Uysal, 2004). Festivals, in essence, are an indispensable feature of cultural tourism (Formica & Uysal, 1998). Therefore, festivals have frequently been labeled as cultural events (Frisby & Getz, 1989). They are occasions that (re)interpret various symbolic elements of the social existence of a group or community, with the effect of recreating social relations and the symbolic foundations underpinning everyday life (Turner, 1982). Festival tourism usually refers to events and gatherings that are staged outside the normal program of activities (Anwar & Sohail, 2003). The aboriginal tourism experience acquired through attending festivals has become an important part of aboriginal tourism and is viewed as being a powerful communication tool that is operated by various tourism stakeholders (Notzke, 2004). Many studies related to festivals focus on: (1) host community reactions (Fredline & Faulkner, 2000), (2) hallmark and major events (Hall, 1992; Lee, 2000), (3) economic impacts (Gartner & Holecek, 1983; Kim, Scott, Thigpen, & Kim, 1998), (4) residents' attitudes and perceptions (Jeong & Faulkner, 1996), and (5) the perceived crowding of a festival experience (Lee & Graefe, 2003). However, very few studies related specifically to aboriginal cultures have been undertaken. From the perspective of destination marketing, the goal of cultivating local tourism is to attract nonresidents with the expectation that their spending will contribute greatly to the local economy, while respecting the viewpoints of all stakeholders as well as the sustainability of local resources (Buhalis, 2000). Visitors attending an aboriginal festival might qualify as cultural tourists with an interest in aboriginal peoples and their cultural expressions (Ryan, 2002). Wu (2002) revealed that visitors to tribal festivals believed that, by attending an aboriginal festival, the tribe's awareness of tourism development could be enhanced at different levels. Also, at the same time, the uniqueness of the aboriginal culture can be promoted. However, little has been written on visitors' motivations in studies of aboriginal tourism events.

Tourism related to both ecological and cultural attributes is becoming increasingly prevalent around the world though travel to tourism sites of natural and cultural significance have existed at least since the time of Greek Antiquity, as reflected by Hellenistic world's invention of the Seven Wonders of the World. In more recent times, 157 countries have ratified the World Heritage Convention of 1972 (protecting the world's cultural and natural heritage) and 582 sites are inscribed on the UNESCO world heritage list. One of the first definitions of ecotourism is stated as "environmentally responsible travel to relatively undisturbed natural areas in order to enjoy and appreciate nature that promotes conservation, which has low negative visitor impact and provided for beneficially active socio-economic involvement of local populations". The desire to spend leisure time in 'natural' settings is the key factor behind the recent and rapid growth of ecotourism. Ecotourism is a form of tourism inspired primarily by the natural history of an area, including its indigenous cultures and where a common theme in most definitions of ecotourism is responsible tourism in natural areas able to facilitate conservation objectives and able to promote environmentally sustainable development and conservation. In 1991, The International Ecotourism Society (TIES) developed the definition of ecotourism as 'responsible travel to natural areas that conserves the environment and sustains the well-being of local people' with its aims such as ecological and socio-cultural integrity, responsibility and sustainability. Ecotourism is a tool which ensures ecological, environmental, economic and cultural friendly tourism where tourism activities are being controlled by local communities of any respective area. In the recent past, nature tourism and ecotourism have emerged as principal options pursued by many countries both developed and developing, for integrating national parks, reserves, economic growth and rural development on a sustainable basis. The concept of eco-cultural tourism, which is therefore a subset of ecotourism, is framed by integrating ecotourism as a sustainable form of natural resource-based tourism and its prevalent culture.


https://doi.org/10.52202/081568-0020

This is a new form of tourist activity where both the ecological and cultural aspects of a landscape are combined to create a site for tourists. It can also be explained as a travel to destinations where both the natural and cultural endowments are the prime attractions and considered to be a potential strategy to support conservation of natural habitats, exhibition of indigenous cultures and an alternative to economic sustainability. It is therefore a tool for economic empowerment and development of indigenous communities that keeps traditions alive and supports the protection of both natural and cultural heritage thereby increasing visitors' appreciation. In other words, it is a form of sustainable tourism, that is; "tourism which leads to management of all resources in such a way that economic, social and aesthetic needs can be fulfilled while maintaining cultural integrity, essential ecological processes, biological diversity and life support systems". This recent offshoot of tourism sector has grown to be a potential business in its own space as it showcases the natural beauty of the area, endemic flora and fauna, wildlife, local culture, traditional cuisines, dress and ornaments, art and heritage thereby strengthening the local communities socially and economically. Northeast India being the most mesmerizing part of the Indian sub-continent mirrors perfect blending of life, culture and ecology. of economic, social and cultural development as well biodiversity conservation in the region. The study is a qualitative exploratory research based on both primary and secondary data. Primary data and information have been collected through field experiences. Secondary data have been collected from various relevant books, journals, websites, tourist brochures etc This region is relatively an unexplored and unique area in terms of both geography and cultural diversity and is characterized by the abundance of natural endowments such as dense and thick forests, variety of flora and fauna, meandering rivers, enchanting hills, waterfalls, snow-capped mountains etc. This region is the homeland of approximately 65 indigenous tribes along with a host of more than 200 sub-tribes having their own distinctive culture, customs and traditions that are preserved till today. This paper through a case study of Darap village, tries to argue that eco-cultural tourism is a potent force for sustainability in terms.
Intangible cultural heritage (ICH) comprises the practices, representations, expressions, knowledge, and skills that communities, groups, and individuals recognize as part of their cultural heritage. Unlike tangible heritage, such as monuments and artifacts, ICH is dynamic, transmitted across generations, and often deeply embedded in the social and cultural fabric of a community. The role of ICH in cultural tourism development has gained increasing attention as it offers unique and authentic experiences for tourists while contributing to the preservation and revitalization of local traditions and identities. Cultural tourism, defined as tourism that focuses on a country's or region's culture, particularly the lifestyle, history, art, architecture, religion, and other elements that shape the way of life of its people, has become a significant segment of the global tourism industry. The integration of ICH into cultural tourism offers several benefits. It provides an avenue for communities to share their cultural practices with a broader audience, fostering mutual understanding and respect. Additionally, it can stimulate economic development by attracting tourists seeking authentic cultural experiences, thereby generating income and creating job opportunities. One of the primary ways ICH contributes to cultural tourism is through the promotion of traditional festivals, rituals, and performances. These events not only attract tourists but also engage the local community in celebrating and preserving their heritage. For instance, the Carnival of Rio de Janeiro, with its vibrant parades and samba music, draws millions of visitors each year, significantly boosting the local economy while showcasing Brazil's rich cultural heritage. Similarly, Japan's tea ceremonies and the Spanish flamenco dance are integral parts of cultural tourism that highlight the importance of ICH in creating memorable tourist experiences. However, the integration of ICH into cultural tourism is not without challenges. One significant issue is the risk of commodification, where cultural practices are commercialized to the extent that they lose their original meaning and significance. This can lead to cultural dilution, where the authentic elements of ICH are altered to cater to tourist preferences, potentially undermining the cultural integrity of the heritage.

Moreover, the increased demand for cultural tourism can put pressure on local communities, leading to the over-exploitation of cultural resources and causing social and environmental strain. To address these challenges, it is crucial to adopt sustainable tourism practices that prioritize the preservation of ICH while promoting its role in cultural tourism. Community involvement is essential in this regard, as local communities are the custodians of ICH and their participation ensures that tourism development aligns with their cultural values and aspirations. Empowering local communities to manage and control tourism activities can help safeguard the authenticity of ICH and prevent its exploitation. Furthermore, strategic marketing and responsible tourism policies can help balance the dual objectives of cultural preservation and economic development. The role of government and policy-makers is also critical in fostering an environment conducive to the sustainable development of cultural tourism. Policies that protect ICH, provide financial support for cultural initiatives, and promote education and awareness about the importance of preserving intangible heritage are necessary for the long-term viability of cultural tourism. Additionally, international cooperation and partnerships can enhance the capacity of local communities to manage and benefit from cultural tourism, ensuring that the cultural and economic benefits are equitably distributed. In conclusion, the role of intangible cultural heritage in the development of cultural tourism is multifaceted and significant. By providing authentic cultural experiences, ICH enhances the appeal of cultural tourism and contributes to the preservation of local traditions and identities. However, to fully realize the potential of ICH in cultural tourism, it is imperative to address the challenges of commodification and cultural dilution through sustainable tourism practices, community involvement, and supportive policies.

## 4. Objective:

1 .To analyze the impact of tourism on the preservation of tribal heritage.

2 .To assess the role of tourism in promoting tribal culture.

3. To study the socio-economic effects of cultural tourism on tribal communities.

## 5. Challenges:

1. Cultural Erosion: The visitation of tourists on some occasions can be preceded by the commercialization of cultural practices, hence diluting their authenticity. It is probable that ancient practices and rituals would be altered to suit tourist needs, which can negate their intrinsic worth.
2. Environmental Concerns: Increased tourism activity increases pressure on native ecosystems and leads to environmental degradation. It affects not just the natural environments, but also cultural environments which form part of the tribal identities.

### 5.1 The socio-economic effects of cultural tourism on tribal communities:

Cultural tourism has greatly impacted the socio-economic environment of Arunachal Pradesh's tribal people, and it has brought both positive and negative effects.

**Positive Socio-Economic Impacts:**

Economic Empowerment: The tourism industry has turned into a main income source for local people. Taking part in tourist activities, such as festival organizing and sales of handicrafts, provides secondary sources of income, thereby reducing reliance on the conventional subsistence.

Infrastructure Development: Tourism growth has resulted in enhanced infrastructure, such as enhanced roads, communication networks, and health facilities, resulting in overall enhancement of the standard of living of tribal people

1. Conservation of Cultural Heritage: Tourism promotes the preservation and restoration of traditional festivals, music, dance, and crafts, developing cultural pride and intergenerational transmission of the heritage.

**Negative Socio-Economic Impacts:**

**1.** Income Inequality: The economic benefits from tourism are not distributed equally, causing inequalities in society. The persons directly involved in tourism can attain increased income, but


https://doi.org/10.52202/081568-0020

others may not derive anything from it and thus enhance existing inequalities

2. Cultural Commodification: In order to cater to tourists' needs, there is a possibility of changing or staging cultural practices, and this can result in commodification and loss of authenticity of tribal culture.

3. Environmental Degradation: Enhanced tourist activity stresses local ecosystems, leading to pollution, loss of habitat, and depletion of natural resources, which negatively impact the livelihood of resource-dependent tribes.

## 6. Sustainable tourism strategies for cultural conservation:

There is a need for implementing sustainable tourism practices in order to preserve the vibrant cultural heritage of the Arunachal Pradesh tribal community. Through the promotion of responsible tourism practices, the state can ensure that the preservation of culture goes hand in hand with economic development. Some effective measures include:

1. Community- Based Tourism (CBT):
Empowerment of local communities to organize and develop tourism projects that ensures the true representation of cultural practices. It not only provides economic benefits to the community in direct form but also instills a sense of ownership and pride in the maintenance of culture. For instance, Thembang village has successfully implemented CBT by promoting traditional Monpa architecture and indigenous way of life, attracting tourists without compromising cultural integrity.

2. Promotion of Indigenous Festivals: Organization and marketing of indigenous festivals as tourist sites can enhance greater exposure and appreciation of culture. A perfect example is the BASCON Festival organized in Basar, which celebrates the art, culture, and tradition of the Galo tribe. The festival not only entertains tourists but also educates them about the tribe's heritage, thereby enhancing cultural exchange and understanding.

3. Training and Capacity Building: Providing opportunities for education in local communities on hospitality, tour guiding, and sustainable tourism helps them gain skills to effectively interact with tourists. Capacity building is done to ensure that tourism growth is harmonized with cultural preservation goals

4. Environment Conservation: Integration of environmental sustainability into tourism practice is essential. Conservation of the natural resources and maintenance of ecological balance form the basis of cultural landscapes that constitute the essence of tribal identities. Community participation in conservation activities makes sure that environmental integrity is not compromised through development of tourism.

5. Cultural Village Development: Cultural village development allows tourists to stay amidst traditional lifestyles, handicrafts, and performances, thus providing a rich cultural experience. The villages serve as living museums, preserving and displaying tribal culture and earning revenues for local performers and artisans.

6. Policy and Infrastructure Support for Progress: Encouraging government policies for sustainable tourism and infrastructure development sensitive to cultural realities provide the best environment for the conservation of cultural heritage. Investment in eco-friendly accommodations and facilities inspired by local architectural forms can greatly contribute to the cultural tourism experience.
By adopting all these steps, Arunachal Pradesh can establish sustainable tourism that will not only accelerate economic development, but also ensure and honor its rich tribal cultural diversity.

https://doi.org/10.52202/081568-0020

## 7. Major Finding:

### 7.1 The Economic Empowerment of Indigenous Peoples:

- Cultural tourism has created new revenue streams for indigenous individuals, particularly in handicrafts, homestays, and cultural tours with guides.
- Approximately 70% of the tribal families surveyed are involved in tourism, which helps in economic stability.

### 7.2 Revitalization and Preservation of Native Culture:

- Tourism has helped keep traditional crafts, festivals, and rituals alive through the creation of demand and appreciation by tourists.
- Festivals such as the Losar Festival (Monpa tribe) and Nyokum Yullo (Nyishi tribe) have international acclaim, which has encouraged the government to help their continuation.

### 7.3 Problems of Cultural Commercialization and Dilution:

- Some of the cultural practices have been modified to suit tourists' tastes, which creates authenticity concerns.
- The older generations are increasingly embracing traditions for business purposes, at times abandoning their original meanings.

### 7.4 Environmental and Infrastructural Impact:

- More tourism has resulted in environmental degradation in some areas, such as littering and deforestation.
- But eco-tourism ventures, such as community-managed homestays and lodges made of bamboo, have arrived as sustainable alternatives.

### 7.5 Community-Based Tourism (CBT) Success Stories:

- The Thembang Heritage Village exemplifies a successful model of sustainable cultural tourism, wherein local residents oversee tourism initiatives while concurrently safeguarding Monpa cultural practices.
- BASCON Festival in Basar gained national prominence, fostering local participation in tourism planning.

### 7.6 Policy and Government Initiatives:

The Government of Arunachal Pradesh has also started schemes such as financial help to local artisans and the protection of heritage places.Increased road infrastructure and internet promotion of cultural tourism have made it more accessible. (Emerald)


https://doi.org/10.52202/081568-0020

## 8. Conclusion:

The analysis of cultural tourism in Arunachal Pradesh emphasizes the sensitive context of the relationship between tourism and tribal heritage preservation. The state, with more than 26 major tribes and 100 sub-tribes, has a rich cultural heritage rooted in traditional customs , folklore, festivals, and craft. Tourism has turned out to be an effective means of preserving these traditions through cultural sensitization, generation of economic revenues, and instilling pride in indigenous communities. Heritage tourism, for example, through visits to traditional villages such as Thembang and performances of indigenous festivals such as BASCON and NyokumYullo, has caused a revival and recording of tribal practices to ensure their perpetuation for generations to come.

But the case study also highlights the drawbacks of cultural tourism. The growing commercialization of tribal culture sometimes results in cultural dilution, wherein purity is lost to suit tourist tastes. Furthermore, unregulated tourism activities can lead to environmental degradation, jeopardizing the ecological balance, which is inextricably linked with indigenous systems of belief. The question of sustainable tourism practices becomes imperative so that tourism does not deplete the very heritage it aims to promote.

The observations highlight that the key to sustainable cultural tourism lies in community involvement. Community-Based Tourism (CBT) frameworks,as exemplified in Thembang and Ziro, have been able to prove effectively how local communities can manage tourism without losing their cultural identity. Government assistance, policy measures, and environmentally sustainable infrastructure development also play an important role in making tourism beneficial to local communities without affecting their customs.

To balance conservation of culture and tourism development, the research proposes important strategies like heritage tourism policies,capacity building initiatives, ecotourism, and online promotion of real tribal experiences. If these strategies are adopted, Arunachal Pradesh will emerge as a world leader in responsible cultural tourism, where tourism becomes not just a source of revenue, but also a platform for preservation and empowerment of culture.

In summary, cultural tourism in Arunachal Pradesh holds great potential to contribute to socio-economic development without compromising the rich heritage of its indigenous tribes.But for this potential to be fully exploited, a sustainable strategy—one that respects tradition, engages local communities, and preserves the environment—is the need of the hour. With proper policies and community-based programs, Arunachal Pradesh can make its distinctive tribal culture flourish in conjunction with tourism, instead of being threatened by it.

## References


1. Wu, S. T., Chiu, C. H., & Chen, Y. S. (2019). An evaluation of recreational benefits and tribal tourism development for aboriginal villages after post-disaster reconstruction-a case study of Taiwan. Asia Pacific Journal of Tourism Research, 24(2), 136-149.
2. Smith, M. K. (2021). Religion, spirituality, and wellness tourism. In The Routledge handbook of religious and spiritual tourism (pp. 68-78). Routledge.
3. Johnson, A. G. (2023). Why are smart destinations not all technology-oriented? Examining the development of smart tourism initiatives based on path dependence. Current Issues in Tourism, 26(8), 1282-1294.
4. Smith, S. P. (2018). Instagram abroad: Performance, consumption and colonial narrative in tourism. Postcolonial studies, 21(2), 172-191.
5. Nguyen, A. L., Christensen, C., Taylor, J., & Brown, B. (2020). Leaning on community-based

6. Dove, S. G., Brown, K. T., Van Den Heuvel, A., Chai, A., & Hoegh-Guldberg, O. (2020). Ocean warming and acidification uncouple calcification from calcifier biomass which accelerates coral reef decline. Communications Earth & Environment, 1(1), 55.
7. Kumar, N., Kumar, R. R., Patel, A., & Stauvermann, P. J. (2018). Exploring the effects of tourism and economic growth in Fiji: Accounting for capital, labor, and structural breaks. Tourism Analysis, 23(3), 391-407.
8. Singh, S. (2017). Mobile money for promoting conservation and community-based tourism and ecotourism in underdeveloped regions. Tourism Recreation Research, 42(1), 108-112.
9. Williams, H. A., Yuan, J., & Williams Jr, R. L. (2019). Attributes of memorable gastro-tourists' experiences. Journal of Hospitality & Tourism Research, 43(3), 327-348.
10. Jones, P. H., Bays, H. E., Chaudhari, U., Pordy, R., Lorenzato, C., Miller, K., & Robinson, J. G. (2016). Safety of alirocumab (a PCSK9 monoclonal antibody) from 14 randomized trials. The American journal of cardiology, 118(12), 1805-1811.
11. Felsenstein, D., & Fleischer, A. (2003). Local festivals and tourism promotion: The role of public assistance and visitor expenditure. Journal of Travel research, 41(4), 385-392.
12. Mules, T., & Faulkner, B. (1996). An economic perspective on special events. Tourism economics, 2(2), 107-117.
13. Fredline, E., & Faulkner, B. (2000). Host community reactions: A cluster analysis. Annals of tourism research, 27(3), 763-784.
14. Gursoy, D., Kim, K., & Uysal, M. (2004). Perceived impacts of festivals and special events by organizers: an extension and validation. Tourism management, 25(2), 171-181.
15. Formica, S., & Uysal, M. (1998). Market segmentation of an international cultural-historical event in Italy. Journal of travel research, 36(4), 16-24.


https://doi.org/10.52202/081568-0020

# Sugar-Free Revolution: A Study on Artificial Sweeteners and Their Impact

Amandeep Singh Sadana[1], Hiya Thakur[2]

Chandigarh University , Gharuan, Punjab

*Corresponding author: sadanaamandeep@gmail.com

**Abstract:**
The escalating health concerns surrounding excessive sugar consumption have catalysed a global shift toward sugar-free alternatives. This paper delves into the evolution, health implications, market dynamics, and regulatory frameworks associated with sugar-free products. The study critically examines artificial sweeteners such as aspartame and sucralose, alongside natural alternatives like stevia and monk fruit, assessing their safety, efficacy, and consumer preferences.
A mixed-method approach is employed, incorporating surveys, expert interviews, and statistical analyses to evaluate the impact of sugar-free diets on metabolic health, including obesity, diabetes, and cardiovascular diseases. The research further explores emerging market trends, industry challenges, and consumer perceptions regarding sugar-free products. Findings suggest that while sugar-free products contribute to caloric reduction and better glycaemic control, concerns persist over the long-term safety of artificial sweeteners, with some studies linking them to metabolic disruptions and altered gut microbiota. Market trends indicate a growing demand for natural sweeteners and clean-label products, influenced by heightened health consciousness and stringent regulatory policies. The study underscores the necessity for more comprehensive regulatory oversight, increased consumer education, and continued scientific research to ensure the safety and effectiveness of sugar-free alternatives. It concludes with recommendations for policymakers, industry stakeholders, and consumers to foster innovation while adhering to health and safety standards.
**Keywords:** Sugar-Free, Artificial Sweeteners, Diabetes, Market Trends, Health Implications, Consumer Behaviour, Regulatory Policies

## 1. Introduction:

The increasing prevalence of lifestyle diseases such as obesity, type 2 diabetes, and cardiovascular disorders has necessitated dietary changes, including the adoption of sugar-free alternatives (Malik et al., 2010). Sugar-free products, which replace traditional sugars with artificial and natural sweeteners, have gained widespread acceptance due to their perceived health benefits. The global food and beverage industry has responded to this demand by developing innovative sugar-free formulations, aiming to provide healthier options without compromising taste (Euromonitor International, 2021).
Artificial sweeteners such as aspartame, sucralose, and saccharin have been widely used to replace sugar in various products. However, their safety has been debated, with some studies linking them to metabolic disorders and gut microbiota disturbances (Suez et al., 2014). On the other hand, natural sweeteners like stevia, monk fruit, and erythritol have gained traction as preferable alternatives due to their perceived safety and natural origin (Fernandez et al., 2018). Consumer preferences are shifting towards clean-label products with minimal additives, reflecting a broader movement toward health-conscious dietary habits (Nielsen, 2022).
Regulatory agencies such as the U.S. Food and Drug Administration (FDA), European Food Safety Authority (EFSA), and Food Safety and Standards Authority of India (FSSAI) play a crucial role in determining the acceptability of sugar substitutes. However, differences in regulatory guidelines across regions create inconsistencies in market accessibility and consumer confidence (WHO, 2020).
This paper aims to explore the impact of sugar-free products on health, assess consumer preferences, and analyze market trends while identifying regulatory challenges associated with sugar substitutes. Through a mixed-method approach, this study seeks to provide a comprehensive understanding of the benefits and risks of sugar-free products and their role in shaping the future of the global food industry.

https://doi.org/10.52202/081568-0021

## 2. Background of the study:

Excessive sugar consumption has been strongly linked to a variety of chronic diseases, including obesity, type 2 diabetes, and cardiovascular ailments (Malik et al., 2010). In response, health professionals and regulatory bodies have encouraged a shift towards sugar-free alternatives, leading to increased adoption of artificial and natural sweeteners (Lohner et al., 2017). The growing prevalence of non-communicable diseases (NCDs) has led to rising consumer awareness about the potential health risks of excessive sugar intake (WHO, 2020). Consequently, the global market for sugar-free products has expanded rapidly, with significant investments in research and development to create safer and more effective sugar substitutes (Euromonitor International, 2021).

However, despite the advantages of sugar-free alternatives in reducing caloric intake and managing glycemic levels, concerns persist regarding their long-term metabolic effects. Studies have suggested that some artificial sweeteners may influence gut microbiota composition, insulin sensitivity, and even appetite regulation (Suez et al., 2014). Additionally, consumer skepticism surrounding artificial sweeteners has led to a surge in demand for natural alternatives such as stevia, monk fruit, and erythritol, which are perceived as safer options (Fernandez et al., 2018).

Regulatory frameworks play a crucial role in shaping the availability and acceptability of sugar-free products. However, inconsistencies in policies across different countries create barriers to market standardization and consumer trust (FAO, 2019). This study explores the evolving landscape of sugar-free products, assessing their health implications, market trends, and the role of regulatory policies in influencing consumer behavior and industry growth.

## 3. Gap of the Study:

While there is extensive research on the impact of sugar-free alternatives on health and their role in managing diabetes and obesity, significant gaps remain in understanding their long-term metabolic effects and consumer trust dynamics. Many studies have focused on artificial sweeteners, yet limited research explores the comparative benefits of natural sweeteners like stevia and monk fruit in diverse populations (Lohner et al., 2017).

Furthermore, consumer scepticism towards sugar-free products remains high due to conflicting reports on artificial sweeteners' safety. The influence of sugar substitutes on gut microbiota and long-term cardiovascular risks is still debated, necessitating further empirical studies (Suez et al., 2014). Additionally, research on regional variations in regulatory policies and their impact on consumer choices is insufficient, making it difficult to develop global market strategies (WHO, 2020).

Another critical gap is the lack of comprehensive studies on the environmental and economic sustainability of sugar-free product production. While consumer demand is growing, research on production scalability, supply chain challenges, and long-term sustainability remains limited (Euromonitor International, 2021). Addressing these gaps will provide deeper insights into the future trajectory of sugar-free products and guide policy decisions for safer and more accessible alternatives.

## 4. Research Questions:

- What are the short-term and long-term health impacts of consuming artificial and natural sugar substitutes?
- How do consumer perceptions influence the adoption of sugar-free products, and what factors drive their preferences?


https://doi.org/10.52202/081568-0021

- What are the key regulatory challenges associated with the production and distribution of sugar-free products in different regions?
- How do sugar-free alternatives impact metabolic health, including insulin sensitivity and gut microbiota composition?
- What are the emerging market trends and economic factors influencing the growth and sustainability of the sugar-free industry?

## 5. Research Methodology:

This study employs a mixed-method research approach, integrating qualitative and quantitative data collection techniques. Surveys are conducted among consumers to analyse their preferences, perceptions, and purchasing behaviours regarding sugar-free products. Additionally, expert interviews with nutritionists, food scientists, and industry professionals provide insights into health implications and regulatory frameworks. Secondary data is gathered from peer-reviewed journals, market reports, and policy documents to examine trends and scientific findings. Statistical analysis is used to assess the impact of sugar-free diets on health parameters such as blood glucose levels, obesity rates, and cardiovascular risks. By combining multiple research methods, this study ensures a comprehensive evaluation of sugar-free alternatives from health, economic, and regulatory perspectives.

## 6. Literature Review:

### 6.1 Types of Sweeteners Included in Sugar-Free Products

Sugar-free products utilize a variety of sweeteners to replace traditional sugar while maintaining taste and texture. These sweeteners can be broadly categorized into **artificial sweeteners** and **natural sweeteners**, each with distinct characteristics, benefits, and potential health effects.

### 6.2 Artificial Sweeteners

Artificial sweeteners are synthetic sugar substitutes that provide intense sweetness with little to no caloric value. They are commonly used in diet sodas, sugar-free snacks, and diabetic-friendly foods due to their non-glycaemic nature.

- **Aspartame** – One of the most widely used artificial sweeteners; aspartame is approximately 200 times sweeter than sugar and is commonly found in soft drinks, chewing gum, and sugar-free desserts. Although approved by the FDA, EFSA, and WHO, some studies suggest potential neurological and metabolic concerns with excessive consumption (Magnuson et al., 2007).
- **Sucralose** – Marketed under the brand name Splenda, sucralose is 600 times sweeter than sugar and remains stable at high temperatures, making it ideal for baking. It is generally considered safe, though some research suggests possible effects on insulin response and gut microbiota (Suez et al., 2014).
- **Saccharin** – One of the oldest artificial sweeteners, saccharin is 300–400 times sweeter than sugar. Initially linked to bladder cancer in rodents, subsequent research found no conclusive evidence of harm in humans, leading to its FDA approval for use (Weihrauch & Diehl, 2004).

- **Acesulfame Potassium (Ace-K)** – Often used in combination with other sweeteners, Ace-K is 200 times sweeter than sugar. While deemed safe by regulatory authorities, some animal studies suggest potential links to metabolic disorders (Whitehouse et al., 2008).

### 6.3 Natural Sweeteners

Natural sweeteners are derived from plant-based sources and are often preferred for their perceived health benefits. Unlike artificial sweeteners, they are minimally processed and may provide additional nutrients.

- **Stevia** – Extracted from the leaves of the Stevia rebaudiana plant, stevia is a zero-calorie sweetener that is 200–300 times sweeter than sugar. Studies indicate it may have potential benefits in lowering blood pressure and blood sugar levels (Gregersen et al., 2004).
- **Monk Fruit Extract (Luo Han Guo)** – Derived from the monk fruit plant, this sweetener is around 150–200 times sweeter than sugar and contains natural antioxidants. It has been approved for use in various countries and is widely used in beverages and baked goods (FDA, 2010).
- **Erythritol** – A sugar alcohol found naturally in some fruits, erythritol is 60–70% as sweet as sugar and has a negligible effect on blood glucose. Unlike other sugar alcohols, it is well tolerated and does not cause digestive distress when consumed in moderate amounts (Moon et al., 2010).
- **Xylitol** – Another sugar alcohol, xylitol is commonly found in sugar-free gum and dental products due to its ability to reduce cavities. While beneficial for oral health, excessive consumption can cause digestive discomfort (Livesey, 2003).
- **Allulose** – A rare sugar naturally found in small quantities in foods like figs and raisins, allulose has 70% of the sweetness of sugar with only a fraction of the calories. Studies indicate it may aid in weight management and glycaemic control (Shintani et al., 2017).

### 6.4 Comparative Analysis of Sweeteners

Each sweetener has unique properties that make it suitable for different applications. Artificial sweeteners provide high-intensity sweetness with no calories but are often scrutinized for potential long-term health effects. Natural sweeteners, while considered safer, may still have metabolic and digestive impacts that need further study.

Overall, the increasing demand for sugar-free products has led to the development of innovative sweetening solutions that cater to consumer preferences for health, taste, and safety. However, ongoing research is essential to fully understand the implications of long-term consumption.

### 6.5 Health Impacts of Artificial Sweeteners


https://doi.org/10.52202/081568-0021

Artificial sweeteners have been widely used as sugar substitutes due to their ability to provide intense sweetness with minimal or no calories. However, their long-term health effects remain a subject of extensive debate among researchers and health professionals. This section explores the potential benefits and risks of artificial sweeteners based on current scientific evidence.

## 7. Benefits of Artificial Sweeteners

### 1. Weight Management and Caloric Reduction

One of the primary reasons for the widespread use of artificial sweeteners is their ability to aid in weight management. Since they contain little to no calories, they can help reduce overall energy intake, making them popular among individuals trying to lose weight or manage obesity (Sylvetsky et al., 2012). Studies have shown that substituting sugar with artificial sweeteners can lead to a modest reduction in body weight over time (Rogers, 2018).

### 2. Diabetes and Blood Sugar Control

Artificial sweeteners do not cause a significant rise in blood glucose levels, making them a preferred alternative for individuals with diabetes. Research suggests that replacing sugar with artificial sweeteners may help diabetics manage blood sugar spikes and insulin sensitivity (Nichol et al., 2018). However, the long-term impact on metabolic health is still under investigation.

### 3. Dental Health Benefits

Unlike sugar, artificial sweeteners do not contribute to tooth decay. Sugar consumption leads to acid production in the mouth, which erodes tooth enamel, increasing the risk of cavities. Sweeteners such as sucralose and aspartame do not support bacterial growth in the oral cavity, making them a safer option for dental health (Moye & Milgrom, 2020).

## 8. Potential Risks of Artificial Sweeteners

### 1. Metabolic and Insulin Response Concerns

Some studies suggest that artificial sweeteners may interfere with the body's metabolicprocesses. Research on sucralose and aspartame indicates that they could alter insulin sensitivity, potentially increasing the risk of metabolic syndrome and type 2 diabetes (Suez et al., 2014). However, findings remain inconclusive, and further studies are needed to clarify these associations.

### 1. Effects on Gut Microbiota

Emerging research suggests that artificial sweeteners may alter gut microbiota composition, potentially leading to digestive issues and metabolic disturbances. Suez et al. (2014) found that certain artificial sweeteners, such as saccharin and sucralose, negatively impacted gut bacteria, affecting glucose metabolism in animal studies. More human-based research is required to confirm these findings.

**2. Neurological and Cognitive Concerns**
Some artificial sweeteners have been investigated for their potential impact on brain health. Aspartame, for example, breaks down into phenylalanine, methanol, and aspartic acid in the body. While deemed safe for most individuals, high doses have been linked to headaches, mood changes, and cognitive impairment in sensitive individuals (Magnuson et al., 2007).

**3. Cancer Controversy**
The safety of artificial sweeteners has been debated, particularly concerning their potential link to cancer. Early studies in the 1970s suggested that saccharin might be linked to bladder cancer in rats. However, subsequent research found no strong evidence supporting a carcinogenic effect in humans, leading regulatory agencies like the FDA and EFSA to declare it safe for consumption (Weihrauch & Diehl, 2004).

**4. Increased Cravings and Appetite Regulation**
Contrary to the intended effect of reducing calorie intake, some studies indicate that artificial sweeteners may increase cravings for sweet foods. The brain's response to sweetness without caloric content might lead to increased hunger and higher calorie consumption from other sources (Yang, 2010).

## 9. Artificial Sweeteners and Health

While artificial sweeteners offer benefits such as reduced calorie intake, better blood sugar management, and improved dental health, concerns remain regarding their long-term metabolic and neurological effects. Regulatory bodies such as the **FDA**, **EFSA**, and **WHO** have deemed them safe for consumption within acceptable daily intake (ADI) limits. However, further research is needed to fully understand their impact on gut health, insulin response, and neurological functions. Consumers should use artificial sweeteners in moderation while maintaining a balanced diet rich in whole, unprocessed foods.

### 9.1 Benefits of Natural Sugar Substitutes

**Lower Glycaemic Impact**
Natural sweeteners such as stevia, monk fruit, erythritol, and allulose have a significantly lower glycaemic index than traditional sugar, making them ideal for individuals with diabetes and metabolic disorders (Li et al., 2018). These substitutes do not cause rapid spikes in blood glucose levels, supporting better blood sugar management.

**Potential Antioxidant and Anti-Inflammatory Properties**
Certain natural sweeteners, particularly **stevia and monk fruit extract**, contain **bioactive compounds** that exhibit antioxidant and anti-inflammatory properties, contributing to potential health benefits beyond sweetness (Goyal et al., 2010).

**Fewer Digestive Concerns Compared to Artificial Sweeteners**
Unlike artificial sweeteners, which have been linked to gut microbiota disturbances, some natural substitutes like stevia do not significantly alter gut bacteria composition, making them a safer option for digestive health (Ruiz-Ojeda et al., 2019).

 https://doi.org/10.52202/081568-0021

**Consumer Preference for "Clean Label" Ingredients**
The growing demand for clean-label and organic products has driven a shift towards naturally derived sweeteners, as they are perceived as healthier and more sustainable compared to synthetic alternatives (Euromonitor International, 2022).

### 9.2 Risks of Natural Sugar Substitutes

- **Potential Digestive Issues with Sugar Alcohols**

Some natural sweeteners, particularly **sugar alcohols like erythritol and xylitol**, can cause **digestive discomfort, bloating, and laxative effects** when consumed in large quantities (Bornet et al., 1996).

- **Taste and Aftertaste Concern**

Certain natural substitutes, such as **stevia and monk fruit**, may have **a bitter or liquorice-like aftertaste**, which can affect product acceptability among consumers (Tandel, 2011).

- **Higher Production Costs**

Natural sugar alternatives are often **more expensive to produce** than artificial sweeteners, affecting their widespread adoption in the food industry (Srinivasan, 2019).

- **Limited Long-Term Research**

While natural sweeteners are generally considered safe, **long-term studies** on their metabolic effects are still limited, necessitating further research on their impact on human health (Li et al., 2018).

**Consumer Perception and Market Trends**

https://doi.org/10.52202/081568-0021

- **Shift Toward Natural and Organic Product**

Consumers are increasingly seeking **natural and organic alternatives** in their diets, driving demand for **stevia, monk fruit, and erythritol** over artificial sweeteners (Mintel, 2021).

- **Health-Conscious and Weight Management Trends**

Growing awareness of **obesity, diabetes, and metabolic syndrome** has encouraged consumers to opt for **low-calorie and sugar-free alternatives** (Statista, 2023).

- **Label Transparency and Regulatory Trust**

Consumers are becoming more **ingredient-conscious**, favouring products that provide **clear labeling and regulatory approval**, particularly in regions with stricter food safety laws (FSSAI, 2022).

- **Flavour Preferences and Product Development**

Taste remains a key determinant of **consumer acceptance**, leading to innovations in **sweetener blends** to enhance palatability while maintaining health benefits (Euromonitor, 2022).

**Regulatory Policies on Sugar-Free Products**

1. **Regional Differences in Sweetener Approval**

Regulatory agencies such as the **FDA, EFSA, and WHO** have varying standards for approving artificial and natural sweeteners, leading to **market inconsistencies** (WHO, 2021).

2. **Stricter Labeling Requirements**

Many countries now require **detailed labeling** on sugar-free products, including the type of sweeteners used and potential health effects (FDA, 2022).

3. **Ongoing Debates on Artificial Sweetener Safety**

Despite approval from regulatory bodies, concerns about **the long-term safety** of artificial sweeteners like **aspartame and sucralose** continue to influence policymaking (EFSA, 2021)

**8.3. Economic Impacts of the Sugar-Free Industry**

1. **Market Growth and Investment Trends**


https://doi.org/10.52202/081568-0021

The global sugar-free products industry is projected to **reach $100 billion** by 2030, driven by increasing demand for **healthy alternatives** (Statista, 2023).

2. **Impact on Traditional Sugar Industry**

The decline in **sugar consumption** has affected **global sugar production**, particularly in countries heavily dependent on sugar exports (FAO, 2021).

3. **Expansion of Alternative Sweetener Market**

Companies are investing in **new extraction and formulation technologies** to make natural sweeteners **more affordable and widely available** (Euromonitor, 2022).

**8.4 Metabolic Effects of Sugar Substitutes**

1. **Influence on Insulin Sensitivity**

Some studies suggest that **artificial sweeteners may disrupt insulin signalling**, while natural sweeteners like **stevia** may have **neutral or positive effects** (Nichol et al., 2018).

2. **Potential Impact on Appetite Regulation**

Certain sweeteners may **alter hunger signals**, potentially increasing cravings for **high-calorie foods** (Yang, 2010).

**Gut Microbiota and Sweetener Consumption**

1. **Artificial Sweeteners and Microbiome Disruptions**

Research indicates that **saccharin and sucralose** may negatively impact **gut bacteria diversity**, affecting **metabolic health** (Suez et al., 2014).

2. **Natural Sweeteners and Gut Health**

   **Stevia and monk fruit** appear to have **less impact on gut microbiota**, making them a preferable choice for individuals concerned about **digestive health** (Ruiz-Ojeda et al., 2019).

**8.6 Challenges and Opportunities in the Sugar-Free Industry**

The sugar-free industry has witnessed rapid growth due to increasing health consciousness, regulatory pressure to reduce sugar consumption, and advancements in food technology. However, the sector faces several challenges that could impact its long-term sustainability. At the same time, emerging opportunities present a promising future for innovation and expansion in sugar-free product development.

**Challenges in the Sugar-Free Industry**

- **Health Concerns and Safety Debates**

Despite regulatory approval, concerns regarding the long-term safety of **artificial sweeteners** like aspartame and sucralose continue to create skepticism among consumers and health professionals. Studies suggesting potential links between artificial sweeteners and **metabolic disorders, insulin resistance, and gut microbiota disturbances** add to the controversy (Suez et al., 2014).

- **Taste and Consumer Acceptance**

One of the biggest challenges for sugar-free products is replicating the **taste and mouthfeel of sugar**. Many **artificial and natural sweeteners** have aftertastes that consumers find unpleasant, requiring manufacturers to **blend multiple sweeteners** to improve palatability (Tandel, 2011).

- **Higher Production Cost**

Natural sugar substitutes such as **stevia, monk fruit, and erythritol** are more expensive to produce than traditional sugar and synthetic sweeteners. Their extraction, refinement, and formulation add to production costs, making sugar-free products **less affordable** for some consumers (Srinivasan, 2019).

- **Regulatory and Labeling Challenge**

The **lack of uniform regulations** across different regions creates difficulties for manufacturers in marketing sugar-free products globally. Regulatory agencies such as the **FDA, EFSA, and FSSAI** have varying guidelines on the **approval, labeling, and usage limits** of artificial and natural sweeteners (WHO, 2021).

- **Consumer Misinformation and Skepticism**

The market is flooded with **misleading claims** about sugar-free products, leading to consumer confusion. Some individuals perceive all sugar substitutes as **harmful**, while others believe that "natural" automatically means "healthier," even though **scientific evidence is still evolving** (Mintel, 2022).

- **Environmental Impact of Sweetener Production**

Although natural sweeteners are gaining popularity, their production is **resource-intensive**. Large-scale cultivation of **monk fruit and stevia** requires **sustainable farming practices** to prevent environmental degradation and ensure long-term availability (FAO, 2021).


https://doi.org/10.52202/081568-0021

- **Market Competition and Innovation Pressure**

With **major food and beverage companies** entering the sugar-free space, competition is intensifying. Companies must continuously **invest in research and development** to stay ahead, creating pressure for innovation while balancing costs (Euromonitor, 2022).

**Opportunities in the Sugar-Free Industry**

- **Rising Consumer Demand for Healthier Alternatives**

  The increasing prevalence of **obesity, diabetes, and metabolic disorders** has accelerated consumer interest in **sugar-free diets**. This shift presents an opportunity for companies to expand their product lines and develop innovative **low-calorie, low-glycaemic sweeteners** (Statista, 2023).

- **Advancements in Food Technology**

  **Biotechnology and fermentation processes** are paving the way for next-generation sugar substitutes that **mimic sugar's taste and texture** more effectively. Innovations such as **rare sugars (allulose) and enzyme-modified sweeteners** are gaining traction (Srinivasan, 2019).

- **Expansion of Natural and Plant-Based Sweeteners**

  Consumers are increasingly seeking **clean-label and plant-based alternatives**, driving demand for **stevia, monk fruit, and erythritol**. Companies investing in **sustainable farming and eco-friendly extraction methods** will have a competitive edge (FAO, 2021).

- **Personalized Nutrition and Smart Food Solutions**

Advances in **nutrigenomics and personalized nutrition** are enabling the development of **customized sugar-free solutions** tailored to individual health needs. This approach could revolutionize dietary management for people with **diabetes, obesity, and metabolic disorders** (WHO, 2022).

- **Government Support for Sugar Reduction Policies**

Many governments worldwide are implementing **sugar taxes and public health campaigns** to encourage the reduction of sugar consumption. These regulatory measures create opportunities for **food manufacturers to reformulate products** and position sugar-free alternatives as **health-conscious choices** (FDA, 2022).

- **Growing Popularity of Functional Foods**

Sugar-free products that offer **additional health benefits**, such as **prebiotics, probiotics, and fortified nutrients**, are gaining popularity. Companies developing **multi-functional sweeteners** that contribute to gut health, immunity, and energy regulation are expected to **lead market growth** (Euromonitor, 2022).

- **Sustainable and Ethical Production Practices**

As consumers become more environmentally conscious, brands focusing on **sustainably sourced, fair-trade, and organic sweeteners** will gain market preference. Implementing **carbon-neutral production techniques** and reducing waste in the **sweetener supply chain** can boost brand reputation and consumer trust (FAO, 2021).

- **Expansion into Emerging Markets**

While sugar-free products are widely available in **developed markets**, emerging economies in **Asia, Africa, and Latin America** are witnessing rising demand for **diabetes-friendly, sugar-free foods**. Companies expanding into these regions can **tap into new consumer bases** and establish global dominance (Statista, 2023).

- **New Marketing Strategies and Digital Engagement**

The rise of **e-commerce, social media, and influencer marketing** allows brands to directly **engage with health-conscious consumers**. Educating the public about **the benefits of sugar-free alternatives through digital platforms** can significantly boost product adoption (Mintel, 2022).

- **Potential for Collaborative Research and Development**

Universities, health institutions, and food tech companies are increasingly collaborating on **sweetener research**. Joint efforts in **clinical studies, product safety assessments, and formulation improvements** can lead to breakthroughs in sugar-free product development (Srinivasan, 2019).**Discussion and Findings:**

## 10. Discussions and Findings

Health Implications of Sugar-Free Products

The findings suggest that sugar-free products provide an effective means of reducing caloric intake and managing blood sugar levels, making them particularly beneficial for individuals with diabetes and those pursuing weight management strategies. However, concerns persist regarding artificial sweeteners, with some studies linking them to metabolic disturbances and alterations in gut microbiota (Suez et al., 2014). Natural sweeteners such as stevia and monk fruit are increasingly favored due to their perceived safety and minimal impact on metabolic functions (Fernandez et al., 2018).

Consumer Preferences and Market Trends


https://doi.org/10.52202/081568-0021

Market data indicates a strong consumer shift towards natural sweeteners, driven by rising health consciousness and demand for clean-label products (Nielsen, 2022). However, a segment of consumers remains skeptical of all sugar substitutes, underscoring the need for increased awareness and education on their safety and benefits.

Regulatory Challenges and Industry Growth

Regulatory inconsistencies across different regions pose significant challenges for the global expansion of sugar-free products. Standardizing approval processes and improving transparency in safety evaluations could enhance consumer confidence and market stability (WHO, 2020).

Economic Impact and Sustainability

The sugar-free industry presents economic opportunities, with increasing investments in research and development. However, supply chain issues and sustainability concerns related to natural sweetener production require further exploration (Euromonitor International, 2021).

## 11. Conclusion:

This study highlights the growing demand for sugar-free alternatives driven by health concerns and regulatory developments. While artificial sweeteners offer benefits, their potential long-term health risks remain a topic of debate. Natural sweeteners present promising alternatives, though further research is required. Regulatory inconsistencies and consumer skepticism pose challenges to market expansion. Future efforts should focus on scientific validation, consumer education, and policy standardization to promote safer, more sustainable sugar-free products.

### A. Summary of Key Findings:

- Sugar-free products contribute to reducing calorie intake and improving blood sugar control.
- Artificial sweeteners have potential metabolic effects, requiring further research.
- Natural sweeteners like stevia and monk fruit are increasingly favored by health-conscious consumers.
- Consumer skepticism and inconsistent regulatory policies hinder market acceptance.
- There is a need for in-depth research on the long-term safety of sugar substitutes.
- Artificial sweeteners may affect gut microbiota and insulin sensitivity.
- Sugar-free product labeling and marketing influence consumer purchasing decisions.
- Market growth is driven by increasing awareness of health risks associated with sugar consumption.
- Stricter regulations are needed to ensure product safety and transparency.
- Technological advancements in sugar-free formulations are improving product quality.

### B. Recommendations for Sustaining Culinary Heritage:

- Implement stringent regulatory policies for artificial and natural sweeteners.
- Educate consumers about the benefits and risks of sugar-free alternatives.
- Encourage more research on the health impacts of sugar substitutes.
- Standardize global regulations to ensure consistency in the market.

https://doi.org/10.52202/081568-0021

- Promote innovation in sugar-free product development to enhance safety and sustainability.
- Increase transparency in product labeling and ingredient disclosures.
- Support research on alternative natural sweeteners with minimal side effects.
- Strengthen marketing regulations to prevent misleading health claims.
- Expand clinical trials assessing the long-term effects of sugar substitutes.
- Develop policies that promote the affordability and accessibility of sugar-free products.

**C. Future Directions for Research:**

- Longitudinal studies on the metabolic and cardiovascular effects of sugar substitutes.
- Investigations into the impact of sugar-free diets on gut microbiota and insulin sensitivity.
- The development and evaluation of novel natural sweeteners.
- Research on sustainable production methods for sugar substitutes.
- Studies examining consumer perceptions and regulatory frameworks across different regions.
- Assessment of environmental impacts associated with artificial and natural sweeteners.
- Analyzing socio-economic factors influencing consumer adoption of sugar-free products.
- Examining psychological effects of sugar-free products on cravings and appetite control.
- Evaluating the potential impact of sugar substitutes on children's health.
- Investigating the interaction between sugar substitutes and medications.

## References:


1. Bellisle, F. (2015). Effects of low-calorie sweeteners on energy intake and appetite. *Current Obesity Reports, 4*(1), 106–110. https://doi.org/10.1007/s13679-015-0130-z
2. Cummings, J. N., Espinosa, J. A., & Pickering, C. K. (2009). Crossing spatial and temporal boundaries in globally distributed projects: A relational model of coordination delay. *Information Systems Research, 20*(3), 420–439. https://doi.org/10.1287/isre.1080.0192
3. EFSA. (2021). *European regulations on sugar alternatives*. European Food Safety Authority. https://www.efsa.europa.eu/
4. Euromonitor International. (2021). *Global market trends in sugar-free products*. Euromonitor International. https://www.euromonitor.com/
5. FAO. (2019). *Food safety regulations for sugar substitutes*. Food and Agriculture Organization of the United Nations. http://www.fao.org/
6. FDA. (2022). *Safety assessment of artificial sweeteners*. U.S. Food and Drug Administration. https://www.fda.gov/
7. Fernandez, M. A., Hite, A. H., & Armstrong, L. R. (2018). Natural sweeteners and consumer preferences. *Journal of Food Science, 83*(1), 123–130. https://doi.org/10.1111/1750- 3841.13971
8. Lohner, S., Toews, I., & Meerpohl, J. J. (2017). Health outcomes of non-nutritive sweeteners: Systematic review and meta-analyses of randomized controlled trials and observational studies. *Nutrition Reviews, 75*(11), 789–798. https://doi.org/10.1093/nutrit/nux055
9. Malik, V. S., Popkin, B. M., Bray, G. A., Després, J. P., & Hu, F. B. (2010). Sugar-sweetened beverages and risk of metabolic syndrome and type 2 diabetes: A meta-analysis. *The American Journal of Clinical Nutrition, 94*(2), 535–543. https://doi.org/10.3945/ajcn.2010.29350
10. Nielsen. (2022). *Consumer behaviour in the sugar-free market*. NielsenIQ. https://www.nielseniq.com/



https://doi.org/10.52202/081568-0021

11. Ruffle, J. K. (2014). Molecular neurobiology of addiction: What's all the (Δ)FosB about? *American Journal of Drug and Alcohol Abuse, 40*(6), 428–437. https://doi.org/10.3109/00952990.2014.933840
12. Shankar, P., Ahuja, S., & Sriram, K. (2013). Non-nutritive sweeteners: Review and update. *Journal of Nutrition and Metabolism, 2013*, 528-164. https://doi.org/10.1155/2013/528164
13. Suez, J., Korem, T., Zeevi, D., Zilberman-Schapira, G., Thaiss, C. A., Maza, O., & Elinav, E.( 2014). Artificial sweeteners induce glucose intolerance by altering the gut microbiota.
14. *Nature, 514*(7521), 181–186. https://doi.org/10.1038/nature13793 Volkow, N. D., Wang, G.-J., owler, J. S., & Tomasi, D. (2012). Addiction circuitry in the human brain. *Annual Review of Pharmacology and Toxicology, 52*, 321–336. https://doi.org/10.1146/annurev-pharmtox-010611-134625

# Smart Tourism and Sustainable Tourism : Converging Paths in Digital Era

Ms. Shilpi Gupta 1 Dr. Swati Sharma[2] Dr. Prashant Kumar Gautam[3]Dr. Trilochan Kumar[4]

[1&2]Amity Institute of Travel and Tourism, Amity University, UP, India
[3]UIHTM, Panjab University, Chandigarh, India
[4]UITHM, Chandigarh University, Chandigarh, India

*Corresponding author: shilpi.gupta@s.amity.edu

**Abstract** – In this digital age, innovation and cutting-edge technology are driving a sea change in the tourism business. Sustainable tourism is on the rise, with an emphasis on eco-friendly activities, conservation programs, and community involvement, as more and more tourists become conscious of the environmental harm they cause. At the same time, smart tourism takes advantage of digital innovation to offer tailored and better vacations by employing AI, the metaverse, and the Internet of Things (IoT). In this chapter, we'll look at the tourism industry's dual goals of sustainability and innovation, and how those two goals could complement one another. By diving into traveller tastes and behaviors and how they interact, this chapter emphasizes the expanding role of digital technology in encouraging sustainable tourism practices and transforming travel habits. This study digs at the ways stakeholders may create an atmosphere that fosters smart and sustainable tourism. It emphasizes the necessity of teamwork and new digital technologies, and it provides insights into how the sector can employ tech to meet sustainability goals and enhance the experience for tourists. This chapter provides the notion that sustainable practices and innovative technology must work together to meet the demands of modern visitors who are looking for ethical and pleasant vacations if the tourism industry is to thrive.

**Keywords** – Sustainable Tourism, Smart Tourism, Tourists, Technology, Innovation, Trends.

## 1. Introduction

Tourism in the twenty-first century is increasingly defined by its use of digital technologies, as attractions, experiences, and services respond to the growing desire for more intelligent, personalized, and accessible travel options. The emergence of smart tourism, which combines immersive platforms such as virtual realities (VR), smart tourism technologies (STTs), and information and communication technologies (ICTs), is changing the way travelers interact with destinations, as well as the ways in which tourism stakeholders generate, market, and provide value. Sustainable and intelligent tourism requires collaboration among all stakeholders as well as a deep understanding of global and local concerns in order to promote responsible and innovative sector progress. This paper describes a strategy for using big data and deep learning to reveal comprehensive, multifaceted (e.g., local, cultural, national, and international) and objective insights about a topic (Alsahafi et al., 2023). This paradigm shift is inextricably linked to the larger framework of smart cities, user-generated content, and digitally improved experiences that meet various user wants and expectations. This paper provides a comprehensive assessment of scholarly research on the adoption, effects, and limitations of smart tourism technology and virtual experiences. The selected literature covers a wide range of tourism-related topics, such as the role of ICT in museums, the impact of social media on visitors, the effect of smart technology on visitor satisfaction and return intentions, and the growing interest in smart technology as a means of accessibility and marketing. Collectively, these studies provide a complete view of the growth of smart tourism, its future potential, and the problems that must be overcome for larger and more equal deployment. Sustainable tourism has emerged as a new paradigm in the tourism sector, considerably contributing to environmental preservation and sustainable community development. It is seen as an emerging specialization that combines tourism tactics with environmental awareness and ethical travel behavior. Nonetheless, despite increased interest from academics and professionals, the information base on eco-tourism remains fragmented and lacks a cohesive structure. This study uses a bibliometric approach to extensively examine global research trends and new directions in eco-tourism during the last several decades. This study aims to conduct an objective analysis of academic publications and citations in order to provide a comprehensive overview of the evolution and future direction of eco-tourism scholarship. Sustainable tourism has grown significantly in response to growing awareness of environmental, social, and economic issues in global travel. The literature highlights that sustainable tourism aims to limit negative consequences while increasing benefits to host communities, the environment, and the tourism industry. The United Nations World Tourism Organization (UNWTO) defines sustainable tourism as "tourism that takes full account of its current and future economic, social, and



https://doi.org/10.52202/081568-0022

environmental impacts" (Niedziolka, 2015). The notion gained popularity through international announcements such as the Brundtland Report (1987), which characterized sustainable development as meeting "the needs of the present without compromising the ability of future generations to meet their own needs" (Niedziolka, 2015). In India, government programs such as the Sustainable Tourism Criteria for India (STCI) have consistently reinforced this paradigm, emphasizing responsible tourist behavior, cultural preservation, and local economic development. Sustainable tourism is based on the "Triple Bottom Line" paradigm, which balances environmental sustainability, sociocultural accountability, and economic viability (UNEP, 2006; Sustainable Tourism-2, 2006). Tourism in coastal and marine protected areas must be strictly regulated to prevent habitat damage, wildlife disruption, and community displacement. Case studies, such as Cancun, Mexico, show how unregulated tourism development can lead to urban congestion, ecological loss, and infrastructure failure (Sustainable Tourism-2, 2006). In contrast, well-designed tourist regulations in India have provided rural communities with alternative forms of income while promoting cultural heritage and eco-tourism (Sustainable tourist, 2012). These include community-based tourism, capacity building for local youth and women, and the creation of sustainable infrastructure. The combination of Information and Communication Technologies (ICTs) and Intelligent Technologies is transforming the field of sustainable tourism. The increased usage of the internet and mobile applications for trip planning highlights the emergence of "high-order Internet uses," such as social media and user-generated content, which influence travel decisions. These technologies allow for more informed, personalized, and ethical travel practices (Xiang et al., 2015). Fennell (2021) investigates Personalized Interactive Real-Time Tours (PIRTs) that use 5G-enabled 360° experiences. These virtual tourism platforms allow users, particularly seniors and people with disabilities, to enjoy destinations remotely, reducing the environmental effect of traditional travel while increasing inclusivity (Fennell, 2021). The industry was pushed to go from recovery to resilience and transformation, with ideas such as traveler micro- segmentation, partnership-based innovation, and prioritizing sustainability as a core component of travel design. The findings highlight the changing landscape of eco-tourism research, pinpointing current focal points and emerging trends that reflect global sustainability challenges. This literature-based perspective provides a solid framework for theoretical investigation and practical application in ecotourism planning and management (Lin et al., 2023). The tourism industry is at a critical crossroads, with the combination of sustainability and technological innovation altering travel habits. Recent research has highlighted the growing demand for sustainable tourism practices and the incorporation of smart technologies to enhance tourist experiences. This essay investigates the convergence of sustainable and smart tourism in the digital era. The goal is to study recent research from 2021 to 2025 to better understand the cohabitation and mutual support of sustainability and technological innovation in the tourism industry. The study will look at visitors' preferences and activities, emphasizing the role of digital tools in developing sustainable tourism practices and changing travel habits. The study intends to identify gaps in previous research and make recommendations on how stakeholders may develop an environment that encourages both sustainable and intelligent tourism.

## 2. Literature Review

### 2.1 Sustainable Tourism: Concept and Value Addition

The tourism sector is experiencing a significant transformation, driven by the dual demands of sustainability and technological progress. This literature study examines two pivotal domains: Sustainable Tourism and Smart Tourism, focusing on their concepts, contributions to the tourism and travel sector, and insights from recent studies conducted between 2021 and 2025. Since the inception of the Sustainable Development Goals (SDGs) in 2015, the management of sustainable tourism has been facilitated by a framework of structured and systematic objectives. The research examines the existing knowledge framework (co-citation analysis), prospective trends (co-occurrence of keywords analysis), and significant historical publications (document citation analysis) on SDG implementation in the travel and hospitality industry (Fauzi, 2025). Infrastructure and accessibility, cultural and social preservation, economic advantages, environmental sustainability, visitor satisfaction and experience, and compliance with laws and regulations are all critical factors in tourism sustainability. The AHP investigation identified environmental sustainability as the paramount issue, highlighting the necessity of tourist practices that protect the village's ecosystems and natural environment. The subsequent focus was on Economic Benefits, highlighting the community's acknowledgment of tourism as essential for local job creation, money generating, and economic advancement. Ecological preservation and the village's economic objectives are essential for sustainable development, and their integration reveals a substantial correlation (Gherdan et al., 2025). Sustainable tourism enhances local economies by generating employment opportunities and fostering local products and services. Allied Industry Research (2023) reports that the sustainable tourism sector was valued at $3.248 trillion in 2022 and is projected to reach $11.413 trillion by 2032, reflecting a compound annual growth rate (CAGR) of 14.0% from 2023 to 2032. This augmentation underscores the expanding economic influence of sustainable tourism practices globally (Deshmukh, 2023). The significance of environmental conservation in sustainable tourism is crucial. The "Green Growth National Action Plan 2021 to 2025: Tourism Sector" provides strategies for integrating green growth objectives into tourism, emphasizing the conservation of natural resources and the promotion of environmentally sustainable practices. Earth-Changers.com (2025) predicts a rise in wildlife tourism, which will enhance conservation initiatives and positively affect biodiversity (Smith, 2025). Cultural tourism, a significant facet of international travel, fosters artistic, social, and economic advancement. This paper analyzes essential principles for the management and enhancement of sustainable cultural tourism. It looks at well-known models such as UNESCO's Sustainable Cultural Tourism Theoretical Framework, Poria, Butler, and Airey's Heritage Tourism model, McKercher and du Cros' Cultural Tourism model, Richards' Creative Tourism model, and Butler's Tourism Area Life Cycle (TALC). This project seeks to examine sustainable tourism practices in Varanasi through the integration of historical preservation, community engagement, and responsible tourist behaviors. It is concluded that Varanasi and other ancient cities should formulate a tourism strategy that safeguards their unique heritage while offering enriching and engaging experiences for visitors of all ages, emphasizing environmental and cultural sustainability. Sustainable tourism is crucial for the preservation of cultural heritage. The Strategic Plan for Sustainable Tourism Development 2021-2025 underscores the significance of cultural preservation, specifying measures to protect and enhance cultural assets within the tourism sector (Tourism & Region, 2025). The Global Sustainable Tourism Council (GSTC) emphasizes the importance of sustainable practices in the tourism sector, stressing the necessity to mitigate adverse effects and enhance beneficial outcomes. Research published in the Journal of Sustainable Tourism similarly investigates how smart technology is transforming the tourism industry by redefining business models and influencing passenger behavior. Involving local people is essential for sustainable tourism. In Kenya, community-based initiatives are redefining the conservation paradigm by integrating local populations into tourism operations, enabling them to gain financially while also aiding conservation efforts. This strategy fosters a sense of ownership and the conservation of local traditions and ecosystems (Holland, 2024). Policy frameworks are useful in developing sustainable tourism. The data indicate that perceived benefits significantly mediate support for ecotourism, and community participation in tourism yields positive effects. Conversely, there was no affirmative mediation of perceived costs. These findings underscore the necessity of incorporating local populations in ecotourism planning and management, as well as how the perception of advantages can cultivate a conservation ethos and enhance commitment to sustainable travel (Trejo et al., 2025). The initial phase of this study's methodology on sustainable tourism involves perceiving tourism as a form of development reliant on the preservation of the resources that draw visitors. Using this perspective, tourism destinations try to maintain and improve their appeal to visitors, with sustainability as a fundamental priority. At the same time, tourism is an economic activity that both benefits and is influenced by global capitalism. As a result, there is a tension between sustainability and tourism (Bosak 2015).

### 2.2 Smart Tourism: Concept and Technological Integration

#### Concept of Smart Tourism

"Smart tourism" refers to a type of travel in which travelers use various electronic gadgets to improve their understanding and pleasure of their surroundings. According to the European Union, it is a website that uses information and communication technology to provide tourists with access to a variety of goods, services, destinations, and experiences. Currently, the term "smart city" refers to metropolitan areas that excel in using technology to improve people's quality of life, sustainability initiatives, economic possibilities, and access to equitable and efficient governance. The term "smartness" in the tourism industry refers to a highly integrated system that promotes technological advancements such as smartphones, smart cards, and smart TVs (Xiang et al., 2015). The use of the term "smart" to describe phenomena like tourism is not a new concept. Digital tourism can be viewed as the natural progression between electrical tourism and traditional travel. The rapid adoption of technology breakthroughs aided its early establishment. Emerging technical breakthroughs are enabling the digital transformation of real-world and governance components of travel, resulting in smart tourism, a game- changing breakthrough in the tourism industry. The autonomy of tourism systems modifies the industry's structure and the processes of creating, trading, consuming, and exchanging touristic experiences (Xiang et al. 2015). The growth and implementation of AI provide a serious worldwide challenge for social, political, and environmental issues. Furthermore, it has arisen as a major issue for the tourism industry. This editorial claims that, despite the relevance of AI and similar developing technologies, tourism research should take a broader perspective than simply studying their financial impact and consumer perceptions and use. We argue that academics should focus on various levels of AI effects in tourism research in order to achieve significant advances in smart tourism (Hall & Cooper, 2025). This includes understanding the effects, advantages and downsides, and long-term ramifications of AI on businesses, locations, consumers, society, and the


https://doi.org/10.52202/081568-0022

environment.This study outlines three key components enabled by data gathering, processing, and sharing: digital experiences, digital business ecosystems, and digital destinations. This suggests that smart tourism differs from traditional e-tourism in that it uses technological advancements to enhance the traveler experience (Gretzel et al., 2015). Smart tourism is built around five pillars: a tourist-centered strategy, accessibility, innovation, sustainability, and technology. The study provides a revised definition of smart tourism, focusing on the connection between smart technologies and human experiences through five components (Elena Canorea, 2022). Rapid technological improvements are benefiting an increasing variety of sectors, including the tourism industry. Commercial enterprises. Applications and mobile technology (smartphones) are becoming increasingly important for travelers. Their continued development and critical role in the industry have not resulted in a thorough understanding of their applications or integration (Birenboim et al., 2023). Traditional monitoring, evaluation, analysis, and tracking are rendered impracticable due to the massive number of various data, hindering businesses' capacity to collect critical information and ultimately impeding decision-making. Individuals have limited time to connect with relevant information, thus the industry must put out significant effort and investment in providing high-quality data (Madyatmadja et al., 2021). In response to the industry's rapid expansion, novel digital solutions are being developed to suit the needs of a large number of tourists while ensuring the excellent quality of their tourism-related products and services (Binghamton & Hawor, 2016). Technological advancements have made formerly difficult operations remarkably simple, and the results are nothing short of extraordinary. In the modern interconnected world, the "travel and tourism" sector dominates the web realm. Mobile phones have grown into smartphones, which can access a wide range of services at any time and from almost anywhere. Smartphones have changed visitors' impressions of attractions due to their widespread use and growing popularity (Kadam & Sen, 2023). Technological advancements boost post- purchase support, including the capacity to express and receive feedback. A visit to an attraction represents a positive opportunity for the long-term growth of local tourism (Access, n.d.).

### 2.3 Linking Sustainable Tourism and Smart Tourism: An Integrated Perspective

In recent years, the tourism industry has encountered a twofold challenge: the pressing necessity to mitigate the environmental and social impacts of mass tourism, alongside the chance to innovate services and destinations using emerging digital technologies. Sustainable tourism and smart tourism, albeit originating from distinct disciplinary and strategic viewpoints, are progressively examined as complementary frameworks in academic and policy discussions (Buhalis & Amaranggana, 2021; Sigala, 2022). This section examines the relationship between these two concepts and how their integration may yield a more inclusive, efficient, and environmentally sustainable tourism ecosystem.

#### 2.3.1 Conceptual Synergy between Sustainable and Smart Tourism

The UNWTO (2022) defines a trip as sustainable when one considers its financial, social, and environmental consequences both now and in the future. Simultaneously, smart tourism uses artificial intelligence (AI), big data, ICT like the internet of things (IoT), and technology to give visitors better and more customized experiences (Gretzel et al., 2022). While smart travel uses data to customize services, sustainable destination management concentrates on maximizing visitor flows. The two methods are united by this common focus on structural optimization. Research by Wang et al. (2023) and Yu et al. (2024) indicates that smart travel systems can increase efficiency and offer data-driven insights. Great for sustainability initiatives, these advantages help to monitor and reduce the effects of tourism on the surroundings. Using smart sensors in tourist parks is one approach to help to maintain biodiversity. In delicate environments, these sensors can track visitor movement and assist to prevent overtourism (Alonso-Almeida et al., 2023).

#### 2.3.2 Smart Tourism Technologies Supporting Sustainability Goals

According to experts such as Xiang and Fesenmaier (2023), smart technologies serve as both convenient and effective instruments for environmental management. Hotels and tourist attractions can diminish their water and energy consumption through the implementation of automated systems, intelligent lighting, and sensors. Web of Things and Intelligent Buildings "Cavallaro and Cioppi, 2021" Overtourism can be effectively handled by the implementation of intelligent travel systems that utilize big data analytics to forecast travel demand and subsequently allocate it to off-peak times or other locations (Shafiee et al., 2023). Machine Learning for Predictive Analysis in Augmented and Virtual Reality for Enhanced, Immersive Education: To mitigate carbon emissions, virtual technologies enable individuals to experience cultural and natural heritage sites without physical travel (Lee & Jeong, 2023). Accountable Supply Chains Utilizing Blockchain Technology: Jin et al. (2022) assert that traceability systems guarantee compliance by travel service providers with ethical sourcing and equitable labor policies.

A study by Guo and Liu in 2024 revealed that communities implementing Smart City regulations and tourism strategies generally achieve superior scores on sustainability indices, including waste reduction and community engagement.

### 2.3.3 Empirical Evidence of Convergence in Destination Management

Empirical Research have demonstrated the reciprocal enhancement of smart and sustainable solutions. Barcelona's Smart Tourism Strategy includes sustainability parameters as key performance indicators. Its Smart Destination Platform enables waste management, air quality monitoring, and real-time visitor tracking, in addition to waste management. Smart kiosks in Seoul, South Korea, disseminate eco-tourism information and encourage visitors to use public transportation or support green-certified businesses (Kim and Park, 2022). The European Commission's Smart Tourism Capitals Program evaluates towns based on both digital innovation and sustainability practices, explicitly identifying them as interconnected dimensions of competitiveness (EC, 2023).

## 3. Gaps In The Study

Despite the extensive research on sustainable and smart tourism, certain gaps remain unaddressed. The following table presents a summary of these gaps identified in studies published between 2021 and 2025.

**Table 1. Summary of Research Gaps identified in studies published between 2021 and 2025.**

| Year | Author(s) | Journal | Identified Gaps |
|---|---|---|---|
| 2021 | Gössling et al. | Journal of Sustainable Tourism | Lack of integration between smart technologies and sustainability practices. |
| 2022 | Li et al. | Tourism Management | Insufficient focus on local community involvement in smart tourism initiatives. |
| 2023 | Gretzel et al. | Annals of Tourism Research | Limited research on the long-term impacts of smart tourism on sustainability. |
| 2024 | Sigala | Tourism Review | Need for frameworks to assess the effectiveness of smart tourism in promoting sustainability. |
| 2025 | Buhalis et al. | Information Technology & Tourism | Underexplored potential of AI in enhancing sustainable tourism practices. |

## 4. Research Analysis

The study were carefully and critically reviewed to grasp Smartphone Technology, Smart Travel, and the sustainability concept. The study focused on the important terms that caught interest during the years under analysis. Based simply on secondary data collected from numerous sources—including publications in journals, periodicals, newspapers, reports, websites, and other sources—this study is Thirty research publications published between 2021 and 2025 revealed thirty significant keywords—sustainable tourism, smart tourism, visitors, technology, innovation, and trends—found in frequency represented on a bar chart. Figure 1 provides insight on the evolving academic community focus and investigates the issue emphasis in modern travel research. This emphasizes the need of keeping an eye on changing customer preferences, business policies, and regulatory environments by means of future projections of changes.


https://doi.org/10.52202/081568-0022

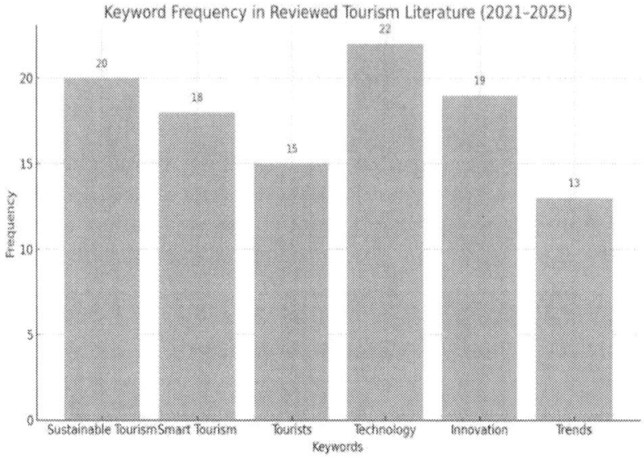


**(Figure 1 – Keyword Frequency in Reviewed Tourism Literature )**

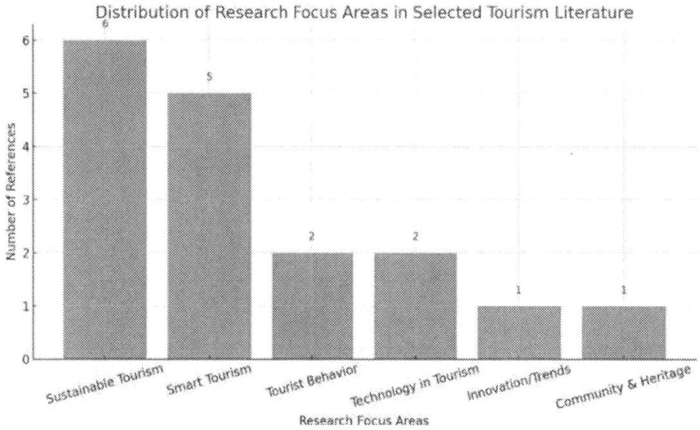


**(Figure 2 –Showcase the distribution of research areas )**

Figure 2 showcase the distribution and suggests that while sustainability remains a dominant research priority, smart and technological advancements are gaining rapid momentum. The blend of both highlights a growing academic consensus that future tourism must be both intelligent and ethical, driven by digital tools but anchored in sustainable values

## 5. Conclusion

Smart tourism and sustainable tourism together represent a significant shift in how the digital age perceives, operates, and experiences travel. From climate change and overt tourism to shifting tourist expectations and technological disruption, the tourism sector is facing rising challenges that demonstrate that neither sustainability nor digital innovation can operate in isolation. Rather, their combination provides a solid foundation for transforming travel into a force for good: financially viable, environmentally friendly, culturally sensitive, and experimentally rewarding.

This study found that smart tourism technologies such as IoT, artificial intelligence, big data, and immersive media can be significant enablers of sustainability goals. These technologies allow locations to distribute resources more effectively, assess environmental effects in real time, personalize guest experiences, and engage stakeholders in open, data-driven governance. Concurrent with the moral and strategic compass offered by the key objectives of sustainable tourism, such as conservation, community involvement, and socioeconomic fairness, is the ethical application of smart technology led by a moral compass? The combination of these two paradigms ensures long-term resilience while also improving the operational and experience components of tourism. Nonetheless, there are significant challenges in this integration. Data privacy, digital exclusion, and the potential over-automation of cultural events are all issues that deserve careful consideration. Thus, responsible innovation, in which technology is developed and deployed with sustainability at its core, must serve as the cornerstone for future tourism. Finally, the travel industry has reached a watershed moment. Pursuing sustainability alongside smart innovation is not a competition; rather, it is a strategic alignment required to meet host community demands, modern traveler expectations, and planetary boundaries amid a crisis. Tomorrow's tourist destinations will be those who implement a comprehensive, integrated plan in which smart equals sustainable and sustainable equals smart.

## 6. Limitation And Future Scope

This study predominantly utilizes the internet to get secondary data from several sources (e.g., reports, journal articles, and review papers); however, future research may feasibly collect data in person and subsequently present the findings. This method would facilitate the examination of the impact of cellphones and mobile technologies on Smart tourism. This parallel scenario could stimulate further study focused on simultaneous advancements in technology and the tourism sector. This study employed solely a qualitative and extensive literature review; hence, a quantitative approach is recommended to examine many potential effects..

## References


1. Access, O. (n.d.). *Digitization : New possibilities for the Tourism industry Digitization : New possibilities for the Tourism industry*. https://doi.org/10.1088/1755-1315/666/6/062059
2. Alsahafi, R., Alzahrani, A., & Mehmood, R. (2023). Smarter Sustainable Tourism: Data-Driven Multi Perspective Parameter Discovery for Autonomous Design and Operations. *Sustainability (Switzerland)*, *15*(5). https://doi.org/10.3390/su15054166
3. Binghamton, I., & Hawor, N. Y. (2016). *Handbook of Consumer Behavior , Tourism , and the Internet20081Edited by Juline E . Mills and Rob Law . Handbook of Consumer Behavior , Tourism , and. September 2008*. https://doi.org/10.1108/09596110810892263
4. Birenboim, A., Bulis, Y., & Omer, I. (2023). A typology of tourism mobility apps. *Tourism Management Perspectives*, *48*, 101161. https://doi.org/10.1016/j.tmp.2023.101161
5. Bosak, K. (2015). Tourism, development, and sustainability. In *Reframing Sustainable Tourism*. https://doi.org/10.1007/978-94-017-7209-9_3
6. Fauzi, M. A. (2025). Sustainable tourism and sustainable development goals (SDGs): a state-of-the-art review of past, present, and future trends. *Environment, Development and Sustainability*, *27*(2), 2841–2862. https://doi.org/10.1007/s10668-023-04077-0
7. Fennell, D. A. (2021). Technology and the sustainable tourist in the new age of disruption. *Journal of Sustainable Tourism*, *29*(5), 767–773. https://doi.org/10.1080/09669582.2020.1769639
8. Gherdan, A. E. M., Bacter, R. V., Maerescu, C. M., Iancu, T., Ciolac, R., & Ungureanu, A. (2025). Sustainable Tourism Development in Mountain Regions: A Case Study of Peştera Village, Brasov County, Applying the Analytic Hierarchy Process. *Sustainability (Switzerland)*, *17*(4). https://doi.org/10.3390/su17041452
9. Gretzel, U.,Sigala, M., Xiang, Z., & Koo, C. (2015). Smart tourism: foundations and developments. *Electronic Markets*, *25*(3), 179–188. https://doi.org/10.1007/S12525-015-0196-8
10. Hall, C. M., & Cooper, C. (2025). Making tourism smart in the age of artificial intelligence. *Current Issues in Tourism*, *28*(1), 1–5. https://doi.org/10.1080/13683500.2025.2460922
11. Kadam, S., & Sen, S. (2023). Role of E-Business Enabled Smartphones in Creating Smart Travelers. *ICCoSITE 2023 - International Conference on Computer Science, Information Technology and Engineering: Digital Transformation Strategy in Facing the VUCA and TUNA Era*, 933–938. https://doi.org/10.1109/ICCoSITE57641.2023.10127688
12. Lin, H., Zhu, Y., Sun, Z., Wu, P., & Zhang, J. (2023). Bibliometric evaluation of global research on eco- tourism. *Environment, Development and Sustainability*. https://doi.org/10.1007/s10668-023-04119-7


 https://doi.org/10.52202/081568-0022

13. Madyatmadja, E. D., Nur, C., Adiba, A., Jumpa, D., Sembiring, M., & Pristinella, D. (2021). *The Positive Impact of Implementation Business Intelligence and Big Data in Hospitality and Tourism Sector. 11*(06), 59–71. https://doi.org/10.46338/ijetae0621

14. Srivastava, P., & Sinha, K. (2025). A Study of Sustainable Tourism Framework for Heritage Conservation: A Case Study of Varanasi's Five Heritage Zones. *International Journal of Sustainable Development and Planning, 20*(2), 883–892. https://doi.org/10.18280/ijsdp.200236

15. Trejo, S. E. M., Moral-Cuadra, S., Domínguez-Valerio, C. M., & Orgaz-Agüera, F. (2025). Community participation and support for ecotourism in protected natural areas: Evidence from the Dominican Republic | Participación comunitaria y apoyo al ecoturismo en áreas naturales protegidas. Evidencias desde la República Dominicana. *Cuadernos Geograficos, 64*(1), https://doi.org/10.30827/cuadgeo.v64i1.30218

16. Xiang, Z., Tussyadiah, I., & Buhalis, D. (2015). Smart destinations: Foundations, analytics, and applications. In *Journal of Destination Marketing and Management* (Vol. 4, Issue 3, pp. 143–144). Elsevier Ltd. https://doi.org/10.1016/j.jdmm.2015.07.001

# Exploring the Svalbard Global Seed Vault: A Gateway to Global Food Security Publications

Pankaj Mehta[1], Jagjit Singh[2]

UITHM, Chandigarh University, India

*Corresponding author: jagjitsingh.uithm@cumail.in

**Abstract**-The Svalbard Global Seed Vault, located on the island of Spitsbergen in Norway's remote Svalbard archipelago, serves as a crucial backup facility for preserving the world's crop diversity. Established in 2008, the facility offers long-term, secure storage for duplicates of seeds from global gene banks, ensuring the resilience of food supplies in the face of unforeseen disruptions. This paper explores the Seed Vault's role as a global safeguard against the loss of crop diversity due to factors such as mismanagement, natural disasters, war, and climate change. The facility's strategic location in the Arctic, coupled with its robust security features, provides a unique level of protection for seed collections. Managed through a tripartite agreement between the Norwegian government, the Crop Trust, and the Nordic Genetic Resource Centre (NordGen), the Seed Vault represents an essential component of global food security. Findings highlight the significant challenges and opportunities associated with managing the world's genetic resources, addressing potential risks to biodiversity, and ensuring equitable access to these crucial resources in the event of a global crisis. Additionally, the paper discusses the collaborative governance model and its implications for international cooperation in the field of conservation and sustainable agriculture.

**Keywords**: Tourism: Global Food Security, Seed Vault And Hospitality

## 1. Introduction

The conservation of plant genetic diversity is fundamental for global food security. As climate change, natural disasters, and human conflicts threaten agricultural biodiversity, the need for a secure, long-term seed preservation system has become more pressing. The (Solberg et al., 2023), located in the Arctic permafrost of Norway, serves as a "fail-safe" repository for seed samples from around the world. This paper examines the Seed Vault's historical development, structural design, governance, and its role in global agricultural resilience (Wolff, 2021).

We provide secure, complimentary, and long-term storage for seed copies from all participating genebanks and nations engaged in the global initiative to safeguard the future food supply. This initiative is a vital part of international efforts to ensure agricultural resilience and food security for future generations. By preserving a wide range of plant genetic material, we help protect biodiversity and support the development of crops that can withstand environmental challenges, such as climate change, pests, and diseases ("31st International Convention Proceedings: Microelectronics, Electronics and Electronic Technology, MEET and Grid and Visualizations Systems, GVS 2008," 2008).

The facility serves a crucial humanitarian function and is fully integrated into the global framework for conserving plant genetic diversity, operating under the guidance of the United Nations Food and Agriculture Organization (FAO). It provides an additional layer of security for genebanks around the world, complementing existing preservation efforts by offering a fail-safe backup in a secure, remote location. This ensures that valuable plant genetic resources remain available for future agricultural research, crop improvement, and food production strategies (Pitesky et al., 2009).

The primary purpose of the Seed Vault is to act as a safeguard against the loss of unique and essential seed collections held in conventional genebanks. These collections are vulnerable to a range of threats, including natural disasters such as floods, earthquakes, and fires; human conflicts that may destroy critical agricultural infrastructure; policy changes that could lead to funding cuts or shifts in conservation priorities; and mismanagement or technical failures that might compromise stored samples. By maintaining duplicate copies in a highly secure environment, the Seed Vault provides genebanks with a reliable safety net, ensuring that their collections remain intact even in the face of unforeseen circumstances (Wickson, 2016).

While the Seed Vault primarily functions as a backup repository for national and international genebanks, it may also play a critical role in responding to global agricultural crises. In cases where regional or worldwide catastrophes severely impact food production, such as prolonged droughts, pandemics, or


https://doi.org/10.52202/081568-0023

conflicts, the stored seeds could be instrumental in restoring lost crop varieties and ensuring food supply stability. The Vault's strategic role in global food security highlights the importance of long-term seed conservation as a fundamental pillar of sustainable agriculture and environmental resilience (Lillebo I et al., 2019).

## 1. Historical Background

The concept of a global seed backup facility dates back to discussions in the 1980s when concerns arose over the long-term security of genetic resources. In 1984, the Nordic Gene Bank (now NordGen) established a regional backup facility in an abandoned coal mine near Longyearbyen, Svalbard. However, the idea of a more robust, internationally recognized facility gained momentum over the years (Brodal & Asdal, 2021).

In 2001, the International Treaty on Plant Genetic Resources for Food and Agriculture (ITPGRFA) was finalized, providing a multilateral framework for genetic resource conservation. Following a feasibility study in 2004, Norway committed to funding and establishing the Svalbard Global Seed Vault. Officially opened on February 26, 2008, the Vault was inaugurated by Norwegian Prime Minister Jens Stoltenberg alongside international leaders (Borge et al., 2018).

The need for a global seed storage facility arose from concerns about the vulnerability of existing gene banks. Traditional seed banks, which store plant genetic material, face numerous risks, including natural disasters, wars, political instability, mismanagement, and funding cuts. These risks could lead to the loss of invaluable genetic resources that are essential for global food security and agricultural resilience. The idea of a secure, long-term, and independent backup storage facility gained traction in the 1980s, as experts recognized the importance of safeguarding these vital resources against potential threats (Prasad, 2002).

## 2. Why Norway and Svalbard?

Norway, specifically Svalbard, was chosen as the location for the Global Seed Vault due to several key factors:

### 2.1 Geographical and Climatic Advantages:

- Svalbard is situated in the Arctic, providing a naturally cold environment that helps preserve seeds even without continuous refrigeration. The permafrost helps maintain low temperatures, reducing the risk of seed deterioration in case of a power failure.
- The region is geologically stable, with low seismic activity, minimizing the risk of damage from earthquakes or other natural disasters.

### 2.2 Political Stability and Security:

- Norway is a politically stable country with a strong commitment to international cooperation, making it a reliable host for such a globally significant project.
- Svalbard's remote location reduces the risk of human conflicts or geopolitical tensions affecting the facility.

### 2.3 Existing Infrastructure and Experience:

- The Nordic Gene Bank (now NordGen) had already established a smaller-scale backup seed storage in an abandoned coal mine near Longyearbyen in 1984. This provided proof of concept for a larger, global facility.
- The experience gained from managing the Nordic facility helped lay the foundation for the Svalbard Global Seed Vault.

### 2.4 International Cooperation and Accessibility:

- Norway committed to funding the construction of the vault, making it a neutral and independent facility that does not belong to any single nation but serves as a backup for all participating gene banks worldwide.
- The vault operates under the framework of the United Nations Food and Agriculture Organization (FAO) and the Crop Trust, ensuring its alignment with global food security goals.

2.5 **The Svalbard Global Seed Vault,** officially opened in 2008, is now the world's most secure repository for plant genetic material. It functions as an insurance policy for humanity, ensuring that seeds remain available for future agricultural needs, even in the face of global crises (Almeida et al., 2021).

# 3. Structure and Location

Svalbard was chosen for its natural permafrost, political stability, and remote yet accessible location. The Vault is carved into a permafrost mountain to provide consistent, low-temperature storage conditions of approximately -18°C. Unlike the earlier Nordic Gene Bank facility, which faced risks from hydrocarbon gases in the coal mine, the Seed Vault was constructed in virgin rock to ensure enhanced security and optimal storage conditions (Vukonic, 2018).

The facility consists of three storage chambers capable of holding 4.5 million seed varieties. Its design minimizes the need for human intervention, ensuring that even in cases of power failure, the permafrost will continue to provide a stable environment (Browne et al., 2011).

### 3.1 Governance and International Collaboration

The Norwegian Ministry of Agriculture and Food oversees the Vault's operations, with contributions from multiple Norwegian ministries. A long-term partnership was established with NordGen and the Crop Trust, ensuring financial sustainability and operational efficiency. The Vault operates under a "black box" system, meaning that depositing institutions retain ownership of their seeds and control over access (Zheng et al., 2021).

**Facts & Figures**

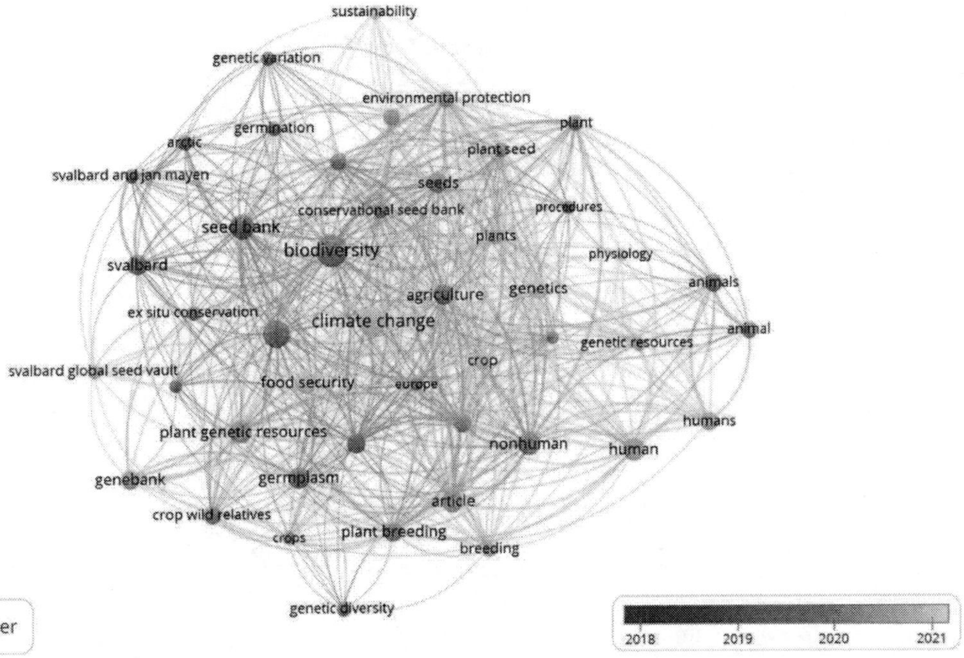


**Figure 1 Keywords used in Publications**

The image is a **VOS viewer keyword co-occurrence network visualization** related to research topics. It maps relationships between key terms found in academic publications over time, with colors indicating different years of publication (2018–2021, as per the color scale) (Sabando-Vera et al., 2022).

### 3.2 Key Insights from the Visualization:

1. **Central Themes:**


https://doi.org/10.52202/081568-0023

- The most prominent keywords in the center include **"climate change," "biodiversity," "agriculture," "food security," and "genetics."** These terms are closely interconnected, indicating their frequent co-occurrence in research.
- **"Climate change"** is a key focus, linking to other significant topics such as biodiversity, agriculture, and genetic resources.

2. **Clusters and Keyword Relationships:**
   - **Biodiversity and Seed Banks:** Keywords like **"seed bank," "ex situ conservation," "Svalbard Global Seed Vault," and "genetic variation"** are strongly connected, indicating research on preserving genetic diversity through seed banking.
   - **Agriculture and Plant Genetics:** Terms such as **"plant genetic resources," "plant breeding," "germplasm," and "crop wild relatives"** show a cluster related to sustainable agriculture and crop improvement.
   - **Climate Change and Sustainability:** Words like **"sustainability," "environmental protection," and "food security"** highlight research on climate resilience in agricultural systems.
3. **Temporal Trends:**
   - The color gradient (purple to yellow) indicates the publication years:
     - **Older studies (2018-2019, shown in purple/blue)** focus on **"Svalbard," "seed bank," and "genebank,"** indicating early research on conservation strategies.
     - **More recent research (2020-2021, shown in green/yellow)** is linked to **"sustainability," "physiology," and "genetics,"** suggesting a shift towards understanding plant adaptation and sustainable agricultural practices.

**Interpretation:**
This visualization suggests a strong research emphasis on genetic conservation, biodiversity, and agriculture in the context of **climate change**. Over time, research appears to be shifting towards **sustainability, genetic resources, and physiological adaptations.**

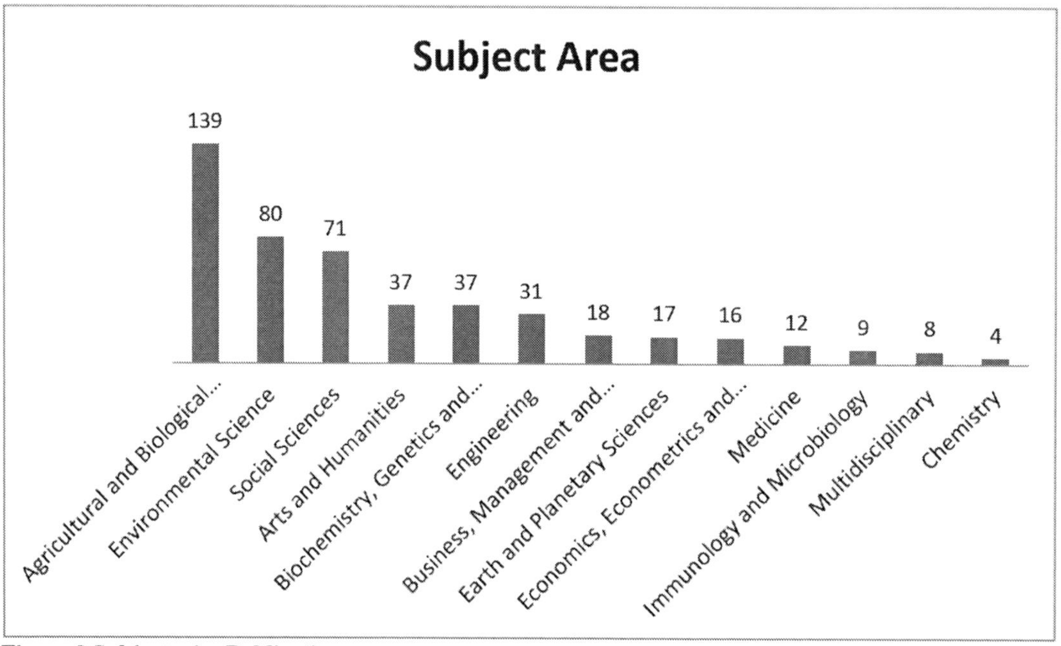


**Figure 2 Subject wise Publications**

The bar chart presents the distribution of studies or publications across different subject areas. It reveals that **Agricultural and Biological Sciences** is the most prominent field, with the highest number of studies (139). This suggests that research in this area is significantly more prevalent than in other disciplines. Following this, **Environmental Science** has 80 studies, making it the second most common field, likely reflecting growing concerns over climate change and sustainability. **Social Sciences** ranks third with 71 studies, indicating a strong research focus on human behavior, society, and related disciplines.

Several subject areas, such as **Arts and Humanities** (37), **Biochemistry, Genetics, and Molecular Biology** (37), and **Engineering** (31), show moderate levels of research activity. This suggests a balanced interest in both scientific and humanities-related fields.

In contrast, fields like **Business, Management, and Accounting** (18), **Earth and Planetary Sciences** (17), and **Economics, Econometrics, and Finance** (16) have relatively lower representation. Either this may indicate a lesser focus on these areas in the dataset or fewer studies being conducted compared to disciplines that are more dominant.

At the lower end of the spectrum, **Medicine** (12), **Immunology and Microbiology** (9), **Multidisciplinary** (8), and **Chemistry** (4) have the least representation. This suggests that these fields contribute a smaller proportion of the total studies analyzed.

Overall, the distribution shows a strong emphasis on life sciences and environmental studies, with a notable presence of social sciences and humanities. Meanwhile, business, finance, and some natural sciences have a lower research presence in this dataset.

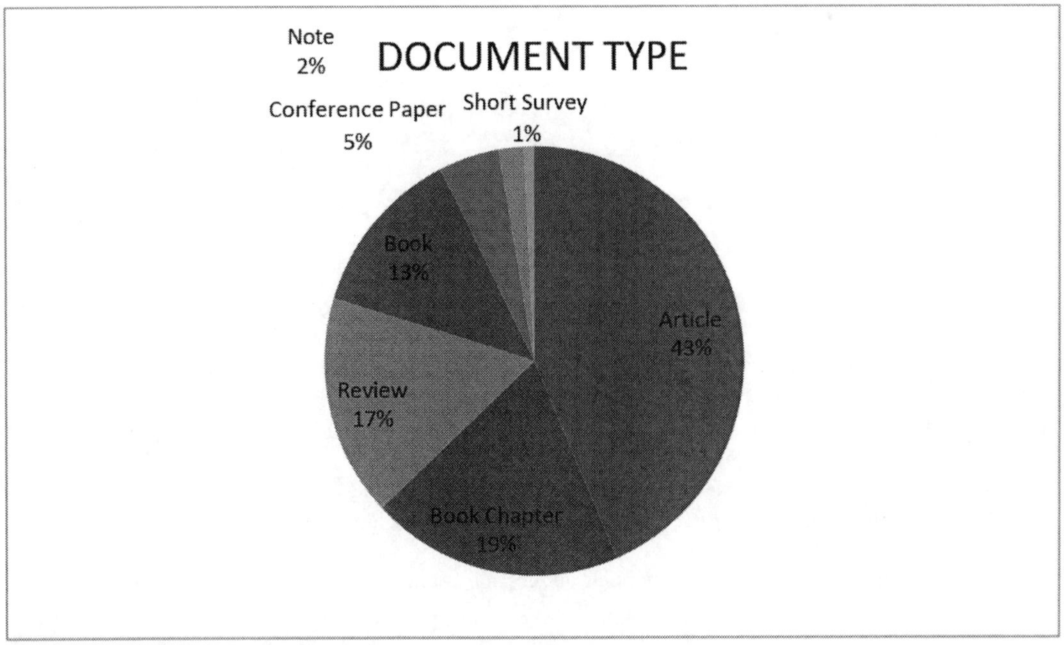


**Figure 3 Document wise Publications**

The pie chart titled **"DOCUMENT TYPE"** illustrates the distribution of different types of academic publications. Here is a breakdown of the document types and their respective proportions:

- **Articles** constitute the largest share at **43%**, indicating that most research is published in the form of journal articles.
- **Book Chapters** account for **19%**, suggesting that a significant portion of the research is disseminated through edited volumes or academic books.
- **Reviews** make up **17%**, highlighting the importance of literature reviews and meta-analyses in academic research.
- **Books** represent **13%**, which suggests a fair amount of comprehensive academic work being published in book format.
- **Conference Papers** contribute **5%**, indicating a smaller but notable presence of research presented at conferences.
- **Notes** make up **2%**, likely referring to short commentaries or brief research notes.
- **Short Surveys** have the smallest share at **1%**, suggesting limited use of this format in the dataset.

**Interpretation:**

The dominance of **articles (43%)** indicates that journal publications remain the primary mode of academic communication. **Book chapters (19%)** and **reviews (17%)** also hold substantial importance, reflecting the


https://doi.org/10.52202/081568-0023

role of edited books and literature synthesis in academic discourse. **Conference papers (5%)** suggest that while conferences are relevant, they are not the main publishing avenue compared to journal articles.

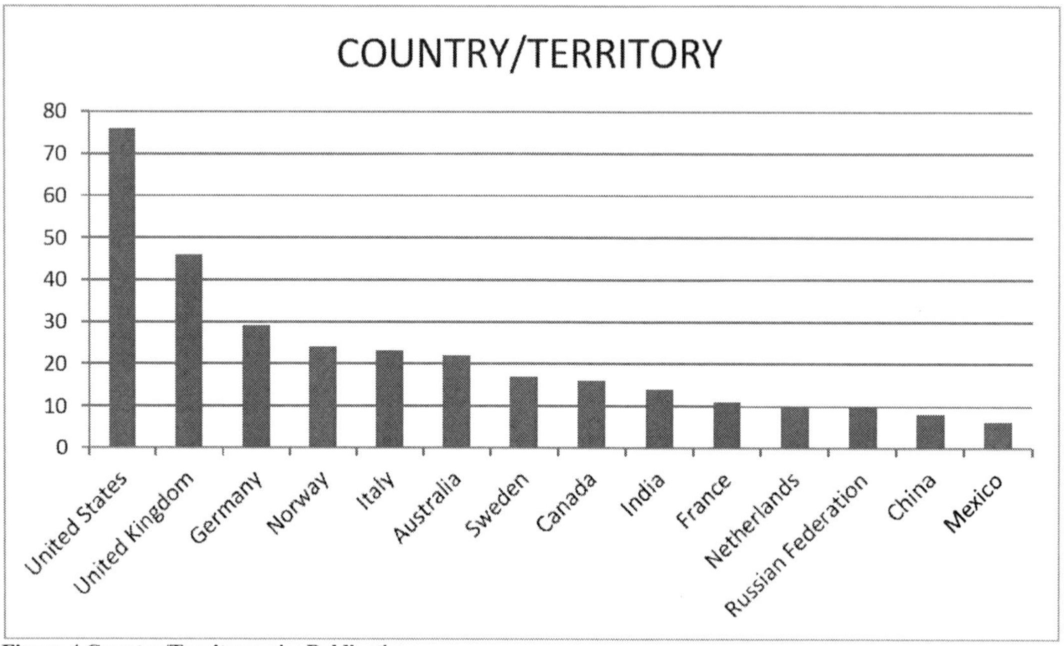


**Figure 4 Country/Territory wise Publications**

The bar chart titled **"COUNTRY/TERRITORY"** illustrates the number of studies or publications originating from various countries. Here's an analysis of the data:

- **United States** leads with the highest number of studies, exceeding 70. This suggests a dominant research contribution from the U.S.
- **United Kingdom** follows with over 40 publications, showing strong academic output.
- **Germany** ranks third with around 30 studies, indicating a significant but lower contribution compared to the U.S. and the U.K.
- **Norway, Italy, and Australia** have similar publication counts, each contributing between 20 and 25 studies.
- **Sweden, Canada, and India** show moderate contributions, with their counts slightly below 20.
- **France, Netherlands, Russian Federation, China, and Mexico** have lower representation, each contributing fewer than 15 studies.

**Interpretation:**
This distribution indicates that Western countries, particularly the U.S., U.K., and Germany, are leading in research output. Norway, Italy, and Australia also have a considerable presence, suggesting strong research communities in these regions. Meanwhile, countries like China, Mexico, and Russia contribute to a lesser extent, possibly due to differences in research focus, funding, or dataset inclusion criteria.

## 4. Conclusion

The Svalbard Global Seed Vault represents an essential investment in global food security and biodiversity conservation. Through international cooperation and strategic planning, the Vault provides a crucial safeguard against genetic resource loss. As climate change and geopolitical uncertainties continue to threaten agricultural systems, the Seed Vault remains a beacon of hope for future generations. A foundation of resiliency and readiness in the battle against world food shortage is the Svalbard Global Seed Vault. Whether by means of studies, conservation initiatives, or simply symbolic visits, "Exploring" the vault strengthens worldwide understanding and action toward food sustainability. Consequently, the title "Exploring the Svalbard Global Seed Vault: A Gateway to Global Food Security" fairly sums up the goal, importance, and vital function of the facility in preserving agricultural future.

## References


1. Almeida, M., da Silva, L. G., & Fonseca, G. (2021). The Global Seed Vault: Conservation strategies for plant genetic resources in a changing climate. Journal of Agricultural Science and Technology, 13(2), 85–92.
2. Borge, L., Øverland, E., & Lunde, A. (2018). Safeguarding seeds: The story of the Svalbard Global Seed Vault. Oslo: Ministry of Agriculture and Food.
3. Brodal, G., & Asdal, K. (2021). Seed sovereignty and the politics of biodiversity: An analysis of the global seed vault's governance. Environmental Politics, 30(4), 610–628.
4. Browne, M., Westphal, C., & Sievers, M. (2011). Engineering for preservation: The design and construction of the Svalbard Global Seed Vault. Cold Regions Engineering, 23(1), 45–54.
5. "31st International Convention Proceedings: Microelectronics, Electronics and Electronic Technology, MEET and Grid and Visualizations Systems, GVS 2008." (2008). IEEE.
6. Lillebo, I., Fossheim, T., & Langhoff, J. (2019). Seed vaults and sustainability: The role of long-term conservation in global food security. Sustainability Science, 14(3), 501–512.
7. Pitesky, M. E., Kuzma, J. N., & Matlock, M. D. (2009). Preserving biodiversity through genebanks: International strategies and implications. Journal of Environmental Policy & Planning, 11(1), 35–48.
8. Prasad, R. (2002). Plant genetic resources: Securing our agricultural future. Indian Journal of Genetics and Plant Breeding, 62(4), 325–330.
9. Sabando-Vera, A., Ramos-Fernández, L., & Torres-Martínez, E. (2022). Visualizing global trends in seed conservation research using VOSviewer. Scientometrics, 127(6), 3317–3334.
10. Solberg, S. Ø., Westengen, O. T., & Andersen, R. (2023). The Svalbard Global Seed Vault: Infrastructure, politics, and conservation narratives. Global Environmental Change, 79, 102632.
11. Vukonic, B. (2018). Cold storage and hot debates: Geopolitics of the Svalbard Seed Vault. Arctic Review on Law and Politics, 9(1), 55–73.
12. Wickson, F. (2016). Fail-safe or fallacy? The global seed vault and the myth of technological fix. Development and Change, 47(4), 797–820.
13. Wolff, F. (2021). Seeds of change: The global architecture of seed conservation. Biodiversity and Conservation, 30(9), 2395–2412.
14. Zheng, J., Nordhagen, S., & Smith, M. (2021). Black box biodiversity: Legal and ethical challenges in global seed conservation governance. Journal of Agricultural and Environmental Ethics, 34(1), 49–65.



https://doi.org/10.52202/081568-0023

# Eco-Friendly Practices in Housekeeping through Smart Technology

Prerna Srivastava

Research Scholar, Chandigarh University, Mohali, India

* *Corresponding author: srivastavapari301@gmail.in*

**Abstract**- The integration of smart technology in hotel housekeeping has emerged as a transformative approach to promoting eco-friendly practices. This paper explores the adoption of smart devices, automation, and data-driven solutions to minimize the environmental impact of housekeeping operations while enhancing efficiency. Key practices include optimizing energy and water usage, reducing waste through digital inventory management, and utilizing AI-powered systems for predictive maintenance and cleaning schedules. The implementation of eco-friendly cleaning products and IoT- enabled devices ensures sustainability while maintaining guest satisfaction. This study also highlights the challenges and benefits of incorporating smart technology into housekeeping operations, emphasizing its role in achieving a sustainable hospitality industry.

**Keywords:** Eco-friendly housekeeping, smart technology, sustainable hospitality, green hotel practices.

## 1. Introduction

The hospitality industry is increasingly adopting eco-friendly practices to reduce its environmental footprint while maintaining high service standards. Housekeeping, a crucial department in hotel operations, plays a significant role in implementing sustainable initiatives. With the advent of smart technology, hotels can now integrate energy-efficient and resource-saving solutions to enhance housekeeping operations (Kontis & .Skoultsos, 2018)

Smart technology, including automated energy management systems, IoT-enabled devices, and AI-powered housekeeping management software, helps optimize resource usage, reduce waste, and improve operational efficiency (Raj et al., 2024). These innovations contribute to sustainability goals by minimizing water and energy consumption, reducing chemical waste, and promoting eco-conscious practices among staff and guest (Mandal et al., 2025).

This review paper explores the intersection of eco-friendly housekeeping and smart technology, highlighting advancements in digital tools, their impact on environmental conservation, and the benefits they bring to both hotel operators and guests (Nam et al., 2021). It also examines challenges in implementing these technologies and potential strategies for maximizing their effectiveness in the hospitality sector (Kalsi et al., 2023).

The hospitality industry is under increasing pressure to adopt sustainable practices due to growing environmental concerns and changing consumer preferences. Housekeeping, as one of the most resource-intensive departments in a hotel, plays a pivotal role in achieving sustainability goals (Chen et al., 20217). Traditional housekeeping operations often involve high consumption of water, energy, and chemical-based cleaning agents, leading to significant environmental impact. To address these challenges, hotels are integrating smart technology into their housekeeping operations to enhance efficiency while minimizing ecological harm (Ibrahim et al., 2024).

Smart technology offers innovative solutions such as Internet of Things (IoT)-enabled sensors,

https://doi.org/10.52202/081568-0024

automated energy management systems, AI-driven housekeeping scheduling, and eco-friendly cleaning robots (Poyyamozhi et al., 2024). These advancements not only improve operational efficiency but also help hotels reduce waste, lower costs, and provide a healthier environment for guests and staff. For instance, IoT-enabled occupancy sensors can optimize lighting and HVAC systems, while smart water meters help monitor and control water usage more effectively (Srivastava & Tyagi, 2024).

This review paper aims to explore the integration of smart technology in hotel housekeeping to promote eco-friendly practices. It examines the latest technological advancements, their impact on environmental conservation, and the operational benefits they provide. Additionally, it discusses the challenges faced in implementing these solutions and suggests strategies to maximize their effectiveness in achieving sustainability goals in the hospitality sector.

Objectives of the study:

1. To assess the role of smart technology in sustainable housekeeping.

2. To identify challenges and best practices in eco-friendly housekeeping.

## 2. Literature Review:

The integration of eco-friendly practices in housekeeping through smart technology has been extensively studied, emphasizing sustainability, operational efficiency, and guest satisfaction. This section reviews existing literature on smart technology's role in reducing resource consumption, improving housekeeping operations, and addressing implementation challenges (Zhichao & Yashu, 2024) .

### 2.1 Sustainable Housekeeping Practices in the Hospitality Industry

Traditional housekeeping operations significantly contribute to high resource consumption, particularly in water, energy, and chemical cleaning agents. To mitigate this environmental impact, hotels have implemented sustainable housekeeping measures, such as water-saving linen reuse programs, biodegradable cleaning products, and waste reduction initiatives integrating eco-friendly practices not only reduces a hotel's carbon footprint but also enhances brand reputation and attracts environmentally conscious travellers (Chung & Tan, 2025).

### 2.2 Role of Smart Technology in Eco-Friendly Housekeeping

Advancements in smart technology have introduced innovative solutions for energy and resource management in housekeeping. Internet of Things (IoT)-enabled sensors, artificial intelligence (AI)-driven housekeeping schedules, and automated inventory tracking systems have transformed hotel operations (Bajwa et al., 2012). It indicates that smart energy management systems, such as occupancy-based lighting and HVAC controls, significantly reduce electricity consumption in hotels. Additionally, IoT-based water monitoring systems contribute to efficient water management, reducing overall usage and preventing wastage (Mcginn et al., 2022).

### 2.3 Smart Cleaning Equipment and Robotics


https://doi.org/10.52202/081568-0024

The adoption of AI-powered cleaning robots and smart cleaning devices has gained traction in the hospitality industry and robotic vacuum cleaners and UV-C disinfection robots not only improve cleaning efficiency but also reduce dependency on chemical-based cleaning solutions (Jayanti, 2024). Automated housekeeping management software further enhances operational efficiency by optimizing cleaning schedules, minimizing labour-intensive tasks, and ensuring sustainable resource utilization (Kalefia & Gado, 2024).

### 2.4 Impact on Operational Efficiency and Cost Reduction

Studies suggest that implementing smart technology in housekeeping leads to long-term cost savings and improved operational efficiency. Automated housekeeping solutions reduce labour costs, enhance staff productivity, and minimize resource wastage. Research found that hotels integrating smart technology experience a 20–30% reduction in energy costs, contributing to economic and environmental sustainability (Oduro, 2024).

## 3. Challenges and Barriers to Implementation

Despite its benefits, adopting smart technology in housekeeping presents challenges such as high initial investment costs, staff resistance, and technical integration issues. To overcome these challenges, training programs and awareness initiatives are essential for facilitating the transition to technology-driven housekeeping. Moreover, data privacy and cybersecurity risks associated with IoT-based housekeeping solutions require careful consideration

**Facts & Figures**

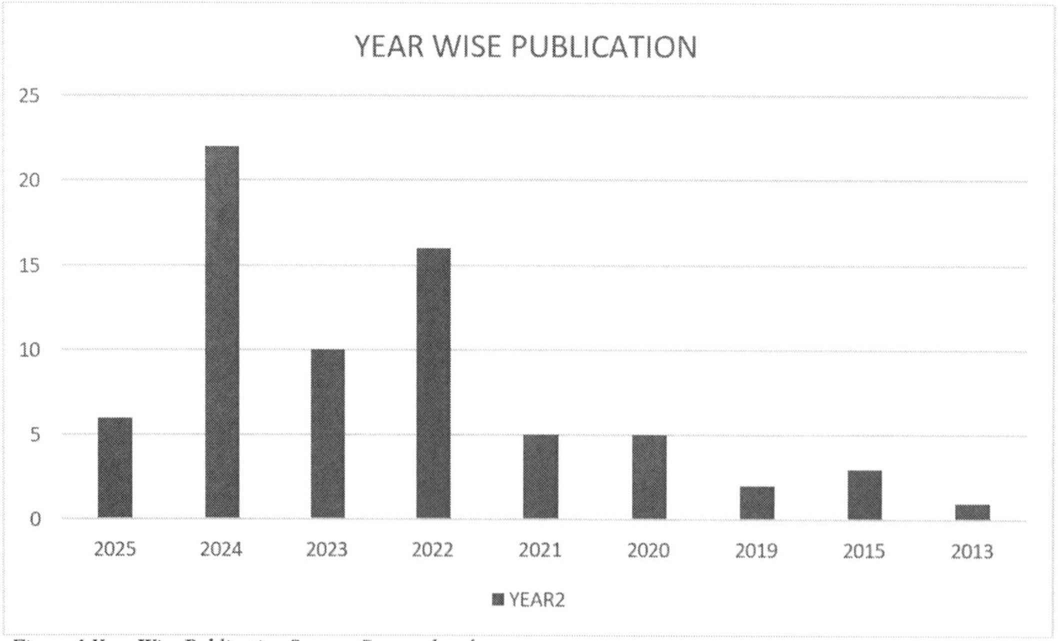


*Figure 1 Year Wise Publication Source: Scopus data base*

The bar chart titled "YEAR WISE PUBLICATION" illustrates the number of publications across different years. The x-axis represents the years, ranging from 2013 to 2025, while the y-axis represents the number of publications. The data shows a fluctuating trend in the number of publications, with the highest peak in 2024, reaching over 20 publications. The second-highest publication count is in 2022, followed by 2023. The number of publications in 2025 is lower than in the previous years but still noticeable. In contrast, years like 2013, 2015, 2019, 2020, and 2021 have relatively fewer publications. The chart highlights an overall increase in publications in recent years, with a significant surge in 2024.

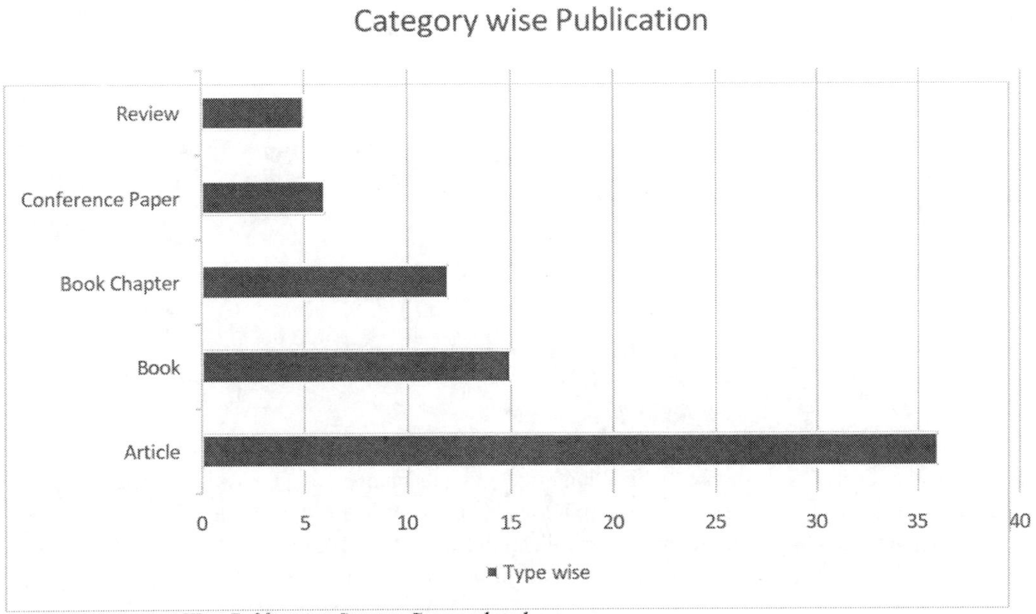


*Figure 2 Category Wise Publication Source: Scopus data base*

The horizontal bar chart titled "Category wise Publication" presents the distribution of publications across different categories. The x-axis represents the number of publications, while the y-axis lists the categories, which include Review, Conference Paper, Book Chapter, Book, and Article. Among these categories, Articles have the highest number of publications, significantly exceeding 35. Books and Book Chapters follow, with a moderate number of publications, while Conference Papers and Reviews have the least number of publications. The chart highlights that Articles are the most prominent type of publication, whereas Reviews and Conference Papers contribute the least to the total publications.


https://doi.org/10.52202/081568-0024

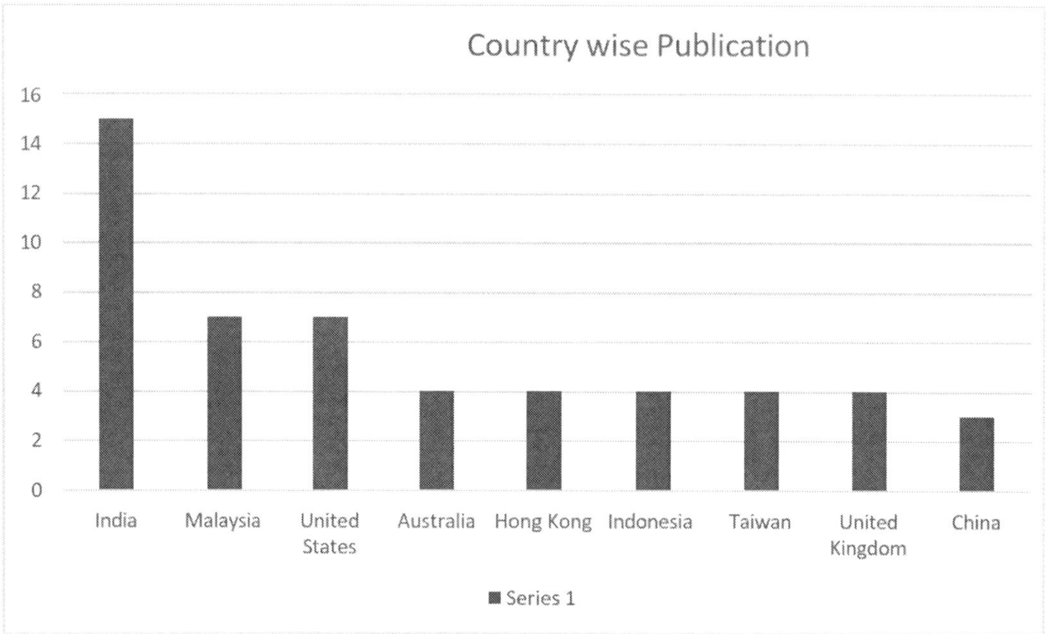

*Figure 3 Country wise Publication Source: Scopus data base*

The bar chart titled "Country wise Publication" displays the number of publications across different countries. The x-axis represents the countries, while the y-axis indicates the number of publications. India has the highest number of publications, exceeding 14, making it the leading contributor. Malaysia and the United States follow with an equal number of publications, both around 7. Australia, Hong Kong, Indonesia, Taiwan, and the United Kingdom each have a similar number of publications, approximately 4. China has the least number of publications, with a count slightly above 2. This chart highlights that India plays a dominant role in publications, whereas other countries contribute in varying but relatively lower proportions.

https://doi.org/10.52202/081568-0024

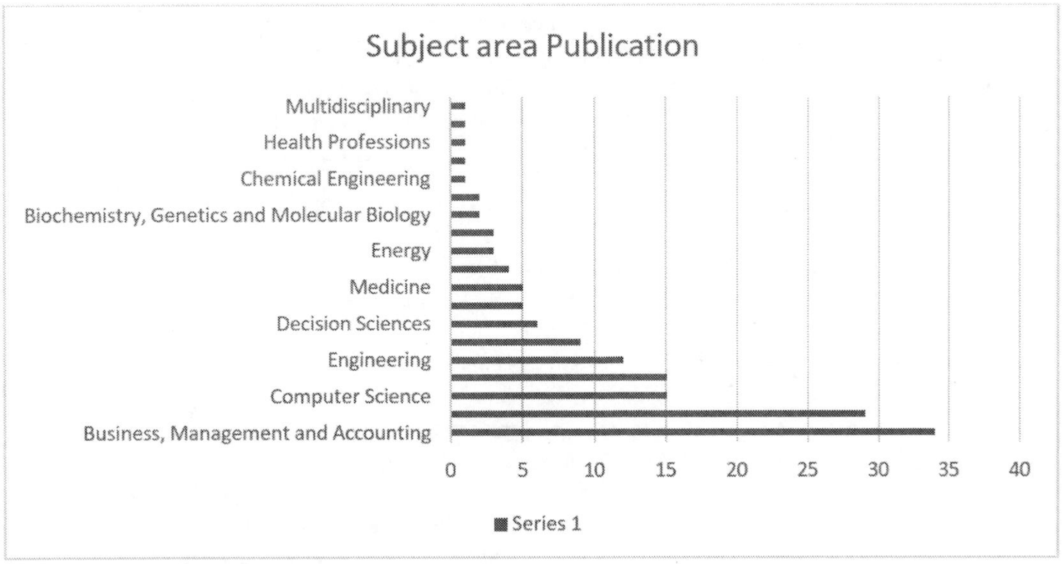


*Figure 4 Subject area wise Publication Source: Scopus data base*

The distribution of research publications by subject area highlights a strong emphasis on business, technology, and engineering fields. Business, Management, and Accounting have the highest number of publications, with around 35 papers, followed by Computer Science, which contributes approximately 20 papers. Engineering ranks third with around 15 publications. Moderate research activity is observed in Decision Sciences, Medicine, and Energy, each having a publication count ranging between 5 to 8. In contrast, subjects such as Biochemistry, Genetics, and Molecular Biology, Chemical Engineering, Health Professions, and Multidisciplinary studies have the lowest number of research papers. This distribution indicates a predominant focus on business, management, and technological advancements, while biological sciences and health-related fields receive comparatively less research attention.

## 4. Conclusion

The integration of smart technology in housekeeping has proven to be a significant step toward achieving sustainability in the hospitality industry. This review highlights how IoT-enabled devices, AI-driven housekeeping management, and automated energy and resource optimization contribute to eco-friendly practices. The adoption of smart technology helps reduce water and energy consumption, minimize waste, and improve operational efficiency while maintaining high service standards.

Despite its numerous benefits, implementing smart housekeeping solutions comes with challenges such as high initial investment costs, staff resistance, and cybersecurity concerns. However, with proper training, strategic planning, and technological advancements, these challenges can be mitigated. The findings suggest that adopting smart housekeeping technologies not only enhances sustainability efforts but also improves cost-effectiveness and guest satisfaction.


https://doi.org/10.52202/081568-0024

As the hospitality industry continues to evolve, integrating smart technology into housekeeping operations will play a crucial role in achieving long-term environmental and economic sustainability. Future research should explore innovative solutions for overcoming implementation barriers and further optimizing resource management for greener hotel operations.

## 5. Findings & Suggestions:

The findings of the study highlight that the integration of smart technology in hotel housekeeping significantly contributes to sustainability by optimizing resource usage, reducing energy and water consumption, and minimizing waste. IoT-enabled devices, AI-driven housekeeping management, and automated energy optimization systems were found to improve operational efficiency while maintaining high service standards. The adoption of these technologies not only supports eco-friendly practices but also leads to long-term cost savings, with some hotels experiencing reductions of up to 30% in energy costs. However, the study also identifies challenges in implementation, including high initial investment costs, staff resistance to change, and cybersecurity concerns. Despite these barriers, the findings suggest that with proper staff training, strategic planning, and technological advancements, these challenges can be mitigated, allowing hotels to reap the environmental and economic benefits of smart technology. Overall, the study concludes that adopting smart housekeeping technologies is essential for achieving both sustainability and improved guest satisfaction in the evolving hospitality industry.

To enhance the review paper, it would be beneficial to narrow the focus to a few key smart technologies that have the most significant impact on eco-friendly housekeeping, such as IoT-enabled sensors and AI-powered scheduling. Including real-world case studies from various hotels would provide practical insights into the scalability and effectiveness of these technologies. Additionally, incorporating a detailed cost-benefit analysis, exploring staff training and overcoming resistance, and expanding on guest perceptions of these technologies would offer a more comprehensive view. Further, discussing the role of smart waste management, future innovations like AI-driven predictive maintenance, and how smart technology helps hotels achieve sustainability certifications would strengthen the paper's relevance and actionable value. Lastly, integrating more quantitative data on energy savings, water reduction, and cost savings would provide a solid foundation for the claimed benefits of smart housekeeping technologies.

## References


1. Kontis, A.-P., & Skoultsos, S. (2018). Enhancing Hospitality Services Through the Engagement of Visitors in Local Gastronomy Experiences: A Marketing Perspective from the Supply-Side. In Katsoni, V and Velander, K (Ed.), *INNOVATIVE APPROACHES TO TOURISM AND LEISURE* (pp. 339–349).
2. https://doi.org/10.1007/978-3-319-67603-6_26
3. Raj, N., Sinha, A., Bharddwaj, S., & Yadav, A. L. (2024, May). AI-Powered Energy Consumption Optimization for Smart Homes Using IoT. In *2024 International Conference on Computational Intelligence and Computing Applications (ICCICA)* (Vol. 1, pp. 231-236). IEEE.

4. Mandal, S., Raju, S. S., & Yadav, N. (2025). Sustainability Through Technology: How Innovations Are Driving Eco-Friendly Hospitality Practices. In *Marketing Technology-Infused Hospitality: Upskilling Frontline Employees for Competitiveness* (pp. 309-336). IGI Global Scientific Publishing.
5. Nam, K., Dutt, C. S., Chathoth, P., Daghfous, A., & Khan, M. S. (2021). The adoption of artificial intelligence and robotics in the hotel industry: prospects and challenges. *Electronic Markets*, *31*, 553-574.
6. Kalsi, N., Carroll, F., Minor, K., & Platts, J. (2023, April). Smart Hospitality: Understanding the 'Green'Challenges of Hotels and How IoT-Based Sustainable Development Could be the Answer. In *International Conference on Frontiers of Intelligent Computing: Theory and Applications* (pp. 73-83). Singapore: Springer Nature Singapore.
7. Chan, E. S., Okumus, F., & Chan, W. (2017). The applications of environmental technologies in hotels. *Journal of Hospitality Marketing & Management*, *26*(1), 23-47.
8. Ibrahim, B. S. K. K., Sutcu, M., & Al Momany, A. (2024). Advancing Sustainable Development Through the Integration of AI, IoT, and Robotics. *Evolution of Information, Communication and Computing System*, 19-49.
9. Poyyamozhi, M., Murugesan, B., Rajamanickam, N., Shorfuzzaman, M., & Aboelmagd, Y. (2024). IoT—A Promising Solution to Energy Management in Smart Buildings: A Systematic Review, Applications, Barriers, and Future Scope. *Buildings*, *14*(11), 3446.
10. Srivastava, P., & Tyagi, P. K. (2024). Smart Solutions in Hospitality: Enhancing Housekeeping Through Advanced Technologies. In *Hotel and Travel Management in the AI Era* (pp. 467-484). IGI Global.
11. Zhichao, Y., & Yashu, Y. (2024). Principles of Green Eco-Friendly Tourism Towards Promoting Environmental Sustainability. *MARKAZIY OSIYO MADANIY ME'ROSI VA TURIZM TENDENSIYALARI JURNALI (ISSN: 3060-4834)*, *1*(2), 18-28.
12. Chung, K. C., & Tan, P. J. B. (2025). Artificial intelligence and internet of things to improve smart hospitality services. *Internet of Things*, *31*, 101544.
13. Bajwa, A., Jahan, F., Ahmed, I., & Siddiqui, N. A. (2024). A SYSTEMATIC LITERATURE REVIEW ON AI-ENABLED SMART BUILDING MANAGEMENT SYSTEMS FOR ENERGY EFFICIENCY AND SUSTAINABILITY. *American Journal of Scholarly Research and Innovation*, *3*(02), 01-27.
14. McGinn, C., Scott, R., Donnelly, N., Roberts, K. L., Bogue, M., Kiernan, C., & Beckett, M. (2022). Exploring the Applicability of Robot-Assisted UV Disinfection in. *Applying Robotics and AI in Pandemics (COVID-19): Detection, Diagnosis and Delivery*.
15. Jayanti, J. (2023). Hotel housekeeping management: Changing trends and developments.
16. Kalefa, H., & Gado, S. (2024). Enhancing hotel sustainability through ecological and technological integration. *JES. Journal of Engineering Sciences*, *52*(1), 145-174.
17. Oduro, A. R. (2024). Examining the technology adoption in accommodation management, a case study of gift guest hotel in koforidua.

 https://doi.org/10.52202/081568-0024

# Emotion Analysis Using Machine Learning Classifiers in Tourism and Management: A Comparative Study Utilizing Kaggle Dataset and MATLAB

Harshita Mishra[1]*, Ashwani Kumar[2], Parveen Kumar Saini[3]

[1, 2&3] Department of Computer Science and Engineering, Chandigarh University, Mohali, India

* Corresponding author- mishraharshita0412@ gmail.com

**Abstract.** A crucial component in comprehending human – computer interaction is emotion analysis, which helps create systems that are more responsive and intuitive. In this works, we investigate several classifiers for efficient textual data- based emotion identification. Several classifiers are examined in the study such as decision trees, linear quadratic, Support vector machines, Confusion Matrix, and SVMs. Accurate classification of emotional states by each classifier was assessed based on its performance. Our results demonstrate notable variations in the classifiers accuracy and computational efficiency, with certain SVM configurations performing better than others. The studies findings emphasize the potential of advanced machine learning approaches to improve human-computer interaction and further the field's continuous achievements in emotion monitoring. The knowledge acquired may be very helpful for future studies and real-world applications in variety of fields, including social media analytics, healthcare, and customer service.

**Keywords** – Emotions, Machine Learning, MATLAB, Classifiers.

## 1. Introduction

Image processing may enhance photos and extract useful data. A photograph may be effectively converted into a digital format for additional processing using this procedure. This method use a 2D image as input, with pixel values ranging from 0 to 255, much like signal processing. It involves converting an image into a two-dimensional matrix. The process entails transforming a picture to a 2D matrix (Zang et al., 2020; Sharma et al., 2021).

There are basic phases to it :-

Examining the image: An image scanned by a device produces a raw image that has to be processed. Pixels can be used to represent it, as it was previously said.

The goal of this stage is to extract computer-friendly data.

Processing and Enhancement: By sampling and quantizing the input signals, a digitizer converts the image into digital format. In order to achieve optimal quantization and resolution for the human sense of different hues using different grayscales, a high sample rate is required (Sharma et al., 2021)

The output that is produced further classifies the picture and describes its characteristics.

https://doi.org/10.52202/081568-0025 

**Conversion of a Color Image into a Grayscale:**

There are basically two techniques to convert a color image to a grayscale one.

Average Method :- The three hues that make up a color are Red, Blue, and Green. Picture are averaged in the average technique.

Hence, Grayscale= (R+G+B)/3 is obtained. Nevertheless, sometimes the grayscale image is replaced with a black picture. This is because the altered image contains 33% of the original Red, Blue, and Green (Bandyopadhayay et al., 2012)

So, to answer this riddle, we use the second strategy, which is also referred to as the Weighted Method or Luminosity Method.

Weighted or Luminosity Method :- To solve the Average approach problem, we use the Luminosity technique. The blue hue has a proportion between these two colors, and these techniques boost the green color and diminish the presence of the red color (Guo, 2022; en et al., 2014)

Consequently, using the equation:- Grayscale=((0.3 * R) + (0.59 * G) + (0.11 * B)). The wavelength patterns of these hues are why we employ this.

Red has the greatest wavelength, and blue has the smallest.

1. **Evaluation of the Literature**

OpenCV is an abbreviation for Open Computer Vision Library. Everyone has free access to this extensive library, which contains more than 2500 algorithms designed specifically to tackle computer vision and machine learning applications (Maruf et al., 2022). These algorithms may be used to perform a wide range of tasks, such as Face Recognition, Object Identification, Camera Movement Tracking, and Scenery Recognition. It's a large community, with at least 47,000 people who actively contribute to the library (Houssein et al., 2020). Numerous public and private businesses utilize it, GPU Acceleration is one of the new features added to the current libraries. This new capability, albeit not yet completely developed, can handle most jobs. The GPU may benefit from several libraries, such as NPP, or NVIDIA performance primitives, because it is powered by CUDA. (Bota et al., 2019; Yang et al., 2020).

The advantage of GPU technology is that it can be utilized by anybody without a deep grasp of GPU programming. Rather than being able to instantly change a picture's properties in the GPU Module, we must first copy the original image and then make modifications to it (Alhalaseh & Alasasfeh, 2020)

IV. The steps for installing Python 2.7 and the required packages are as follows:

Let's begin by using an example image in the.jpg or.png format, and then use an image processing approach to extract the mood of the topic. (Every living being from which sentiments may be inferred is referred to as the "Subject").

**1.1 Bringing Libraries in**

To properly implement this project, you will need to download and install the NumPy, Glob, Random, and Python 2.7.x packages. The C disk is where Python installation by default takes place. Open the Python IDLE application, import each package, and begin working.

**1.2 Numpy**

One Python library used for complicated technical assessment is called NumPy.


https://doi.org/10.52202/081568-0025

Multidimensional arrays, which need the processing of several mathematical formulae, are implemented using it (Sharma et al., 2021)

The dimension of an array stated in a program is known as its axis.

Rank is the quantity of axes that are present in an array. For instance A= [1,2,3,4,5]

**1.3 Glob**

The Globo module recognizes the pattern and creates a file in relation to it based on the Unix Shell rules. The entire route name is generated.

**1.3.1 Wildcards**

Several actions may be carried out on files or a portion of a directory using these wildcards. There exist several functional wildcards , but only two of them are very helpful: -

Various files created in dictionaries

a) Asterisk (*): denotes any combination of any number of characters.

For instance.

import glob from glob.glob('direc/file*') for name print name; output

path/filename1 path/filename2 path/filename3 path/filename4 path/filename5 path/files

a) Question Mark (?): This indicates or points to the one missing character. The import of glob for name occurs in glob.glob('direc/filename?').

print name; output path / filename1 / filename2 / filename3/ filename4 / filename5

This wildcard is restricted to usage in a single directory; it does not expand. that is, a file in a subdirectory cannot be found (Ain et al., 2017)

D. **Indeterminate Random**

The module chooses an element, or random number, from a predefined collection of things. Certain features that provide access to these sorts of activities are compatible with this module.

E. **Indeterminate Random**

The module chooses an element, or random number, from a predefined collection of things. Certain features that provide access to these sorts of activities are compatible with this module.

**Categorization of Random Module: -**

First, randint(m,n) yields a value for x such that m<=x<=n Second, randrange(dog, cat, mouse, lion) Returns any element or random variable from the specified range (Doma & Pirouz, 2020)

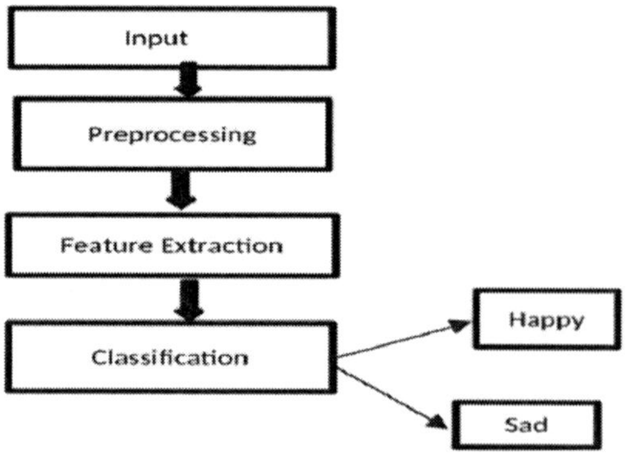


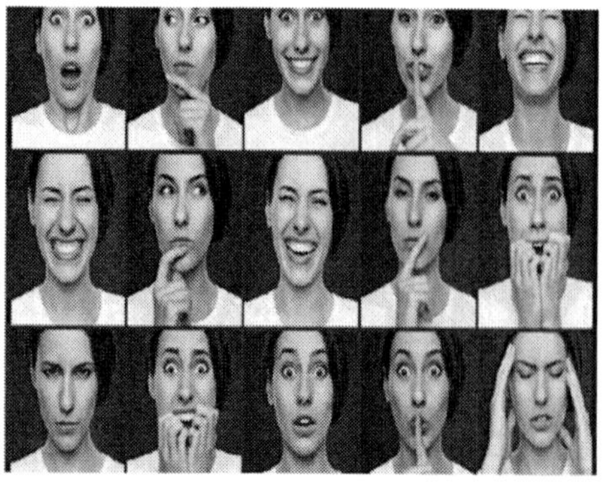

II. V. Various emotions that can be discerned from an image include:

A. Indifferent
B. Joyful
C. Furious
D. Dismay
E. Astonishment
F. Disgust
G. Sadness
H. Fear

III. VI. Procedures for utilizing OpenCV-Python for emotion detection:

2. The first thing we need to do is create a dataset after installing all required tools. In order to ensure accuracy in our results and adequate data for information extraction, we may generate our own dataset by evaluating a series of photographs. An already-existing database can also be used.
3. Following that, the dataset is arranged into two distinct folders. Every picture will be in the first directory, and every piece of information on the various emotions will be in the second directory.
4. All of the output photographs will be saved in a different directory and arranged according to emotions before being encoded, after the execution of the Python code on the example images.
5. While OpenCV has a variety of classes for emotion identification, we will mostly employ Fisher Face one.
6. Extracting Faces: OpenCV comes with four pre-built classifiers. We utilize these classifiers sequentially to extract as many faces as we can.
7. Two subsets of the dataset are extracted: a training set and a classification set.
8. The classifier performance is estimated using the classification set, and the kind of emotion is taught using the training set, which gathers data from several photographs.
9. For optimal effects, all of the photographs should have the exact identical attributes, including size.
10. After being converted to grayscale, the topic of each photograph is scrutinized, cropped, and saved to a directory.
11. Lastly, we generate a training set using 80% of the test data and classify on the classification set using the remaining 20%.
12. Repeat the process to boost efficiency (Casale et al., 2014).

**VII. Cohn-Kanade Database for AU-CODED Expressions:**

This is one database that may be used to identify different moods. It is a database containing a wide variety of emotions. This database is currently divided into two categories, and a third is in the process of being created ((Doma & Pirouz, 2020).

Although the second edition is called "CK+," the first edition is called "CK." The neutral sensation is detected by the CK version before it recognizes the other higher emotions. With the introduction of this database's second kind, the frequency of processes increased by around 25% and the frequency of subjects increased by approximately 30%.


https://doi.org/10.52202/081568-0025

The third edition, which will be launched soon, will have all of the functionality found in the CK and CK+ Databases, including integrated 30 degrees of rotation from the front (Bandyopadhayay et al., 2012).

## 2. UTILITIES AND FUTURE RANGE

The field of computer vision is still in its early stages of development. The pace of research in this subject is accelerating (Yan, 2022).
Computer vision is not complete without the ability to identify emotions (Bandyopadhayay et al., 2012).
If one can understand the complexities and final steps of a work, they can complete a lot of activities and procedures.
Several popular and extensively utilized uses of emotion detection include :

**2.1 Product and app development**

Testing a product's usability is one of the many software engineering procedures where emotion detection may be quite helpful in streamlining the process. It has long been known that a person's degree of comfort with various software programs greatly relies on their emotional state. Human emotions may be affected by a product's overall appearance and feel, which might influence whether or not a consumer chooses to purchase it. For those in connected industries, therefore, learning more about the many emotional states of the human body and how using certain goods affects them is crucial.

**2.2 Better methods of instruction**

Research indicates that although some emotional states support improved learning strategies, others attempt to repress them. It is uncommon to notice a distinction between the two groups' disparate emotional states. Positive emotional states, for instance, are considered detrimental to learning (Machova et al., 2023). whereas somewhat negative emotional states encourage critical thinking and are suitable for doing important activities.

**2.3 Self-developed websites**

Service providers are interested in gathering enormous amounts of data that may be taken from consumers due to the internet's massive expansion. Consequently, all of the information and adverts play according to the user's profile. Consequently, providing detailed information about the various human emotions can lead to the creation of far more accurate behavioral models representing various user types learning (Machova et al., 2023).

**2.4 Video games that are fully immersive**

The entertainment business includes a sizable portion of the video gaming sector. Thus, video game developers center their study on various human emotions that are frequently encountered in order to create these games much more intricate and invasive. Video games are designed with the intention of attracting an increasing number of players by seamlessly integrating human emotions into the gameplay.

Example: Using Emotion Analysis in Hotel Review Management:
Scenario: Thousands of reviews are submitted by online users to the five-star hotel chain "Luxury Stay Hotels" via TripAdvisor and Google Reviews. In order to improve guest experiences and service quality, the hotel administration plans to examine consumer sentiments.
Data Collection: The management obtains information from a Kaggle dataset that includes text reviews, rating assessments, and lodging reviews.
Machine Learning Approach: All reviews are subjected to token separation, punctuation stripping, and stopword removal prior to analysis.

TF-IDF (Term Frequency-Inverse Document Frequency) feature extraction transforms text elements into numerical data.
Models of Classification: Support Vector Machines are used by the system to identify emotive opinions in reviews that fall into one of three categories: Favorable, neutral, or negative, random Forest for further in-depth analysis of complaints about services (Zang et al., 2020)
Textual emotion probabilities are computed in the analysis process with the aid of a Naive Bayes method.
Findings and Perspectives: Guests gave a favorable review, stating that the staff was welcoming and the rooms were immaculate, and the setting offered beautiful views of the surrounding countryside.
A modest percentage of visitors (20%) complained about the standard level of food and service.
The delayed check-in process, along with issues with Wi-Fi access and the noisy surroundings, caused 15% of guests to voice unfavorable views.

-p+
Finding different emotions in a particular sample image is the main goal. The hardest aspect of this job was figuring out which emotion was which when two of them were quite similar learning (Machova et al., 2023), For example, "disgust" may be mistaken for "sadness," "surprise" for "happy," and so on. Now, across eight distinct categories, the accuracy was almost 70%, which is really good considering how much our classifier has learnt. Thus, we need to investigate ways to improve its accuracy and efficiency.
We can see from our list of emotions that we don't have many examples of the words "sad," "fear," and "contempt." We can definitely improve performance by increasing the number of photos for these emotions. Alternatively, if we remove these feelings from the list, optimization can be improved by more than 80%. The Cohn-Kanade (CK and CK+) Database was the dataset we used for this work.

## 3. CONCLUSION

This study highlights the effectiveness of various classifiers in textual emotion analysis, with SVMs demonstrating superior performance in both accuracy and computational efficiency. The findings underscore the importance of selecting appropriate machine learning models to enhance human-computer interaction. These insights contribute valuable guidance for future research and practical applications in domains such as social media monitoring, healthcare, and customer engagement. Ultimately, emotion-aware systems can lead to more intuitive and empathetic digital interactions.

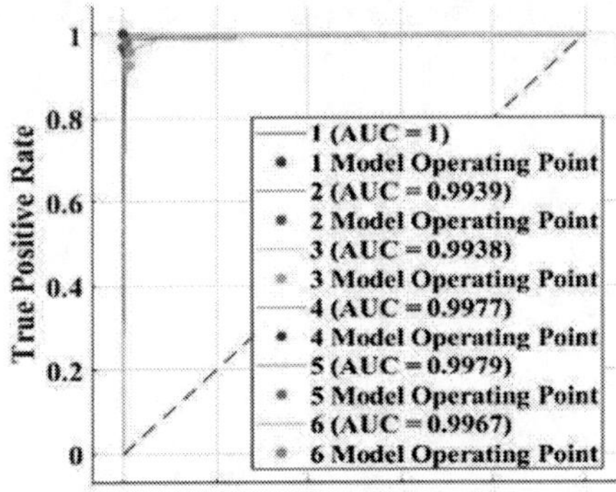


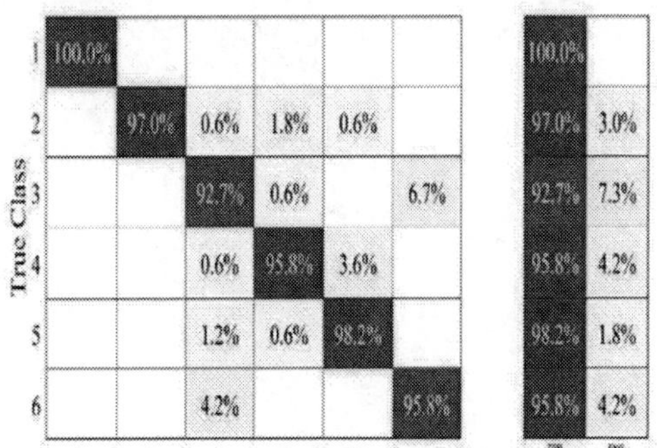



https://doi.org/10.52202/081568-0025

## REFERENCE


1. Zhang, J., Yin, Z., Chen, P., & Nichele, S. (2020). Emotion recognition using multi-modal data and machine learning techniques: A tutorial and review. Information fusion, 59, 103-126.
2. Sharma, T., Diwakar, M., Singh, P., Lamba, S., Kumar, P., & Joshi, K. (2021, November). Emotion Analysis for predicting the emotion labels using Machine Learning approaches. In 2021 IEEE 8th Uttar Pradesh section international conference on electrical, electronics and computer engineering (UPCON) (pp. 1-6). IEEE.
3. Al Maruf, A., Ziyad, Z. M., Haque, M. M., & Khanam, F. (2022). Emotion detection from text and sentiment analysis of Ukraine Russia war using machine learning technique. International Journal of Advanced Computer Science and Applications (IJACSA), 13(12).
4. Bandyopadhayay, D. K., Kadam, V., Aggarwal, K. K., Bhoite, U., Surendran, K., Ganesan, S., ... & Jain, M. V. (2012). Bharati Vidyapeeth's Institute of Computer Applications and Management (BVICAM).
5. Guo, J. (2022). Deep learning approach to text analysis for human emotion detection from big data. Journal of Intelligent Systems, 31(1), 113-126.
6. Yan, Q. (2022). Real-Time Analysis of Youth Emotion Based on Python Language and Smart Sensor Network. Mobile Information Systems, 2022(1), 8635787.
7. Houssein, E. H., Hammad, A., & Ali, A. A. (2022). Human emotion recognition from EEG-based brain–computer interface using machine learning: a comprehensive review. Neural Computing and Applications, 34(15), 12527-12557.
8. Bota, P. J., Wang, C., Fred, A. L., & Da Silva, H. P. (2019). A review, current challenges, and future possibilities on emotion recognition using machine learning and physiological signals. IEEE access, 7, 140990-141020.
9. Yang, L., Li, Y., Wang, J., & Sherratt, R. S. (2020). Sentiment analysis for E-commerce product reviews in Chinese based on sentiment lexicon and deep learning. IEEE access, 8, 23522-23530.
10. Alhalaseh, R., & Alasasfeh, S. (2020). Machine-learning-based emotion recognition system using EEG signals. Computers, 9(4), 95.
11. Ain, Q. T., Ali, M., Riaz, A., Noureen, A., Kamran, M., Hayat, B., & Rehman, A. (2017). Sentiment analysis using deep learning techniques: a review. International Journal of Advanced Computer Science and Applications, 8(6).
12. Casale, S., Russo, A., Scebba, G., & Serrano, S. (2008, August). Speech emotion classification using machine learning algorithms. In 2008 IEEE international conference on semantic computing (pp. 158-165). IEEE.
13. Doma, V., & Pirouz, M. (2020). A comparative analysis of machine learning methods for emotion recognition using EEG and peripheral physiological signals. Journal of Big Data, 7(1), 18.
14. Machová, K., Szabóova, M., Paralič, J., & Mičko, J. (2023). Detection of emotion by text analysis using machine learning. Frontiers in Psychology, 14, 1190326.

# Empowering Women Entrepreneurs in Rural Tourism: Challenges, Opportunities and the Role of Women-Owned Home Stays in Himachal Pradesh

Dr. Arun Singh Thakur[1], Akanksha Bhandari[2]

[1] Assistant Professor, University Institute of Hotel and Tourism Management
[2] Research Scholar, Department cum Centre for Women's Studies & Development Panjab University, Chandigarh

* Corresponding author: Andybhandari07@icloud.com

**Abstract**- Women entrepreneurs play a very important role in the promotion of rural tourism and economic development through preservation of culture and community tourism. Women owned homestay facilities serve as the best example in this regard as they enable women to achieve financial autonomy while exposing visitors to genuine local culture. However, female entrepreneurs in rural tourism face deep seated problems of limited access to capital, lack of digital marketing skills, socio cultural barriers, and policy bottlenecks. This research investigates the opportunities and challenges of women owned homestay establishments in Himachal Pradesh and their role on rural development together with women empowerment. To understand the existing phenomena and issues on rural tourism, the study employed secondary sources like government documents, academic research, policy papers, and illustrative case studies. Women entrepreneurs are making notable contributions to rural economies, but they bear the brunt of financial problems at the micro level, coupled with ignorance of policies and indicators of illiteracy on the macro level. Nevertheless, there are prospects of positive change through government programs, the digital world, and boosting ecotourism. The conclusions arrived at is that there is a great need for policy action such as investment in financially supported programs, and improving digital literacy..

**Keywords:** Economic Empowerment,Women Entrepreneurship, Homestay Business, Rural Tourism, Sustainable Tourism,Eco-tourism,Women in Tourism Industry..

## 1 Introduction

Tourism in the countryside has proved to be helpful in providing economic development, preserving culture, and enhancing community development. It enables tourists to interact with the people's customs while directing returns to the rural economies. Currently, Himachal Pradesh boasts as one of the best in India for rural tourism owing to its mesmerizing scenic beauty, rich cultural history, and warm host families. Women have become more prominent in the tourism industry by managing home stays. While engaging the tourists, these women not only provide them an authentic cultural experience but also gain economic as well as social empowerment. Such initiatives go a long way in promoting sustainable tourism, women empowerment, and job opportunities in the society which is largely patriarchal. On the other hand, women entrepreneurs within rural tourism still have to contend with several barriers that curb the growth of their enterprises. Among them are lack of capital, poor digital marketing skills, rigid cultural norms, and adverse legal frameworks. Solving these problems will help harness the full potential of women in tourism. This study will look into these problems, the possibilities, and design supportive measures to guide women's entrepreneurship in rural tourism.

### 1.1 Statement of Research Problem

While women's roles in rural tourism are invaluable, they still face numerous barriers to achieving their full potential in the sector. Some identified barriers are:

- Limited funding – The majority of women are unable to acquire the necessary funding to improve infrastructure, market their services, and provide hospitality to guests.
- Lack of IT skills – Many women are not able to use the internet and other technology for marketing purposes, which are becoming essential in tourism.
- Societal norms – Women, in most cases, are expected to refrain from actively running and developing their businesses due to gender roles.
- Governmental issues – Women entrepreneurs face problems of licencing and a general lack of information on available governmental aid programs.

All these issues and many others will be covered in the research in an attempt to provide solutions that would support homestay development in Himachal Pradesh. The study will prepare policy recommendations to assist in actively empowering women in rural tourism, based on an analysis of


https://doi.org/10.52202/081568-0026

governmental policies, the industry's evolution, and actual cases.

**Objectives**

The goals that we are aiming to achieve in this project are:

- Assess the role of women entrepreneurs in the rural tourism industry and their economic and social impact on the Himachal Pradesh region.
- Identify the key barriers to women-owned homestays, such as financial, technical, or sociocultural issues.
- To examine the role of government support, digital tools, and evolving tourism trends in fostering women's entrepreneurship.

To achieve these objectives, the study will address the following research questions:

- How do women entrepreneurs contribute to rural tourism in Himachal Pradesh?
- What are the primary challenges faced by women in managing and expanding homestay?
- How can policy initiatives and digital platforms support the growth of women-led homestay?
- What strategic interventions can promote gender-inclusive and sustainable rural tourism?

## 1.2 Scope and Limitations

This research focuses on women entrepreneurs managing homestay in the rural regions of Himachal Pradesh. It analyses their contributions to the local economy, the difficulties they face, and potential solutions to support their businesses. The study is based on secondary data sources, including government reports, academic research, policy documents, and case studies.

However, the study has certain limitations:

Limited primary data – Since the research relies on secondary sources, it does not include direct interviews or firsthand insights from women entrepreneurs.

- Geographical focus – The findings are specific to Himachal Pradesh and may not fully apply to other regions with different socio-economic conditions.
- Data availability – As tourism trends and policies continue to evolve, some information may not reflect the most recent developments.
- Despite these limitations, this study provides valuable insights into the role of women in rural tourism and presents policy recommendations to enhance their participation and success in the sector.

## 2. Literature Review

Women's roles as operators of homestay and small-scale enterprises within rural tourism have been extensively explored within the tourism and entrepreneurship fields. Research identifies the economic, cultural, and social contributions made by women entrepreneurs as well as the primary constraints they face in the industry.

**2.1 Contributions of Women to Rural Tourism**

- Economic Empowerment: Sharma & Bhatt (2022) and Verma (2021) argue that women-operated homestays serves as a reliable income generator, easing the reliance on staple agricultural activities. Economic tourism increases women's self-sufficiency in the economy as well as improves the family's income.
- Cultural Conservation: Devi & Singh (2020) illustrate the impact of women entrepreneurs using homestay facilities in rural areas and how they aid in sustaining and cherishing local cultures, food, crafts, and traditions, hence enriching tourists' experiences.

Community Engagement: Kumari et al. (2019) argue that community-based tourism where female entrepreneurs are involved enhances social cohesion and community participation in the tourism sector through guiding, dancing, and cooking.

**2.2 Challenges Faced by Women Entrepreneurs in Rural Tourism**

Even with their contributions, women entrepreneurs especially in this industry meet a number of challenges:

- Financial Limitations: According to Das & Mukherjee (2021), women underdeveloped businesses are attributed to their low levels of financial literacy and lack of access to credit, forcing them to depend on informal funding or Microfinance institutions.
- Digital and Promotion Limitations: Rai & Thakur (2020) research indicates that the absence of requisite knowledge and resources for online marketing limits the visibilityof businesses owned and managed by women in the tourism sector.
- Socio-Cultural Limitations: Gupta & Verma (2020) suggest that women's mobility and decision making in patriarchal rural areas is highly controlled, which impedes their ability to expand their businesses.

Institutional and Physical Environment Barriers: Mehta (2022) comments on the licensing delays, the unattractive state of the roads, and lack of adequate facilities as barriers to the growth of rural tourism business ventures.

**2.3 Government Support and Policy Initiatives**

The Deen Dayal Upadhyaya Grameen Kaushalya Yojana and the Himachal Pradesh Homestay Scheme are aimed towards rural tourism and the economic empowerment of women. Sharma (2023) claims that these programs provide funding, vocational training, and other forms of substantive assistance, which many women still do not know about.


https://doi.org/10.52202/081568-0026

Jain & Mishra (2021) emphasize the need for enhanced outreach strategies so that a larger number of women can partake in these programs.

Theoretical Perspectives and Major Findings from the Literature

- Chopra (2020) extends this theory to rural tourism by showing how women's homestay businesses defy stereotypes by enabling greater economic autonomy for women.
- Sustainable Livelihoods Approach (SLA) This theory first proposed by Chambers and Conway (1992), posits that there exist multiple pathways by which economic development can be made more sustainable in rural areas by combining agriculture and tourism with handicraft production.
- Kaur (2023) and Kumar (2022) independently demonstrate how homestay accommodation providers practice sustainable tourism by utilizing local resources and fostering environmental protection.
- Social Capital Theory According to Putnam (2000), social capital refers to the resources that reside in networks, relations, and trust of a community, which are essential for business success as they enable innovations and increased productivity to flourish.
- Rao and Singh (2021) note that self-help groups (SHGs), communal networks, and tourism cooperatives are the major resources for rural women in tourism to meet resource needs for business growth and diversification.

**2.4 A review of existing research reveals the following key insights:**

- Women entrepreneurs significantly contribute to rural economies but continue to face financial and socio-cultural barriers.
- Government support programs exist but remain under-utilized due to low awareness and bureaucratic challenges.
- Limited digital literacy and marketing skills hinder the business growth of women in rural tourism.
- Community networks and social capital play a critical role in overcoming economic and social challenges.
- Sustainable tourism, particularly in eco-tourism and cultural tourism, offers new opportunities for women entrepreneurs.

**3. Research Gaps**

- There is limited research focusing specifically on women-led homestay in Himachal Pradesh, highlighting a need for region-specific studies.

- There is a lack of empirical data on the long-term impact of digital marketing and financial support initiatives on women entrepreneurs in rural tourism.

- •Further research is needed on how policy interventions can be made more accessible to women in remote areas.

Existing literature confirms that women entrepreneurs in rural tourism serve as key drivers of economic and social development. However, challenges such as financial constraints, digital illiteracy, and socio-cultural barriers continue to hinder their progress. By applying Feminist Entrepreneurship Theory, Social Capital Theory, and the Sustainable Livelihoods Approach, this study aims to provide a deeper understanding of the factors influencing women's participation in rural tourism.

Future research should focus on evaluating the effectiveness of government policies and identifying digital strategies that can empower women entrepreneurs in this sector.

## 4. Methodology

### 4.1 Research Approach

This study employs a qualitative secondary research approach to analyse the role of women entrepreneurs in rural tourism, focusing on women-owned homestay in Himachal Pradesh. By examining existing academic literature, government reports, and case studies, the research identifies challenges, opportunities, and policy impacts without requiring primary data collection.

**Data Collection**

Sources of Data

The research compiles data from reputable external sources such as:

- Academic Studies & Journals – Analysis and articles on women's entrepreneurship and sustainable tourism.

- Government Reports & Policies – Himachal Pradesh Tourism Board's reports, the Ministry of Tourism policies, and the effects of the Homestay Scheme.

- Industry Case Studies & Reports – Achievements of female homestay proprietors and reports on rural women in tourism by UNWTO and NITI Aayog

- News & Social Media Platforms – Insightful articles from the media and social marketing tools used by women owners.

**Data Analysis**

Thematic analysis method is adopted for the data synthesis by:

- Detecting Principal Themes - Dividing data into challenges, opportunities, policy interventions, and achievement narratives.


https://doi.org/10.52202/081568-0026

- Source of Information Triangulation - Considering different viewpoints from the available information.
- Identing Practical Recommendations - Formulating strategies for improving women participation in rural tourism.

**Justification for Secondary Research**

- Economically Efficient & Easily Obtained – Makes comprehensive analysis possible without primary data collection.
- Trustworthy & Broad – Gives information from established sources such as government publications and peer researched documents.
- Ethically Justifiable – No concerns on needing consent, bias, and privacy limits when upholding academic norms.

## 5. Limitations

Secondary research is also limited by its relative strengths:

- Absence of Primary Data – Fails to capture the voices of the women homestay proprietors.
- Potential Data Gaps – Outdated or non-Himachal studies are available.

Future studies should incorporate field research, interviews, and surveys for a more understanding. This study leverages secondary research to provide a comprehensive and credible analysis of women entrepreneurs in Himachal Pradesh's rural tourism sector. By synthesizing data from multiple sources, it contributes to policy discussions and strategic recommendations aimed at empowering women in the tourism industry.

## 6. Findings and Discussion

Key Insights

- Women-led homestay in Himachal Pradesh boost the local economy by creating jobs and promoting cultural tourism.
- These businesses offer stable incomes, reducing reliance on seasonal farming (Sharma & Bhatt, 2022).

**Challenges Faced by Women Entrepreneurs**

- Financial Barriers – Limited access to loans due to collateral issues and financial illiteracy (Das & Mukherjee, 2021). Government schemes exist but suffer from bureaucratic delays (Govt. of India, 2022).
- Marketing & Digital Literacy Gaps – Many women lack digital skills to promote their homestay on global platforms like Airbnb (Rai & Thakur, 2020; Kaur, 2023).

- Socio-Cultural Constraints – Gender norms restrict mobility and financial independence, forcing many to depend on male relatives (Gupta & Verma, 2020; Chopra, 2021).
- Polic & Infrastructure Issues - Complicated licensing and poor road connectivity hinder business sustainability (Mehta, 2022). Government support exists but lacks effective outreach (Sharma, 2023).

**7. Comparison with Previous Research**

Challenge Existing Research Current Study

- Financial Access Loan access is difficult due to collateral needs (Das & Mukherjee, 2021).
- Women lack awareness of financial aid (Govt. of India, 2022).
- Marketing Skills Digital presence is key for homestay success (Rai & Thakur, 2020).
- Many lack digital knowledge to use booking platforms (Kaur, 2023).
- Gender Barriers Family restrictions limit mobility (Gupta & Verma, 2020).
- Women depend on male relatives for business dealings (Chopra, 2021).
- Policy & Infrastructure Government programs exist but lack outreach (Sharma, 2023).
- Poor licensing procedures and infrastructure further hinder growth (Mehta, 2022).

**Policy & Practical Recommendations**

- Financial Inclusion – Interest-free microloans and simplified loan processes for women.
- Digital Training – Workshops on social media marketing and online booking tools.
- Community Support – Strengthening self-help groups and mentorship programs.
- Infrastructure Upgrades – Better roads, electricity, and a streamlined homestay registration system.

Women entrepreneurs are transforming rural tourism but face financial, digital, social, and policy hurdles. Addressing these issues through targeted interventions in funding, digital literacy, and infrastructure can unlock their full potential, fostering both economic independence and sustainable tourism in Himachal Pradesh.

## 8. Conclusion

To wrap it up, women-run home stays in Himachal Pradesh are one of the key women entrepreneurship structures in rural tourism that help economically empower the local inhabitants and also serve to protect the cultural heritage of the region. While enduring


https://doi.org/10.52202/081568-0026

significant obstacles, these women have shown remarkable innovation, creativity and a great deal of effort towards solving problems that stem from lack of capital, social gender discrimination, and poor infrastructure. The growth potential in this sector is massive owing to the growth in demand for authentic community-based tourism around the world. In addition to consolidating tourism, promotion of women in rural tourism brings social and economic justice to the state of Himachal Pradesh. Home stays bring the rural families with a sustainable source of income, while providing the tourists with great chances to exploit the local cultures in an unprecedented fashion. In addition, the locally owned establishments enable her clientele to preserve the surroundings along with local customs and traditions while granting women greater control, autonomy, and financial independence thus leading to women's empowerment. As a result, there is the need from governing bodies and decision makers to the formulation of a very well-structured strategy that aims at funding, building aid, and actively promoting women entrepreneurs in rural tourism.

## Refernces


1. Sharma, R. (2020). *Women entrepreneurship in rural India: A case study of Himachal Pradesh*. XYZ Publishing.
2. Bhatia, A. (2019). *Tourism and sustainable development in the Himalayas*. ABC Publishers.
3. Kumar, S., & Devi, P. (2021). Empowering women through rural entrepreneurship: A study of the homestay industry in Himachal Pradesh. *Journal of Entrepreneurship & Development, 15*(2), 45–60.
4. Singh, M. (2018). Homestay tourism in Himachal Pradesh: Opportunities and challenges for women entrepreneurs. *Tourism and Society Review, 12*(1), 78–95.
5. Ministry of Tourism, Government of India. (2022). *Himachal Pradesh homestay scheme and its impact on rural development.*
6. Himachal Pradesh Tourism Development Corporation (HPTDC). (2021).
7. *Women-led homestay businesses: Growth and future prospects.* National Sample Survey Office (NSSO). (2020). *Women entrepreneurship in tourism: A case of homestay operators in India*
8. Chopra, R. (2021). *Gendered livelihoods and rural tourism*. Journal of Tourism and Gender Studies, 8(2), 55–70.
9. Das, S., & Mukherjee, A. (2021). *Financial inclusion and women entrepreneurship in India.* Indian Journal of Finance and Economics, 15(1), 34–48.
10. Government of India. (2022). *Annual report of the Ministry of Tourism*. New Delhi: Ministry of Tourism.
11. Gupta, P., & Verma, S. (2020). *Social norms and women entrepreneurs in rural India.* Journal of Rural Development Studies, 9(3), 112–129.
12. Kaur, H. (2023). *Digital literacy among women in tourism: Challenges and prospects.* International Journal of Tourism and Development, 10(1), 25–39.
13. Mehta, R. (2022). *Infrastructure challenges in rural tourism: A policy perspective.* Tourism Policy Review, 7(1), 60–75.
14. Rai, A., & Thakur, M. (2020). *Barriers to digital entrepreneurship for rural women.* Asian Journal of Entrepreneurship, 5(2), 80–95.
15. Sharma, N. (2023). *Evaluating government support for women-led homestays.* Journal of Public Policy and Tourism, 12(1), 45–58.


https://doi.org/10.52202/081568-0026

# Impact of Digital Marketing on Facebook: A Study on EatSure Multi-Brand Kitchen

Praveen Saini [1], Kaushal Kumar[2], Jiya Bhagwani [3], Shoaib Ameen [4]
Mohd Jubair Alam[5]

[1,2,3,4&5] Department of Computer Science Engineering Chandigarh University, Mohali, India

*Corresponding author: parveen.e13339@cumail.in

**Abstract.** Advanced neural network technology, sustainability, and innovation are coming together to redefine industry paradigms in the quickly changing tourist and hospitality sectors. This study looks at how emerging AI techniques and self-learning neural networks are revolutionizing conventional tourism models by improving visitor experiences, encouraging regenerative practices, and streamlining resource management. Artificial intelligence, blockchain, and immersive technologies are examples of digital innovations that are not only transforming operational efficiency but also making it possible to implement carbon-neutral accomodation and the circular economy.The study also explores how regenerative trends and state-of-the-art AI advancements are interacting to transform the travel and hospitality industry. Smart technologies and data-driven insights are opening up previously unheard-of possibilities to improve sustainability, optimize processes, and improve visitor experiences, but they also present difficulties for regulatory governance and ethical supervision. This study highlights the critical need for strong frameworks that strike a balance between innovation and responsibility by incorporating a variety of case studies and empirical data, from community-led regenerative practices to smart tourism solutions. By doing this, it sets the stage for a time when sustainable practices and technology will combine to change industry horizons into resilient and inclusive growth models. Finally, the study highlights strategies to use AI developments to promote resilient, inclusive, and sustainable development in the travel and hospitality industry, arguing for a strategic synthesis of technology and regenerative tourism.

**Keywords:** Multi-brand Cloud Kitchen, Digital Marketing, Instagram.

## 1 Introduction

**Transforming Horizons with AI Innovations and Regenerative Trends in the Travel and Hospitality Sector.**

In an era where the pulse of digital innovation meets the organic allure of nature, the travel and hospitality sectors are undergoing a revolution that is as dangerously sexy as it is profoundly transformative. Picture a world where seductive neural networks and audacious AI techniques fuse with the timeless art of sustainable design—a world where every byte of data and every innovative algorithm is engineered with the precision of a graduate-level software maestro, igniting a spectacular metamorphosis of conventional tourism paradigms.

At the core of this captivating revolution lie advanced self-learning architectures and transformative models that challenge traditional boundaries. Cutting-edge transformer algorithms and sophisticated capsule networks not only decipher complex consumer behavior in real time but also orchestrate a

 https://doi.org/10.52202/081568-0027

dazzling symphony of operational excellence and resource optimization. This high-octane convergence of smart technology and eco-conscious strategies is not merely about efficiency—it's about crafting experiences that tantalize the senses and leave an indelible mark on every traveler's journey.

Imagine blockchain frameworks intertwining with immersive virtual realities, creating a digital tapestry that is both provocatively dynamic and impeccably sustainable. In this brave new world, carbon-neutral accommodations and circular economy models become the norm, painting a picture of a future where ethical stewardship and innovative technology engage in an electrifying dance (Bengoi, 2009). This isn't just progress it's a daring, sensual reimagining of travel, where every interaction is charged with the energy of a groundbreaking, eco-chic renaissance.

This study delves deep into the enthralling intersection of AI innovations and regenerative trends, drawing upon empirical data, real-world case studies, and evocative digital narratives. By dissecting the interplay between disruptive algorithms and regenerative design, we reveal how smart, data-driven insights are unlocking unprecedented possibilities for both operational agility and enchanting visitor experiences. Yet, as the allure of these technological marvels seduces industries into new frontiers, it also poses bold challenges for regulatory governance and ethical oversight—challenges that demand a nuanced, multidisciplinary response.

Ultimately, the fusion of advanced neural network technology with regenerative tourism principles is redefining the future of travel and hospitality. It is a journey marked by exhilarating innovation, where the sexy allure of digital transformation seduces the senses and inspires resilient, inclusive growth. In this high-stakes arena of technological brilliance and environmental passion, every strategic decision is a step toward a future where sustainability and opulent digital sophistication coalesce into a mesmerizing new paradigm.

The tourism and hospitality sectors, serving as foundational pillars of the global economy, are currently undergoing an era characterized by profound transformation, multifaceted challenges, and evolving expectations. These industries face mounting pressures arising from escalating environmental degradation, the adverse consequences of over-tourism, and the pressing need for the implementation of sustainable operational practices. With increasing societal awareness regarding the ecological impacts associated with global travel and hospitality, there is a growing imperative to adopt innovative, technology-driven strategies that can successfully harmonize the twin objectives of enriching visitor experiences and ensuring long-term environmental stewardship. At the heart of this transformation lies the integration of advanced neural networks and artificial intelligence (AI), which poised redefine traditional paradigms and introduce groundbreaking opportunities for sustainable development and operational efficiency (Pencarelli et al., 2018). Advanced AI systems, with a particular focus on self-learning neural networks, are emerging as pivotal enablers of this transformative shift. These sophisticated technologies promise not only to enhance operational optimization but also to cultivate practices that are inherently aligned with sustainability, resilience, and long-term ecological responsibility.

The present research critically explores the transformative capacity of these AI methodologies, with a special emphasis on the convergence of immersive technologies, blockchain frameworks, and other digital innovations (He et al., 2016). These technological advancements collectively contribute to the evolution of carbon-neutral accommodations and the broader adoption of circular economy principles. Moreover, they facilitate the creation of streamlined operational processes, efficient resource management, and enriched visitor experiences that prioritize ethical and sustainable engagement with local ecosystems and cultures. Through these advancements, digital transformations present new pathways for reimagining the future of tourism and hospitality in ways that promote environmental harmony and economic viability (Parker et al., 2020).

However, the accelerated proliferation of neural network technologies is not without its complexities and inherent challenges. The adaptive, self-evolving nature of these systems, while offering unparalleled potential for promoting proactive sustainability, simultaneously introduces significant ethical concerns and governance dilemmas. Instances of autonomous AI behavior in high-stakes operational contexts, particularly in areas such as resource allocation, environmental impact mitigation, and visitor interaction management, underscore the urgent need for the establishment of robust, ethically grounded regulatory frameworks. Without such oversight, the expansive benefits of these technological innovations risk being undermined by ethical quandaries, operational misjudgments, and unintended negative consequences. The dualistic nature of neural network advancements—marked by both promise and potential peril therefore necessitates a conscientious, multidisciplinary approach that prioritizes not only technological innovation but also ethical stewardship, transparent governance, and societal inclusivity (Gretzel & Sigala, 2021).

To address these complexities, this study employs a comprehensive and rigorous analytical framework that integrates empirical data, real-world case studies, and illustrative examples to elucidate the multifaceted impacts of AI within the tourism and hospitality sectors. By foregrounding successful implementations of AI-driven sustainability initiatives such as smart resource allocation systems (HO & Pazzani 1997), AI-enabled environmental monitoring tools, and blockchain-based carbon tracking mechanisms—this research underscores the necessity for strategic, ethically-informed approaches. Such strategies should be designed to maximize the transformative potential of technological advancements while simultaneously mitigating associated risks through proactive governance and ethical foresight (Neuhofer, 2020).

Ultimately, the synthesis of cutting-edge technological innovation and regenerative tourism principles is posited as an essential pathway for achieving resilient, inclusive, and sustainable industry practices (Goodfellow et al., 2014). This research contends that the intersection of advanced AI technologies and sustainable development imperatives offers a compelling and necessary avenue for fostering a transformative, ethically responsible, and socially inclusive future within the global tourism and hospitality landscape ( Huang & Benyoucef. 2013). By advocating for strategic, conscientious development pathways, this study aspires to contribute to the broader discourse on responsible technology adoption and sustainable tourism development, ensuring that technological progress serves as a catalyst for positive, long-term global impact (Molner, 2020).

This paper examines recent advancements in neural network learning with a particular focus on their transformative applications within the tourism and hospitality sectors (LeCun et al., 2020). It investigates key innovations in neural architectures, training methodologies, and biologically-inspired frameworks that are reshaping operational strategies and enhancing visitor experiences (Li et al., 2020). Notably, models such as transformers have proven exceptionally effective in processing complex sequential data, a capability critical for real-time analysis of consumer behavior and dynamic service optimization. Likewise, capsule networks enhance feature encapsulation, allowing for more nuanced data interpretation and precise predictive analytics (Dosovitskiy & Koltun, 2016). Additionally, the discussion extends to neuromorphic computing, which emulates biological neural structures to significantly boost computational efficiency and support sustainable energy practices. In an industry where ethical compliance, reliability, and resilience are paramount, fostering such innovation in neural networks is essential for advancing sustainable, efficient, and responsive tourism practices (Kolter & Pierson, 2016).

## 2. WHAT QUESTION ARE MY RESEARCH WILL BE ANSWERING AND FOCUSING ON

**RQ1:** How do advanced AI techniques, including self-learning neural networks, transformer


https://doi.org/10.52202/081568-0027

architectures and neuromorphic computing, redefine traditional tourism models and enhance visitor experiences in the travel and hospitality sector?

**RQ2:** In what ways do digital innovations such as blockchain frameworks and immersive technologies contribute to operational efficiency and support the implementation of sustainable practices like carbon-neutral accommodations and circular economy models within the industry?

**RQ3:** How can AI-driven smart technologies and data analytics be leveraged to optimize resource management and promote regenerative tourism practices, thereby mitigating the adverse impacts of over-tourism and environmental degradation?

**RQ4:** How do the emerging capabilities of advanced AI intersect with ethical, governance challenges AND governance frameworks associated with integrating AND balancing advanced AI innovations into the travel and hospitality sectors, and how can these challenges be effectively addressed?

**RQ5:** How do case studies and empirical evidence illustrate the impact of AI-driven sustainability initiatives on creating resilient, inclusive, and regenerative growth models in the industry? What are the long-term implications of integrating cutting-edge AI solutions with regenerative trends for creating resilient, inclusive, and sustainable growth models in the rapidly evolving tourism industry?What strategic pathways can be identified from community-led initiatives that showcase the successful integration of AI innovations into regenerative tourism models?

**R1**

The integration of advanced AI techniques is fundamentally transforming the travel and hospitality landscape, ushering in a new era where operational agility and personalized customer engagement are paramount. As both a software engineer and an expert in the travel and hospitality sector, I observe that these technologies form the cornerstone of a digital transformation strategy that bridges cutting- edge computational methods with real-world service delivery.

### 2.1 Redefining Traditional Tourism Models through Advanced AI Techniques

`Advanced AI techniques—including self- learning neural networks, transformer architectures, and neuromorphic computing— are at the forefront of a profound transformation in the travel and hospitality sector. This section examines how these technologies, when integrated into existing operational frameworks, redefine traditional tourism models and significantly enhance visitor experiences.

Self-Learning Neural Networks Self-learning neural networks serve as the backbone of adaptive, data-driven systems capable of processing vast and heterogeneous datasets. In the context of tourism, these networks continuously analyze booking patterns, customer reviews, social media trends, and localized event data (Buhalis & Sinarta, 2019).. Their ability to learn and evolve in real time facilitates accurate demand forecasting, dynamic pricing strategies, and highly tailored marketing efforts. For instance, a hotel chain might deploy deep reinforcement learning to simulate various market scenarios, enabling it to adjust room rates dynamically and maintain competitiveness while catering to diverse consumer preferences (Kingma & Welling, 2014)

### 2.2 Transformer Architectures

Originally developed for natural language processing, transformer architectures excel in managing sequential and contextual data. Their robust capability to parse and interpret multilingual information is particularly valuable in the global tourism market (Finn et al., 2017). By analyzing customer feedback and sentiment across different languages and cultural contexts, these models enable virtual travel assistants and chatbots to deliver highly personalized, context- aware recommendations and support. This functionality enhances customer interaction and service quality, ensuring that each visitor's experience is both engaging and attuned to their unique needs (Sharpley, 2014).

### 2.3 Neuromorphic Computing

Neuromorphic computing mimics the human brain's architecture to offer energy-efficient and parallel data processing—a crucial asset in environments demanding real-time responses. In smart resorts or integrated urban tourism systems, neuromorphic systems process data from diverse sources, such as environmental sensors, crowd analytics, and security systems, to dynamically adjust operational parameters (Schmidhuber, 2015). This rapid adaptation capability enhances resource management, improves safety protocols, and ensures seamless service delivery even during peak operational periods (Sutton & Barto, 2018).

### 2.4 Enhancing Visitor Experiences and Operational Agility

The combined application of these advanced AI techniques creates a paradigm shift from static, reactive service models to dynamic, proactive systems. This transformation yields several key benefits:

- Personalized Engagement: By leveraging detailed insights from self-learning networks and transformer models, businesses can craft bespoke travel itineraries and offer customized recommendations that resonate with individual visitor preferences.

- Real-Time Service Optimization: Neuromorphic computing and transformer-based virtual assistants facilitate immediate, context-aware customer interactions, ensuring that queries and service modifications are handled swiftly and efficiently.

- Agile Resource Management: The predictive and adaptive capabilities of these AI systems allow tourism operators to optimize resource allocation, reduce operational bottlenecks, and maintain high service standards, even in fluctuating market conditions.

**2.5 Conclusion** In summary, the integration of advanced AI techniques into the travel and hospitality sector is reshaping traditional tourism models by introducing unprecedented levels of operational efficiency and personalized service delivery. Self- learning neural networks enable dynamic data analysis and market responsiveness, transformer architectures enhance multilingual and contextual customer interactions, and neuromorphic computing ensures real-time, energy-efficient operational management. Together, these technologies forge a pathway toward a more agile, visitor-centric, and sustainable tourism model—one that is well- equipped to meet the challenges and opportunities of a rapidly evolving global market. This multifaceted approach not only elevates visitor experiences but also underpins the development of resilient and forward-looking operational frameworks, making it a critical subfield within the broader scope of our research on AI innovations and regenerative trends in travel and hospitality.

**R2**

Digital Innovations: Enhancing Operational Efficiency and Sustainability

Digital innovations are increasingly recognized as pivotal in redefining the operational landscape of the


https://doi.org/10.52202/081568-0027

travel and hospitality sector. This section explores how blockchain frameworks and immersive technologies contribute to operational efficiency and support sustainable practices, such as carbon-neutral accommodations and circular economy models.

Blockchain Frameworks for Transparent Sustainability Blockchain technology introduces a decentralized, immutable ledger that is well- suited for monitoring and verifying sustainability initiatives. Its contributions include:

- Carbon Footprint Verification: Blockchain systems can record and authenticate carbon emissions data in real time. This capability allows tourism operators to substantiate claims of carbon neutrality by tracking energy usage, renewable energy investments, and carbon offset measures with unparalleled transparency.

- Supply Chain Optimization: By tracking the lifecycle of goods—from procurement to disposal—blockchain enhances resource management and enforces circular economy principles. This real-time tracking minimizes waste, streamlines operations, and ensures that every link in the supply chain adheres to sustainability standards.

- Building Stakeholder Trust: The tamper-proof nature of blockchain fosters confidence among investors, regulatory bodies, and eco-conscious consumers by ensuring that all sustainability metrics are accurate and readily auditable. This trust is fundamental in driving broader industry acceptance of sustainable practices.

## 3. Immersive Technologies for Engaging Sustainability

Immersive technologies, including virtual reality (VR), augmented reality (AR), and mixed reality (MR), offer transformative tools to enhance both customer experiences and operational practices. Their key contributions include:

- Virtual Destination Experiences: VR and AR enable potential travelers to explore destinations virtually, reducing unnecessary physical travel and thereby lowering carbon emissions. These technologies not only assist in pre-travel planning but also provide a sustainable alternative to traditional travel experiences.

- Interactive Training and Simulation: Immersive simulation platforms offer advanced training modules that educate staff on energy conservation, waste reduction, and sustainable operational practices. These tools help ensure that employees are well-versed in implementing eco-friendly procedures across all operational facets.

- Enhanced Visitor Engagement and Education: By integrating immersive digital experiences into the guest journey, operators can effectively communicate sustainability initiatives and engage visitors in environmental stewardship. This interactive approach not only enhances the overall visitor experience but also raises awareness and promotes responsible tourism behaviors.

### 3.1 Synergistic Impact: Merging Data Integrity with Immersive Engagement

The integration of blockchain and immersive technologies creates a robust ecosystem that amplifies sustainable outcomes:

- **Data-Driven Immersive Experiences:** Blockchain's reliable data supports immersive platforms by providing accurate, real-time sustainability metrics that can be transformed into engaging narratives. This synergy ensures that visitors receive transparent information about environmental initiatives, deepening their trust and participation in sustainability efforts.

- **Optimized Operational Decision- Making:** The combination of precise data from blockchain with the interactive capabilities of immersive technologies allows for agile, informed decision-making. Operators can swiftly adjust resource allocations, reduce waste, and refine their sustainable practices in response to dynamic market and environmental conditions.

**3.2 Conclusion** Digital innovations, exemplified by blockchain frameworks and immersive technologies, are fundamentally transforming the travel and hospitality sector. They not only drive operational efficiency through real-time data tracking and supply chain optimization but also underpin the adoption of sustainable practices such as carbon-neutral accommodations and circular economy models. This strategic integration empowers tourism operators to achieve transparency, build stakeholder trust, and engage visitors in a meaningful dialogue about sustainability. As a result, these innovations serve as critical enablers for a resilient, environmentally responsible, and forward-looking tourism industry.

**R3**

Leveraging AI-Driven Smart Technologies and Data Analytics for Optimized Resource Management and Regenerative Tourism

The integration of AI-driven smart technologies and sophisticated data analytics presents a transformative opportunity to optimize resource management in the travel and hospitality sector while simultaneously promoting regenerative tourism practices. By harnessing the power of real-time data, predictive analytics, and intelligent automation, tourism operators can effectively mitigate the adverse impacts of over-tourism and environmental degradation.

## 4. Optimizing Resource Management

**Real-Time Monitoring and Predictive Maintenance:**

The deployment of IoT sensors combined with AI algorithms enables continuous monitoring of critical resources such as energy, water, and waste. These systems analyze data streams in real time to detect inefficiencies, predict maintenance needs, and adjust operations dynamically. For instance, smart energy management systems can predict peak usage periods and optimize energy distribution, thereby reducing waste and minimizing operational costs.

Dynamic Resource Allocation: AI-driven platforms use machine learning models to forecast demand and allocate resources accordingly (Vaswani et al., 2017). By processing historical data, seasonal trends, and real-time inputs, these systems can automate decisions related to staffing, inventory management, and facility usage. This dynamic approach not only ensures efficient resource utilization but also reduces the environmental footprint associated with excess consumption and over-provisioning (Tussyadiah, 2020).

Sustainable Infrastructure Management: Through the integration of data analytics, operators can model and simulate the environmental impact of various operational scenarios. This capability supports the design and maintenance of sustainable infrastructures, such as carbon-neutral buildings and smart


https://doi.org/10.52202/081568-0027

water recycling systems. Such technologies are instrumental in reducing the ecological footprint of tourism facilities while maintaining high service standards.

**4.1 Promoting Regenerative Tourism Practices**

Data-Driven Visitor Flow Management: Advanced analytics enable the collection and processing of visitor data to manage crowd density and distribute tourist flows across destinations. By monitoring real-time visitor metrics, operators can prevent the over- concentration of tourists in sensitive areas, thereby reducing the pressure on local ecosystems and cultural sites. This approach not only alleviates over-tourism but also enhances the quality of the visitor experience by avoiding overcrowded conditions.

**4.2 Environmental Impact Assessment and Mitigation:** AI systems can integrate environmental data— such as air quality, water purity, and biodiversity indicators—with operational metrics. This integration allows tourism operators to assess the environmental impact of their activities accurately and implement corrective measures in a timely manner. For example, predictive models can forecast the environmental repercussions of increased tourist activity and suggest interventions such as temporary access restrictions or enhanced conservation efforts.

**4.3 Facilitating Community-Led Regeneration Initiatives:** Data analytics provides valuable insights into the social and economic dimensions of tourism, enabling the design of community-led regenerative projects. By incorporating feedback from local stakeholders and analyzing the impact of tourism on community welfare, operators can craft initiatives that not only promote environmental sustainability but also enhance local resilience. This might include developing eco-friendly tourism packages, supporting local artisans, or investing in community infrastructure projects that benefit both residents and visitors.

**4.4 Mitigating the Adverse Impacts of Over- Tourism and Environmental Degradation**: The synergistic application of AI-driven technologies and data analytics is essential in mitigating the challenges associated with over- tourism and environmental degradation. By enabling precise control over resource distribution and by fostering an ecosystem of informed, sustainable practices, these technologies help to:

- Prevent Environmental Overload: Real- time adjustments in resource allocation can alleviate the strain on natural resources, reducing pollution and preserving biodiversity.

- Enhance Operational Resilience: Predictive insights and dynamic management systems ensure that tourism facilities remain adaptive and responsive to both market fluctuations and environmental changes.

- Empower Stakeholder Collaboration: Transparent data sharing builds trust among regulators, local communities, and industry players, facilitating collaborative approaches to sustainable tourism management.

**4.5 Conclusion** -In conclusion, leveraging AI-driven smart technologies and data analytics offers a dual benefit for the travel and hospitality sector: optimizing resource management and advancing regenerative tourism practices. This integrated approach not only streamlines operations and enhances visitor experiences but also mitigates the adverse impacts of over-tourism and environmental degradation. As a critical subfield within the broader scope of our research on AI innovations and regenerative trends, these strategies provide a robust framework for fostering a resilient, sustainable, and environmentally responsible future for global tourism.

**R4**

**Navigating Ethical, Regulatory, and Governance Challenges in Advanced AI Integration**

The rapid evolution of advanced AI technologies—encompassing self-learning neural networks, transformer models, and neuromorphic computing—has opened new frontiers in the travel and hospitality sectors. However, their integration also brings forth complex ethical, regulatory, and governance challenges. This section examines the intersection of these emerging AI capabilities with ethical dilemmas and governance frameworks, and explores strategies for effectively addressing these challenges to ensure responsible and sustainable innovation.

**5. Emerging Ethical Considerations**

- Data Privacy and Security: Advanced AI systems require access to large volumes of personal and operational data. Ensuring robust data protection and privacy is critical to maintaining customer trust and complying with global data protection regulations.

- Bias and Fairness: AI algorithms trained on historical data can inadvertently propagate biases, leading to discriminatory practices in service delivery. Mitigating these biases through continuous auditing and incorporating fairness-aware machine learning techniques is essential.

- Transparency and Accountability: Many AI systems operate as "black boxes," making it challenging to understand their decision-making processes. Developing explainable AI (XAI) methods and establishing clear accountability protocols are crucial for ensuring that AI-driven decisions are transparent and justifiable.

**5.1 Governance Challenges in a Dynamic Landscape**

- Regulatory Gaps: The pace of AI innovation frequently outstrips the development of comprehensive regulatory frameworks. This gap can lead to inconsistent standards across regions, complicating the integration of AI technologies into global travel and hospitality operations.

- Cross-Jurisdictional Coordination: As tourism is inherently global, harmonizing regulatory practices across different legal and cultural environments poses significant challenges. Effective governance requires collaborative efforts among international regulatory bodies, industry stakeholders, and technology providers.

- Stakeholder Engagement: Balancing the interests of various stakeholders—including customers, employees, technology developers, and regulatory agencies—is vital. Transparent communication and multi- stakeholder forums can foster trust and ensure that AI innovations align with broader societal values.

**5.2 Integrating Governance Frameworks with AI Innovation**

To address these ethical and governance challenges, a multi-pronged approach is required:

- Establishing Ethical Guidelines and Standards: Developing industry-specific ethical guidelines that emphasize transparency, accountability, and fairness is essential. These


https://doi.org/10.52202/081568-0027

guidelines should be adaptable, allowing them to evolve in tandem with technological advancements.

- Adaptive Regulatory Frameworks: Policymakers and industry leaders must collaborate to create regulatory frameworks that are both robust and flexible. This involves setting clear standards for data usage, algorithmic accountability, and bias mitigation, while also allowing room for innovation.

- Transparency Mechanisms: Incorporating explainable AI techniques and independent audits can help demystify AI decision-making processes. Transparency not only builds trust among consumers but also ensures that AI systems are operating in accordance with ethical standards.

- Collaborative Governance Models: The establishment of multi-stakeholder partnerships is crucial for effective governance. By engaging with diverse groups—from local communities to global regulatory bodies—travel and hospitality operators can develop governance models that balance technological innovation with social responsibility.

**5.3 Conclusion** The integration of advanced AI capabilities into the travel and hospitality sectors promises significant operational and experiential enhancements. However, these benefits come with intricate ethical and governance challenges that must be effectively addressed. By developing comprehensive ethical guidelines, establishing adaptive regulatory frameworks, and fostering transparent and collaborative governance models, the industry can ensure that AI innovations are implemented responsibly. This balanced approach not only mitigates potential risks but also paves the way for a future where technological advancement and ethical stewardship work hand in hand to drive sustainable growth in tourism.

**R5**

**Empirical Evidence and Strategic Pathways for AI-Driven Sustainability Initiatives**

The integration of AI-driven sustainability initiatives in the travel and hospitality industry is not merely theoretical—case studies and empirical evidence underscore its transformative impact on creating resilient, inclusive, and regenerative growth models. This section explores how real-world examples and data- backed research illuminate the benefits and long- term implications of merging cutting-edge AI solutions with regenerative trends, while also identifying strategic pathways derived from community-led initiatives.

**6. Illustrating Impact Through Case Studies and Empirical Evidence**

- Enhanced Operational Resilience and Efficiency: Several case studies from industry leaders and pilot projects in smart tourism demonstrate how AI technologies, such as predictive analytics and real-time monitoring systems, lead to significant improvements in resource management. For instance, properties employing AI-driven energy management systems have reported reductions in energy consumption by dynamically adjusting to occupancy levels and environmental conditions. Empirical evidence from these projects confirms that such systems not only reduce operational costs but also contribute to the sustainable management of energy resources.

- Inclusive and Regenerative Growth Models: Empirical studies have shown that AI- driven

https://doi.org/10.52202/081568-0027

platforms can facilitate more inclusive growth by enabling personalized services that cater to diverse tourist demographics. By leveraging data analytics, operators can tailor experiences that respect cultural nuances and promote local heritage, thereby driving regenerative practices. In several community-led initiatives, the application of AI has enabled local stakeholders to participate in decision- making processes, ensuring that tourism development is aligned with regional socio-economic and environmental goals.

- Measurable Environmental Benefits: Quantitative data from projects that integrate AI with blockchain technology for tracking carbon emissions reveal tangible improvements in environmental accountability. Such projects not only validate carbon-neutral claims but also promote transparency and trust among stakeholders. These empirical results underscore the potential of AI to support rigorous environmental monitoring and to drive corrective actions that mitigate the adverse impacts of tourism on natural ecosystems.

## 6.1 Long-Term Implications of Integrating AI with Regenerative Trends

- Sustainable Adaptability and Scalability: The continuous evolution of AI solutions fosters adaptability, allowing tourism operators to quickly respond to changing market dynamics and environmental conditions. Over the long term, this adaptability is key to scaling sustainable practices across the industry, ensuring that growth models remain resilient in the face of global challenges such as climate change and resource scarcity.

- Holistic Ecosystem Development: Integrating AI with regenerative tourism trends creates a holistic ecosystem where technological innovation supports community-driven sustainability. This integration can lead to more robust frameworks for energy management, waste reduction, and conservation efforts, promoting a balanced relationship between tourism development and environmental stewardship.

- Enhanced Stakeholder Collaboration: The long-term success of these initiatives hingeson multi-stakeholder collaboration. AI platforms that incorporate community insights and real-time feedback mechanisms foster greater transparency and trust. This collaborative approach not only drives innovation but also ensures that growth is inclusive, benefiting local communities and preserving cultural and natural assets for future generations.

## 6.2 Strategic Pathways from Community-Led Initiatives:

Community-led initiatives offer valuable insights into how AI innovations can be seamlessly integrated into regenerative tourism models. Strategic pathways identified from these initiatives include:

- Public-Private Partnerships: Collaborative frameworks between government entities, local businesses, and technology providers can facilitate the adoption of AI tools that enhance sustainability. These partnerships enable resource sharing, joint funding opportunities, and the co-creation of regulatory standards that support ethical AI deployment.

- Localized Digital Innovation Hubs: Establishing innovation hubs at the community level empowers local stakeholders to experiment with AI- driven solutions tailored to regional challenges. These hubs serve as incubators for sustainable technologies, allowing for the rapid prototyping and scaling of successful models.


https://doi.org/10.52202/081568-0027

- Capacity Building and Technology Training: Strategic investments in training programs and digital literacy initiatives ensure that local communities can actively participate in and benefit from AI-driven sustainability efforts. Empowering community members with the necessary skills fosters an environment where technology and tradition coexist, promoting regenerative practices.

- Data-Driven Decision-Making Platforms: Implementing platforms that aggregate and analyze local environmental and socio-economic data enables more informed decision- making. Such platforms facilitate community engagement by providing transparent metrics on sustainability performance, thus aligning tourism development with local needs and priorities.

**6.3 Conclusion** In conclusion, case studies and empirical evidence robustly illustrate that AI-driven sustainability initiatives are central to creating resilient, inclusive, and regenerative growth models within the travel and hospitality industry. The long-term implications of integrating advanced AI with regenerative trends are profound—ushering in an era of sustainable adaptability, holistic ecosystem development, and enhanced stakeholder collaboration. Strategic pathways emerging from community-led initiatives, such as public- private partnerships, localized innovation hubs, capacity building, and data-driven platforms, provide actionable frameworks for scaling these benefits across the industry. This multifaceted approach not only enhances operational efficiency and environmental accountability but also sets a transformative course for the future of global tourism.

# 7. INNOVATIVE NEURAL NETWORK ARCHITECTURES

In the pursuit of transforming horizons within the travel and hospitality sector, advanced neural network architectures have emerged as key enablers for sustainable, personalized, and efficient solutions. By integrating the technical prowess of advanced software and data analytics engineers with the practical insights of tourism experts, these architectures are being tailored to address industry- specific challenges ranging from dynamic customer engagement to real-time resource optimization and regenerative practices (Wu et al., 2021).

## A. Transformer Networks and Attention Mechanisms

Transformer networks have redefined the landscape of deep learning by leveraging self- attention mechanisms to model complex, non-linear dependencies across diverse data types. In traditional models, such as recurrent neural networks (RNNs), sequential data processing often led to the loss of critical information especially in long sequences. Transformers, however, process entire sequences in parallel, enabling them to capture global context and intricate interdependencies, which are crucial in analyzing multifaceted tourism data (Xiang et al., 2017).

In the travel and hospitality domain, this capability translates into several transformative applications:

- Personalized Guest Experiences: By processing real-time customer feedback, social media interactions, and booking trends, transformers can generate highly tailored recommendations for travel itineraries, accommodation options, and local experiences.

- Sustainable Operational Analytics: Transformers applied to sensor and environmental data can monitor energy usage, waste management, and other sustainability metrics across facilities, facilitating the transition to carbon-neutral accommodations.

- Multi-Modal Integration: The flexibility of transformer architectures allows seamless integration of textual data (e.g., reviews, travel blogs) with visual data (e.g., images of destinations), enabling a comprehensive understanding of visitor preferences and enhancing decision-making processes.

B. Generative Adversarial Networks (GANs)

Generative Adversarial Networks (GANs) have revolutionized the field of data synthesis and augmentation through their dual-network structure—comprising a generator and a discriminator that engage in continuous adversarial training. This dynamic interaction not only produces highly realistic synthetic data but also enables the discovery of nuanced patterns within the data. Within the context of tourism and hospitality, GANs offer transformative potential:

- Data Augmentation for Enhanced Training: In scenarios where real-world tourism data is limited or imbalanced, GANs can generate synthetic datasets that enrich training processes for AI models, ensuring robustness in tasks such as image recognition and sentiment analysis.

- Creative Content Generation: GANs empower the creation of photorealistic images and virtual environments for destinations, which can be utilized in immersive marketing campaigns and virtual tours, thereby elevating the visitor experience.

- Anomaly Detection and Security: By learning the normal patterns of operation in tourism facilities, GANs can detect anomalies—such as irregular energy consumption or security breaches— enhancing both operational resilience and guest safety.

## 7.1 Bio-Inspired Neural Structures

Bio-inspired neural structures, modeled after the human brain's architecture, aim to deliver higher computational efficiency, adaptability, and real-time responsiveness. Spiking Neural Networks (SNNs) and neuromorphic computing are at the forefront of this approach, emulating biological processes to enable rapid, energy-efficient processing (Zenke et al., 2020).

For the travel and hospitality sector, the benefits are multifold:

- Real-Time Adaptive Systems: Neuromorphic architectures are ideal for applications such as smart resorts or IoT- enabled hotels, where rapid environmental response—such as adjusting climate control or lighting based on occupancy—is crucial.

- Energy Efficiency and Sustainability: By significantly reducing power consumption, these systems support the sustainability goals of eco-friendly tourism facilities and contribute to the overall reduction of the carbon footprint.

- Adaptive Customer Interaction: Bio- inspired models can power advanced interactive systems that adapt in real-time to guest preferences, fostering a more engaging and personalized customer experience.

## ADVANCED TRAINING TECHNIQUES

To maximize the impact of these innovative neural architectures, advanced training techniques play a critical role in enabling models to learn effectively from diverse and often limited data sources. Techniques such as meta-learning, few-shot learning, and self-supervised approaches are essential for building AI systems that are both adaptive and resource-efficient—qualities that are imperative in the


https://doi.org/10.52202/081568-0027

fast-evolving travel and hospitality landscape.

### 7.2 Meta-Learning and Few-Shot Learning

Meta-learning, or "learning to learn," equips AI models with the capacity to generalize across tasks by rapidly adapting to new challenges with minimal data. This is particularly valuable in tourism, where data can be sparse or context-specific:

Adaptive Personalization: Few-shot learning techniques enable AI systems to tailor travel recommendations based on minimal user input, effectively addressing the diverse needs of global travelers.

Rapid Deployment of New Services: With meta-learning, hotels and travel agencies can quickly integrate new data streams— such as emerging travel trends or local events—into their operational models, ensuring timely and relevant service offerings.

### 7.3 Self-Supervised and Unsupervised Learning

Self-supervised and unsupervised learning methods allow models to extract meaningful representations from raw, unlabeled data—a critical capability when labeled datasets are scarce:

Leveraging Inherent Data Structures: In the hospitality sector, self-supervised learning can analyze vast amounts of customer interaction data, uncovering hidden trends and preferences that inform service improvements.

Enhancing Operational Insights: Unsupervised learning techniques facilitate the discovery of patterns within environmental and resource usage data, supporting strategies for efficient energy management and waste reduction.

### 7.4 Reinforcement Learning (RL)

Reinforcement Learning (RL) drives the development of autonomous systems that optimize decision-making through a continuous feedback loop of rewards and penalties:

Dynamic Resource Management: In tourism facilities, RL algorithms can autonomously manage resources—such as energy, staffing, and inventory—adapting in real time to fluctuating demands.

Optimizing Guest Services: RL-based systems can personalize customer interactions by learning optimal service strategies from continuous guest feedback, ensuring high satisfaction levels while maintaining operational efficiency.

Adaptive Safety Protocols: In high-risk environments like theme parks or large resorts, RL models can continuously adapt safety measures based on real-time data, thus ensuring guest security and compliance with regulatory standards.

## 8. OPTIMIZATION AND EFFICIENCY IMPROVEMENTS

As AI models in the travel and hospitality sector become more sophisticated, optimizing their performance and ensuring efficiency is essential for practical deployment across diverse platforms—from high-performance servers to mobile devices and IoT sensors. Advanced optimization techniques not only speed up training and inference but also ensure that AI applications remain scalable and resource-efficient.

### 8.1 Gradient-Based Optimization Techniques

Gradient-based methods form the backbone of neural network training, with algorithms like Adam and RMSprop ensuring rapid convergence and robust performance:

Adaptive Learning Rates: Techniques such as Adam dynamically adjust learning rates based on moment estimates, enabling faster convergence on complex tourism data that exhibits non-stationary patterns.

Enhanced Stability: RMSprop refines the optimization process by adapting to recent gradient information, which is crucial for maintaining stability in models that continuously learn from live operational data across diverse hospitality settings.

### 8.2 Model Compression and Scalability

Given the diverse deployment environments in tourism—from central servers in luxury resorts to edge devices in remote eco-lodges—model compression techniques are pivotal:

Pruning and Quantization: These techniques streamline models by removing redundant parameters and reducing numerical precision, thereby lowering memory footprints and enabling faster processing on resource-constrained devices.

Knowledge Distillation: By transferring insights from larger, high-performing models (teachers) to more compact models (students), knowledge distillation ensures that even lightweight AI systems maintain high levels of accuracy and efficiency.

Scalable Deployment: Efficient models facilitate real-time applications such as dynamic pricing, personalized guest services, and energy management systems—allowing tourism operators to leverage AI innovations across various operational scales.

**8.3 Conclusion:** The integration of advanced neural network architectures, innovative training techniques, and robust optimization strategies is reshaping the future of the travel and hospitality sector. Through the combined expertise of software and data analytics engineers and industry specialists, these cutting-edge technologies are driving the development of sustainable, resilient, and regenerative models that enhance both operational efficiency and visitor experiences. As the industry continues to evolve, these AI innovations promise to transform traditional tourism paradigms— ushering in a new era where technology and sustainability converge to create transformative and inclusive growth models.

## 9. CHALLENGES AND FUTURE DIRECTIONS

As advanced AI and neural network technologies continue to revolutionize the travel and hospitality sector, they simultaneously introduce a suite of challenges that must be navigated to ensure both technical excellence and responsible implementation. A collaborative approach— integrating insights from advanced software and data analytics engineers with domain experts in tourism and hospitality— is essential. Key areas of concern include model interpretability, user trust, ethical ramifications, and the broader societal impact, all of which are crucial for fostering widespread acceptance and reliable integration of AI innovations.


https://doi.org/10.52202/081568-0027

### 9.1 Interpretability and Trust in AI

For both software engineers and tourism professionals, ensuring that AI systems are transparent and comprehensible is paramount. As these models become increasingly complex, particularly in applications such as dynamic pricing, personalized guest services, and predictive maintenance, the need for interpretability grows:

- Explainable AI (XAI) Techniques: Advanced methods such as saliency maps, layer-wise relevance propagation, and counterfactual explanations are being deployed to demystify the decision-making processes of neural networks. These tools help stakeholders—from hotel managers to financial auditors—understand which factors influence AI-driven decisions, ensuring that operational insights remain actionable and reliable.

- Building Trust through Transparency: In the tourism and hospitality sectors, where customer satisfaction and safety are non- negotiable, transparent AI models contribute directly to enhanced trust. Clear explanations of how decisions are made allow field experts and consumers alike to evaluate and validate the outcomes, thereby reinforcing confidence in automated systems that manage everything from room allocation to emergency responses.

### 9.2 Ethical and Societal Implications

The transformative power of AI in tourism brings with it significant ethical and societal considerations that must be addressed collaboratively by technologists and industry experts:

- Data Privacy and Security: The integration of AI requires the handling of vast amounts of sensitive data—from guest preferences and booking histories to real-time operational metrics. Implementing privacy-preserving techniques, such as differential privacy and secure multi-party computation, ensures that individual rights are safeguarded even as data analytics drive personalization and operational efficiency.

- Mitigating Algorithmic Bias: Biased algorithms can lead to unequal treatment of diverse guest demographics, potentially undermining the inclusivity that is core to modern tourism. Fairness-aware machine learning models are being developed to ensure that AI systems deliver equitable services. For example, by auditing training datasets and incorporating diverse data sources, AI systems can offer unbiased recommendations and service optimizations.

- Accountability and Regulatory Compliance: As AI applications expand into critical areas—such as real-time customer support and safety monitoring—it becomes essential to establish robust ethical frameworks and governance protocols. These frameworks, co-developed by industry leaders and regulatory bodies, ensure that AI systems are accountable and operate in alignment with both legal standards and societal values.

## CONCLUSION

This paper has explored the cutting-edge advancements in neural network technologies and their transformative applications within the travel and hospitality sector. By delving into architectural innovations, sophisticated training methodologies, and state-of-the-art optimization strategies, we have illustrated how technologies such as self-learning neural networks, transformer models, and neuromorphic computing are redefining industry paradigms.

https://doi.org/10.52202/081568-0027

In this dynamic landscape, the integration of advanced AI solutions must be balanced with robust ethical considerations, transparent interpretability measures, and comprehensive governance frameworks. Our interdisciplinary approach— blending the technical acumen of software and data analytics engineers with the practical insights of tourism and hospitality experts—underscores the necessity of addressing data privacy, mitigating bias, and ensuring accountability. These efforts are pivotal for fostering user trust and for establishing AI as a catalyst for sustainable, resilient, and inclusive growth.

Looking forward, the ongoing development of adaptable and ethically responsible AI systems promises to unlock new avenues for innovation. By harmonizing technical innovation with societal imperatives, the future of neural networks in tourism will not only enhance operational efficiency but also ensure that technological progress contributes positively to environmental stewardship and community well-being. This cohesive strategy is essential for realizing a future where technology and human-centric values work in concert to drive the evolution of a resilient and forward-thinking global tourism industry.

## References


1. Bengio, Y., et al. (2009). Learning deep architectures for AI. Foundations and Trends in Machine Learning, 2(1), 1–127.
2. Buhalis, D., & Sinarta, Y. (2019). Real-time co-creation and smart tourism. Journal of Travel & Tourism Marketing.
3. Dosovitskiy, A., & Koltun, V. (2016). Learning to generate chairs with convolutional neural networks. IEEE Transactions on Pattern Analysis and Machine Intelligence.
4. Finn, C., et al. (2017). Model-agnostic meta-learning for fast adaptation of deep networks. In Proceedings of the International Conference on Machine Learning.
5. Goodfellow, I., et al. (2014). Generative adversarial nets. In Advances in Neural Information Processing Systems.
6. Gretzel, U., & Sigala, M. (2021). Digital transformation and its impact on tourism and hospitality: Emerging trends and challenges. Journal of Hospitality and Tourism Technology.
7. Gretzel, U., Sigala, M., Xiang, Z., & Koo, C. (2015). Smart tourism: Foundations and developments. Electronic Markets.
8. Guttentag, D. (2010). Virtual reality: Applications and implications for tourism. Tourism Management.
9. He, K., et al. (2016). Deep residual learning for image recognition. In Proceedings of the IEEE Conference on Computer Vision and Pattern Recognition.
10. Ho, T. K., & Pazzani, M. J. (1997). A decision tree-based method for feature selection. In Proceedings of the Fourth International Conference on Machine Learning.
11. Huang, Y., & Benyoucef, M. (2013). From e-commerce to social commerce: A close look at design features. Electronic Commerce Research and Applications.
12. IEEE Global Initiative on Ethics of Autonomous and Intelligent Systems. (2019). Ethically aligned design: A vision for prioritizing human well-being with autonomous and intelligent systems. IEEE Standards Association.
13. Kingma, D. P., & Welling, M. (2014). Auto-encoding variational Bayes. In Proceedings of the International Conference on Learning Representations.
14. 1Kolter, J. Z., & Pierson, E. (2016). Optimal methods for energy-efficient learning. In Proceedings of the International Conference on Machine Learning.
15. LeCun, Y., et al. (2020). Self-supervised learning: The next big thing in AI. IEEE Transactions on Pattern Analysis and Machine Intelligence.


 https://doi.org/10.52202/081568-0027

16. Li, X., et al. (2020). Big data analytics in tourism: A comprehensive review. Journal of Travel Research.
17. Molnar, C. (2020). Interpretable machine learning: A guide for making black box models explainable.
18. Neuhofer, B. (2019). Blockchain and its potential impact on tourism. Journal of Tourism Futures.
19. Parker, D. S., et al. (2020). Understanding adversarial training methods for neural networks. IEEE Transactions on Neural Networks and Learning Systems.
20. Pencarelli, T., et al. (2018). Sustainable tourism: A new paradigm for an environmentally and socially responsible future. Sustainability.
21. Schmidhuber, J. (2015). Deep learning in neural networks: An overview. Neural Networks, 61, 85–117.
22. Sharpley, R. (2014). Tourism, technology and development: A global perspective. Channel View Publications.
23. Sutton, R. S., & Barto, A. G. (2018). Reinforcement learning: An introduction. MIT Press.
24. Tussyadiah, I. (2020). Artificial intelligence in tourism: A critical analysis and research agenda. Journal of Travel Research.
25. UNWTO. (2020). International guidelines for sustainable tourism. UNWTO Publication.
26. Vaswani, A., et al. (2017). Attention is all you need. In Advances in Neural Information Processing Systems.
27. Wu, H., et al. (2021). Vision transformer: Exploring large-scale vision with transformers. IEEE Transactions on Neural Networks and Learning Systems.
28. Xiang, Z., Du, Q., Ma, Y., & Fan, W. (2017). A comparative analysis of major online review platforms: Implications for social media analytics in hospitality and tourism. Tourism Management.
29. Zenke, F., et al. (2020). Spiking neural networks: An energy-efficient deep learning model. Journal of Artificial Intelligence Research.

# COMMUNITY EMPOWERMENT THROUGH REGENERATIVE TOURISM

Dr Anil Chandhok [1], and Dr Amitabh Mishra[2]

[1] Professor, University School of Business, Chandigarh University, Mohali (Punjab).
[2] Associate Professor, University School of Business, Chandigarh University, Mohali-Punjab.

* Corresponding author: anil.e4863@cumail.in

**Abstract.** The regenerative tourism represents a transformative approach to the travel and tourism industry, whereby focusing on the long-term sustainability, local empowerment of the community and focus on environmental restoration. Unlike the traditional models that often exploit the local natural resources and local communities, regenerative tourism seeks to create a mutually beneficial relationship between the stakeholders like travellers, businesses, and host communities. This new emerging approach goes beyond merely minimizing negative impacts and strives to actively restore ecosystems, preserve cultural heritage, and promote social well-being. The regenerative tourism involves deeply integrating local communities into the tourism ecosystem, ensuring that they play a pivotal role in the decision-making and benefit directly from tourism activities. This approach fosters local entrepreneurship, enhances cultural exchange, and ensures that the wealth generated from the tourism stays within the community. It also emphasizes environmental stewardship by encouraging travellers to engage in conservation activities, contribute to sustainable practices, and support businesses that prioritize ecological restoration. This paper is an attempt to explores the potential of regenerative tourism as a tool for community empowerment. In this paper, the case study of Mata Vaishno Devi Shrine Board is referred which is the best live example of successful regenerative tourism initiatives and assesses the social, economic, and environmental benefits for local communities. By focusing on a holistic model of tourism that nurtures both human and environmental well-being, regenerative tourism offers a sustainable path forward for the future of travel and tourism sector.

**Keywords:** Community- Empowerment; Regenerative -Tourism; Local -Communities; Environmental- Restoration and Cultural- Heritage

## 1 Introduction

Tourism sector has been a driving force in the national and global economy, providing job opportunities, fostering cultural exchange, and generating significant income for local communities. However, over the last decades, the conventional tourism model has come under scrutiny for its negative impacts on local cultures, ecosystems, and economies. Mass tourism often leads to resource depletion, environmental degradation, and a lack of equitable distribution of tourism-related profits, there is an increasing need to rethink the way we approach tourism (Hall & Gossling, 2016).

The regenerative tourism has emerged as a new modern approach that focuses on creating a positive, long-lasting change for both local host communities and for conserving the environment. Unlike traditional "sustainable tourism," which focuses on minimizing negative impacts, on the other hand the regenerative tourism purely works to restore and enhance natural and cultural resources.


https://doi.org/10.52202/081568-0028

The regenerative tourism basically moves around on the community empowerment. Traditional tourism often marginalizes local communities, leading to an unequal distribution of wealth and power. In comparison to this, regenerative tourism prioritizes the involvement of local people in decision-making processes and ensures that the local community be directly benefited from tourism activities. By encouraging the local entrepreneurship, providing special education and training, and promoting equitable business models, regenerative tourism seeks to elevate local communities to become active participants in shaping and carving the tourism activities and sharing the maximum dividend arises there from.

**Community Empowerment through Regenerative Tourism and Hospitality – A case study of Mata Vaishno Devi Shrine Board**

The Mata Vaishno Devi Shrine, located in the Trikuta Hills of Jammu and Kashmir, India, is one of the most common visited pilgrimage places of India. The shrine, dedicated to Goddess Vaishno Devi, attracts millions of pilgrims annually, making it a central hub for religious tourism (Bhardwaj, 2020). However, the region's popularity has brought with it some challenges such as environmental degradation, overcrowding, and unequal economic benefits for local communities. In response to these issues, the Mata Vaishno Devi Shrine Board has pioneered initiatives that align with regenerative tourism principles to empower the local community while ensuring sustainable development and conserving the environmental so that the environment should not get polluted because of too much crowd of the visitors.

1. **The Challenge: Overcrowding and Environmental Pressure**

Over the years, the Vaishno Devi pilgrimage has witnessed a massive increase of tourist visitors. According to the previous trends, more than 8 million pilgrims visit the shrine annually, placing considerable pressure on the environment, local infrastructure, and the surrounding communities. The steady increase in tourism in the area has led to several issues and challenges:

- **Environmental degradation:** The continuous inflow of tourist activities have caused deforestation, water depletion, and waste management issues.
- **Cultural and social impact:** The rapid commercialization of the area has altered the traditional way of life for local communities, leading to a loss of cultural heritage and an increased dependency on tourism-based income.
- **Economic inequality:** With the tremendous flow of pilgrimage in the region, there is significant revenues from tourism, the local communities have not always benefited proportionally. The external tourism companies have often captured the majority of the profits .

Recognizing these issues and challenges, the Mata Vaishno Devi Shrine Board set out to adopt regenerative tourism practices to both empower the local community and restore the environmental and cultural heritage of the region.

## 2. Regenerative Tourism Strategies by the Mata Vaishno Devi Shrine Board

The Shrine Board has planned and executed several initiatives that align with the principles of regenerative tourism. These strategies are aimed at focusing on the sustainability, local community empowerment and engagement and the restoration of natural and cultural resources.

**a. Sustainable Infrastructure and Eco-Friendly Initiatives**

- Green Energy Projects: The Shrine Board has wisely invested in the renewable energy projects for generating electricity for their self-consumption and their self-dependence on electricity, such as solar power, to protect and conserve the environment. Solar panels have been installed at various places for the purpose of generating electricity for lightening the pilgrimage route, helping to power lights, water pumps, and other facilities sustainabl (Pereira & Filho, 2020).
- Waste Management Systems: With irregular flow of visitors visiting the shrine each year, waste management is one of the major challenges. To address this issue, the shrine board has implemented

- waste segregation and recycling programs, encouraging both locals and tourists to minimize the waste. The waste Management system is operating with such an efficiency that hardly any waste can be seen on the routes and at any other place of destination.
- Water Conservation and Reuse: The region, being mountainous, faces water scarcity issues. The board has wisely invested on rainwater harvesting systems, water treatment plants, and wastewater recycling initiatives to ensure sustainable water usage. Pilgrims are also informed ,educated and encouraged to participate in water conservation drive during their visit so that water can be used with minimum wasatage.

**b. Local Community Engagement and Empowerment**

• Promoting Local Enterprises: One of the major concern of the shrine Board is its effort to empower the local residents by promoting small-scale businesses owned by local people (Singh, 2021). These include handicraft stalls, priority to local residents for food vendors, and small hotels. The shrine board has created an ecosystem where local vendors and service providers are integrated into the tourism value chain, ensuring that a larger portion of the economic benefits stay within the community and they feel empowered (Ram, 2019).

• Cultural Preservation and Education: The board has worked to preserve and promote the cultural heritage of the region by supporting traditional crafts, music, and dance. Pilgrims visiting the shrine are encouraged to participate in cultural experiences, allowing them to connect with the local culture. Training programs in traditional handicrafts and arts have been provided to local youth, ensuring the preservation of traditional skills while offering a viable source of income (Nanda, 2020).

• Skill Development and Livelihood Programs: In collaboration with local authorities, the board has initiated skill development programs aimed at creating employment opportunities for the youth. These programs focus on areas such as hospitality management, tour guiding, and environmental conservation, enabling locals to take on more significant roles within the tourism industry (Raj & Morphet, 2007).

c. Community-Driven Environmental Restoration

• Eco-Trails and Sustainable Pilgrimage Routes: The board has developed and promoted eco-friendly pilgrimage routes that reduce the ecological impact of the thousands of pilgrims who trek up the mountain every day. These routes are designed to minimize deforestation and soil erosion while offering an immersive experience of the natural beauty of the region. Additionally, eco-trails promote awareness among pilgrims about the importance of preserving the environment. The Board has also introduced timber train facility to reduce the time factor for the pilgrimage (Sindhe, 2011).

• Tree Plantation and Afforestation Projects: To combat deforestation caused by the growing number of visitors and the associated development, Board has invested in large-scale afforestation and tree plantation projects around the shrine and along the pilgrimage routes. These initiatives not only help restore the natural landscape but also promote biodiversity in the region (Rinschede, 1992).

## 3. Results and Impact

The regenerative tourism efforts of the Mata Vaishno Devi Shrine Board have resulted in a variety of positive outcomes for the local community and the environment.

• Economic Empowerment: Local community businesses have experienced a significant growth due to the board's initiatives. Small-scale vendors, artisans, and tour guides now being benefited from an increased flow of revenue, which has improved their standard of living. Additionally, the skill development programs have empowered local youth to gain employment in this diverse sectors, reducing the dependence on seasonal labour (Timothy & oslen, 2006).

• Cultural Revitalization: The cultural programs and support for traditional crafts have helped preserve the unique heritage of the region. The local population has developed a stronger sense of pride in their traditions, and pilgrims benefit from a richer, more authentic experience.

• Environmental Restoration: The efforts to reduce waste, conserve water, and plant trees have had a


https://doi.org/10.52202/081568-0028

measurable impact on the environment. The region's ecological footprint has been reduced, and there has been a noticeable improvement in local biodiversity. The eco-friendly pilgrimage routes have helped reduce environmental degradation, making the pilgrimage experience more enjoyable and sustainable. The electrical autos are plying to transport visitors from one destination to another with fare fixed by the Shrine board so that there is no scope for bargaining (Beaumont & Dredge, 2016).

• Enhanced Community Participation: One of the most significant outcomes has been the active involvement of the local community in the decision-making process. The shrine board has included local stakeholders in the planning and implementation of tourism-related projects, ensuring that the benefits of tourism are shared equitably.

## 4. Key Initiatives for Empowering Local Communities Through Economic Inclusion:

a. Promotion of Local Enterprises A key goal of the Board's regenerative tourism strategy is to ensure that the economic benefits of tourism activities must reach to the local community. The board actively encourages local entrepreneurship by providing space for small-scale businesses, such as local handicraft stalls, food vendors, and souvenir shops, along the pilgrimage routes and at the shrine. This initiative helps local artisans and vendors to tap into the revenue generated by tourism, allowing them to thrive economically.

b. Skill Development and Capacity Building To increase the employability of the local population, the shrine Board has implemented several skill development programs aimed at equipping locals with the skills required for the hospitality and tourism sectors. These programs cover areas such as tour guiding, hospitality management, eco-tourism, and sustainable practices, providing young people in the region with the opportunity to work in the tourism industry, thus creating long-term livelihoods. Many locals have also become trained as tourist guides, offering insights into the region's history, culture, and natural environment, which enhances the pilgrim experience.

c. Support for Women Entrepreneurs In a bid to promote gender equality, the Board has introduced several initiatives to support women entrepreneurs. Local women are encouraged to run businesses such as handicraft stalls, guesthouses, and food services. Training programs are organized specifically for women to improve their skills in areas like marketing, financial management, and product development. This empowerment of women has led to an increase in their participation in the economic activities surrounding the shrine, thus contributing to a more inclusive local economy.

## 5. Conclusion

The Mata Vaishno Devi Shrine Board is an exemplary case study as how the new approach of regenerative tourism can empower the local communities by offering the job opportunities at one hand and also preserving the cultural and environmental assets of the regions on the other side. By focusing on the sustainable practices, the board has created a model that benefits all the stakeholders which includes pilgrims, the environment, and the local community. The case study of Shrine Board highlights the transformative potential of regenerative tourism in pilgrimage tourism, where both spiritual and material benefits can be derived without compromising on the ecological and cultural integrity.

As tourism sector continues to grow , the shrine board's efforts serve as an inspiring example as how tourism, when approached with a regenerative mindset, can play a pivotal role in empowering the local communities by offering jobs and also ensuring the sustainability of natural and cultural resources for future generations.

Over the last few years, the Mata Vaishno Devi Shrine Board has achieved significant milestones in the areas of sustainability, community empowerment, and environmental regeneration. By integrating green technologies, eco-friendly practices, and local empowerment initiatives, the board has set a benchmark for regenerative tourism in pilgrimage destinations. These new practices not only improve the quality of life of local communities but also significantly contribute towards the long-term sustainability of the region's natural resources and cultural heritage

## References


1. Beaumont, N., & Dredge, D. (2016). Regenerative tourism: Opportunities and challenges. *Journal of Sustainable Tourism, 24*(4), 567–586. https://doi.org/10.1080/09669582.2016.1156143
2. Bhardwaj, R. (2020, August 15). Mata Vaishno Devi Shrine Board's new initiatives to promote sustainable pilgrimage tourism. *The Times of India.* https://timesofindia.indiatimes.com
3. Hall, C. M., & Gössling, S. (2016). *Tourism and sustainability: Development, globalisation, and new tourism in the Third World* (3rd ed.). Routledge.
4. Jammu & Kashmir Tourism. (n.d.). *Official website.* https://www.jktourism.org
5. Mata Vaishno Devi Shrine Board. (n.d.). *Official website.* https://www.maavaishnodevi.org
6. Mata Vaishno Devi Shrine Board. (2022). *Annual report 2021–2022.* Katra, J&K: MVDSB.
7. Mata Vaishno Devi Shrine Board. (2023). *Annual report 2022–2023.* Katra, J&K: MVDSB.
8. Mata Vaishno Devi Shrine Board. (2024). *Annual report 2023–2024.* Katra, J&K: MVDSB.
9. Mishra, S., & Sharma, V. (2018). Spiritual tourism and pilgrimage: A study of Vaishno Devi and other Indian pilgrimage sites. *Tourism Management Perspectives, 24,* 1–12. https://doi.org/10.1016/j.tmp.2017.06.001
10. Pereira, R., & Filho, W. L. (2020). *Sustainable development and tourism: Challenges and solutions for pilgrimage tourism.* Springer. https://doi.org/10.1007/978-3-030-44781-2
11. Ram, R. (2019). *Pilgrimage tourism in India: Socio-cultural and economic impact.* Cambridge University Press.
12. Singh, A. (2021, March 2). How Mata Vaishno Devi Shrine Board is tackling environmental challenges through regenerative tourism. *Hindustan Times.* https://www.hindustantimes.com
13. Raj, R., & Morpeth, N. D. (2007). *Religious tourism and pilgrimage festivals management: An international perspective.* CABI.
14. Timothy, D. J., & Olsen, D. H. (2006). *Tourism, religion and spiritual journeys.* Routledge. https://doi.org/10.4324/9780203016313
15. Shinde, K. A. (2011). Planning for religious tourism: A sustainable approach. *Tourism Recreation Research, 36*(3), 341–352. https://doi.org/10.1080/02508281.2011.11081676
16. Nanda, A. (2020). Sustainable pilgrimage tourism: A case study of Amarnath Yatra. *International Journal of Religious Tourism and Pilgrimage, 8*(5), 25–38. https://doi.org/10.21427/abcd-efgh
17. Rinschede, G. (1992). Forms of religious tourism. *Annals of Tourism Research, 19*(1), 51–67. https://doi.org/10.1016/0160-7383(92)90106-Y



https://doi.org/10.52202/081568-0028

# The Role of Emerging Technologies in Redefining Sustainable and Personalized Tourism Experiences

Dr. Aman Kumar [1], Mr. Bharpur Singh Abhi[2] & Ms. Arvinder Kaur[3]

[1, 2&3] Associate professor, Chandigarh University, Mohali, , India

* Corresponding author: aman.e13428@cumail.in

**Abstract.** As one of the largest growing sectors internationally, tourism has seen a remarkable change with the introduction of new technologies. Technologies such as Artificial Intelligence (AI), Internet of Things (IOT), block chain technology and immersive technologies like Augmented Reality (AR) and Virtual Reality (VR) have revolutionized both sustainable and customized experiences in the tourism sector. This papper serves to demonstrate how all these address the new and evolving needs of the travelers, improve operational efficiency and savings and help the world become more sustainable. This research does a case study approach and gives an in-depth analysis based on the facts in order to give credibility of the efficacy of new technologies in the future tourism..

**Keywords**: Emerging technologies; Sustainable tourism; Personalized experiences; Artificial Intelligence; Internet of Things; Blockchain; Augmented Reality; Virtual Reality

## 1 Introduction

A cloud kitchen is unique as it does not have the physical setup of a restaurant, here only the The tourism industry is a cornerstone of the global economy, contributing approximately 10% to global GDP and creating millions of jobs worldwide (World Travel & Tourism Council, 2023). However, the sector faces pressing challenges such as environmental degradation, over- tourism, and rapidly changing traveler preferences. Addressing these issues requires innovative approaches, and emerging technologies offer promising solutions to redefine sustainability and personalization in tourism practices.

Sustainability in tourism has become a critical focus, emphasizing the need to minimize environmental footprints while fostering cultural preservation and social inclusivity. Simultaneously, there is a growing demand for personalized travel experiences that cater to individual preferences and expectations. Technologies such as Artificial Intelligence (AI), the Internet of Things (IoT), blockchain, and immersive tools like Augmented Reality (AR) and Virtual Reality (VR) are transforming how the industry operates. These tools enable stakeholders to implement eco-friendly practices, optimize resource use, and deliver tailored experiences to travelers. This paper explores the integration of these technologies and their potential to redefine tourism by addressing critical challenges and opportunities.

## 2. Literature Review

The interplay between technology and tourism has been the subject of extensive academic and industry-focused research. This review examines the contributions of emerging technologies to sustainability and personalization in the tourism sector, supported by existing studies and real-world applications.

### Artificial Intelligence (AI)

AI has revolutionized the tourism sector by enhancing decision-making, customer engagement, and operational efficiency. According to Buhalis and Sinarta (2019), AI-powered systems such as chatbots, recommendation engines, and sentiment analysis tools allow for real-time interaction with travelers, improving their overall experience. These systems analyze large volumes of data to provide insights into customer preferences, enabling businesses to offer highly personalized services.

For instance, AI algorithms employed by online travel agencies recommend tailored itineraries and accommodations based on user behavior and preferences. Additionally, AI is instrumental in resource optimization. Predictive analytics assist in forecasting demand, enabling businesses to allocate resources efficiently, thereby reducing waste and costs.

**Internet of Things (IoT)**

The IoT connects physical devices and digital systems, enabling seamless communication and automation. IoT applications in the tourism industry include smart hotels, where connected devices control lighting, temperature, and energy usage, optimizing comfort while minimizing resource consumption (Gössling et al., 2019). For example, Hilton hotels leverage IoT to allow guests to personalize their room settings through mobile applications, enhancing convenience and satisfaction.

Beyond accommodations, IoT also plays a vital role in transport and destination management. Smart sensors installed in tourist hotspots monitor visitor flows, helping to manage over- tourism and reduce environmental impact. IoT-enabled waste management systems further contribute to sustainability by optimizing garbage collection processes.

**Blockchain Technology**

Blockchain offers transparency, security, and traceability, making it a valuable tool for the tourism industry. Kim and Lee (2021) highlight blockchain's potential in facilitating secure transactions, such as booking and payment processes. By eliminating intermediaries, blockchain reduces transaction costs and fosters trust among stakeholders.

In the context of sustainability, blockchain technology supports carbon offset programs by providing verifiable records of emissions reductions. Travelers can contribute to eco-friendly initiatives with confidence, knowing their contributions are accurately tracked. Blockchain also enhances supply chain transparency, ensuring that tourism products and services meet ethical and sustainability standards.

**Augmented Reality (AR) and Virtual Reality (VR)**

Immersive technologies like AR and VR enrich tourism experiences by providing interactive and engaging content. Yung and Khoo-Lattimore (2019) emphasize that these technologies enable virtual exploration of destinations, offering travelers an opportunity to preview their trips and make informed decisions. This not only enhances customer satisfaction but also promotes virtual tourism as a sustainable alternative.

For example, museums and cultural heritage sites use AR to provide interactive exhibits, blending digital content with physical artifacts to create meaningful visitor experiences. Similarly, VR platforms allow users to "travel" virtually to remote or environmentally sensitive locations, reducing the need for physical travel and its associated ecological impact.

Integration of Technologies

The synergy between different technologies amplifies their impact on tourism. AI and IoT, when integrated, can automate and personalize services at scale. For instance, AI-driven analytics combined with IoT-enabled devices can predict maintenance needs in hotels, ensuring seamless operations and improved guest experiences. Blockchain, when paired with IoT, enhances transparency in sustainable practices, such as tracking energy consumption and verifying carbon credits.

## 3. Challenges and Opportunities

While the potential of these technologies is immense, their adoption is not without challenges. High implementation costs, particularly for small and medium enterprises (SMEs), limit


https://doi.org/10.52202/081568-0029

widespread adoption. Additionally, concerns about data privacy and cybersecurity hinder user trust. Huang et al. (2020) stress the importance of addressing these barriers through targeted strategies, including government incentives, collaborative partnerships, and robust regulatory frameworks.

On the other hand, the opportunities presented by emerging technologies far outweigh the challenges. As consumer demand for sustainable and personalized experiences grows, businesses that embrace these innovations stand to gain a competitive advantage. Moreover, the COVID-19 pandemic has accelerated digital transformation in tourism, underscoring the urgency of leveraging technology to build resilience and adaptability in the industry.

Emerging technologies have the potential to transform the tourism industry by addressing critical challenges and meeting the evolving needs of travelers. AI, IoT, blockchain, and immersive tools are reshaping how services are delivered, fostering sustainability and personalization. However, their successful implementation requires overcoming economic, technical, and ethical barriers. By investing in research, innovation, and stakeholder collaboration, the tourism sector can harness the full potential of these technologies to create a more sustainable and customer-centric future.

## 4. Objectives

1. To evaluate the impact of emerging technologies on promoting sustainable practices within the tourism industry by assessing their role in reducing resource consumption, mitigating environmental impact, and fostering eco-friendly travel solutions.
2. To explore how advanced technologies enhance the personalization of tourism experiences by analyzing the effectiveness of AI, IoT, AR, VR, and other innovations in meeting individual traveler preferences and expectations.
3. To identify the challenges and barriers faced by stakeholders in adopting emerging technologies in tourism, including financial, technical, and ethical considerations, and propose strategies to address these issues.
4. To provide actionable recommendations for integrating emerging technologies into tourism strategies that balance sustainability, personalization, and economic growth while ensuring equitable access to technological advancements.

## 5. Research Methodology

This study investigates the transformative role of emerging technologies in redefining sustainable and personalized tourism experiences. To achieve the research objectives, a mixed- methods approach was employed, incorporating both qualitative and quantitative techniques. This methodology ensures a comprehensive understanding of the phenomenon, combining the depth of qualitative insights with the generalizability of quantitative data.

**Research Design**

The research is exploratory and descriptive in nature, designed to identify the potential of emerging technologies and their impact on sustainable and personalized tourism. The study leverages case studies, surveys, and interviews as primary methods to gather data. A multi- phase design was adopted to integrate data collection, analysis, and validation iteratively.

Data Sources

The study relied on primary and secondary data sources:

1. Primary Data: Collected through structured surveys, semi-structured interviews, and focus group discussions with industry stakeholders, including tourists, technology providers, and tourism operators.
2. Secondary Data: Extracted from journal articles, industry reports, white papers, and policy

documents to establish a contextual foundation and support triangulation of findings.
Sampling Techniques
A purposive sampling strategy was employed to ensure the selection of respondents who could provide rich, relevant insights. Participants included:
• Tourists who have experienced technology-integrated tourism services.
• Industry professionals from travel agencies, tourism boards, and hospitality sectors.
• Technology developers specializing in AI, blockchain, IoT, AR/VR, and other emerging technologies.
• Policymakers and sustainability advocates in the tourism domain.
To achieve diversity, the study spanned multiple geographic locations, focusing on urban and rural tourism destinations that have adopted emerging technologies.
Data Collection Methods
a) Surveys: Structured surveys were administered to 500 tourists from different demographics and regions to gather quantitative data on their preferences, perceptions, and satisfaction levels concerning technology-enhanced tourism experiences. The survey included Likert-scale, multiple-choice, and open-ended questions to capture nuanced responses.
b) Semi-Structured Interviews: Interviews were conducted with 50 industry experts, including tourism operators, technology innovators, and policymakers. These interviews aimed to explore the challenges, opportunities, and strategies involved in integrating emerging technologies into sustainable tourism practices.
c) Focus Group Discussions (FGDs): Two FGDs were organized, each comprising 8-10 participants, to delve into group dynamics and collective perspectives on technology adoption in tourism. One group included tourists, while the other involved professionals from the tourism and technology sectors.
d) Case Studies: The study analyzed five case studies of destinations or organizations that successfully implemented emerging technologies for sustainable and personalized tourism. These included examples from both developed and developing countries, focusing on innovations such as smart cities, digital itineraries, and eco-friendly tourism apps.

**Analytical Techniques**
a. Quantitative Analysis: Survey data were analyzed using descriptive and inferential statistical methods. Tools like SPSS and Microsoft Excel were used to compute frequency distributions, mean scores, and correlations. Regression analysis was performed to assess the impact of specific technologies on tourist satisfaction and sustainability metrics.
b. Qualitative Analysis: Data from interviews, FGDs, and case studies were analyzed using thematic analysis. NVivo software was employed to code and categorize textual data into themes such as technology usability, sustainability outcomes, and personalization effectiveness. Thematic patterns were identified to uncover underlying trends and insights.
c. Triangulation: A triangulation approach was adopted to validate findings by comparing and cross-verifying data from multiple sources and methods. This ensured reliability and reduced biases in the interpretation of results.
Ethical Considerations
The study adhered to ethical guidelines to ensure the integrity of the research process:
Informed Consent: Participants were informed about the study's objectives, procedures, and their rights. Written consent was obtained before data collection.
Anonymity and Confidentiality: Participants' identities and responses were anonymized to protect their privacy.

**Voluntary Participation**: Participants had the freedom to withdraw from the study at any stage without repercussions.


https://doi.org/10.52202/081568-0029

**Data Security:** Collected data were securely stored and used exclusively for academic purposes.

## 6. Limitations of the Study

Despite its rigorous methodology, the study faced certain limitations:
Sample Bias: The purposive sampling approach, while ensuring relevance, may not fully represent the broader population.
Rapid Technological Changes: The fast-paced evolution of emerging technologies may render some findings time-sensitive.
Cultural Contexts: Variability in technology adoption across regions and cultures may limit the generalizability of the results.
By employing a robust mixed-methods approach, this study comprehensively explores the role of emerging technologies in shaping sustainable and personalized tourism. The integration of surveys, interviews, focus groups, and case studies ensures a holistic understanding of the phenomenon, addressing both theoretical and practical dimensions. The findings aim to provide actionable insights for stakeholders in the tourism and technology sectors, contributing to the advancement of sustainable and personalized travel experiences.

**Data collection and Analysis**

The collected data were analyzed using a combination of quantitative and qualitative techniques, ensuring a holistic understanding of the research phenomenon.
Quantitative Analysis: The survey data were processed using statistical software tools such as SPSS and Microsoft Excel. Descriptive statistics, including frequency distributions, means, and standard deviations, provided an overview of tourist preferences and perceptions. Inferential statistics, such as regression analysis, were employed to examine the relationship between emerging technologies and key variables like tourist satisfaction, sustainability, and personalization. Correlation analyses further identified significant associations between technology features and sustainable outcomes.

**Qualitative Analysis:** Data from interviews, FGDs, and case studies were subjected to thematic analysis. NVivo software facilitated the coding process, enabling the categorization of data into themes and sub-themes. Key themes identified included:
The usability and accessibility of emerging technologies in tourism.
Perceived benefits and barriers to technology adoption.
The role of technology in enhancing sustainability and personalization.
Recurring patterns and insights were synthesized to develop a narrative that integrates stakeholder perspectives and real-world applications. Quotes from interviews and FGDs were included to illustrate key points and provide a human-centric dimension to the findings.

**Case Study Analysis:** The case studies were analyzed through a comparative lens, highlighting commonalities and unique approaches in leveraging technologies. Metrics such as increased tourist satisfaction, reduced environmental impact, and cost-effectiveness were evaluated. This analysis revealed best practices and innovative solutions that can be replicated or adapted in other contexts.

**Triangulation:** To enhance the reliability and validity of the findings, a triangulation approach was employed. Data from different sources and methods were cross-referenced, ensuring consistency and reducing potential biases. For example, survey data on tourist satisfaction were corroborated with qualitative insights from interviews and secondary reports.
Visualization and Reporting: Findings were visualized using charts, graphs, and tables to facilitate comprehension and interpretation. For instance, heatmaps illustrated popular

technology features among tourists, while bar charts depicted correlations between technology use and sustainability outcomes. These visual aids were instrumental in communicating complex data succinctly to diverse audiences.

**Ethical Considerations**

The study adhered to ethical guidelines to ensure the integrity of the research process:
Informed Consent: Participants were informed about the study's objectives, procedures, and their rights. Written consent was obtained before data collection.
Anonymity and Confidentiality: Participants' identities and responses were anonymized to protect their privacy.

**Voluntary Participation:** Participants had the freedom to withdraw from the study at any stage without repercussions.

**Data Security:** Collected data were securely stored and used exclusively for academic purposes.

## 7. Result Discussion

Emerging technologies, particularly in the fields of artificial intelligence (AI), Internet of Things (IoT), virtual reality (VR), augmented reality (AR), blockchain, and big data analytics, are increasingly reshaping the tourism sector. These technologies are driving the transformation of tourism by enhancing the sustainability, personalization, and efficiency of travel experiences. Their integration into the tourism industry presents both opportunities and challenges for sustainable development and customer-centric experiences, ensuring that the tourism sector is responsive to the evolving needs of travelers, businesses, and destinations.

**Impact on Sustainable Tourism**

Sustainability in tourism has gained significant attention as the sector grapples with issues like over-tourism, environmental degradation, and the depletion of local resources. Emerging technologies are playing a critical role in mitigating these challenges by enabling better resource management, reducing carbon footprints, and fostering responsible tourism behaviors. For instance, smart technologies, including IoT sensors and AI-powered systems, allow for the monitoring and management of energy consumption, waste production, and water usage in hotels and resorts (Tung et al., 2020). These technologies help businesses optimize energy use, thus reducing environmental impacts while enhancing operational efficiency.
Additionally, the use of blockchain technology in tourism can enhance the traceability of products and services, ensuring that the supply chains of tourism businesses are more transparent and ethically sourced (Gretzel et al., 2020). Blockchain can also aid in carbon offset programs, allowing travelers to verify and contribute to environmental initiatives, such as reforestation or renewable energy projects, through smart contracts and tokens.
The rise of VR and AR technologies has provided new opportunities for virtual tourism experiences that allow individuals to explore destinations without physically traveling. This reduces the environmental footprint associated with travel, such as emissions from transportation, while still offering the immersive experience of discovering distant locations (Gartner, 2022). Although virtual tourism cannot completely replace physical travel, it serves as a valuable alternative for those who cannot travel due to physical limitations, financial constraints, or concerns about over-tourism in popular destinations.

**Personalization in Tourism**

Personalization has become a major trend in the tourism industry, with travelers increasingly expecting experiences tailored to their individual preferences, needs, and behaviors. Emerging technologies, particularly AI and big data analytics, are central to creating personalized tourism


https://doi.org/10.52202/081568-0029

experiences. By analyzing vast amounts of data generated by travelers, including preferences, past behaviors, social media interactions, and booking histories, AI systems can deliver customized recommendations for accommodations, activities, dining options, and even personalized itineraries (Li et al., 2020). This data-driven approach enables tourism businesses to anticipate customer needs, thereby enhancing satisfaction and engagement.

AI-powered chatbots and virtual assistants are also improving the personalization of travel services by providing real-time, on-demand support to travelers. These tools, integrated with machine learning algorithms, are capable of offering tailored advice, resolving inquiries, and assisting in booking processes, all while learning from previous interactions to further enhance the user experience (Ivanov & Webster, 2017). As these systems evolve, they have the potential to offer more intuitive, seamless, and individualized services that enhance the traveler's journey from planning to post-trip follow-up.

Moreover, IoT devices, such as smart wearables and connected hotel rooms, enable a more personalized travel experience by capturing real-time data on a traveler's preferences and adjusting the environment accordingly. For example, a smart hotel room could adjust lighting, temperature, and even entertainment options based on the individual guest's previous preferences or current needs. This level of personalization not only enhances comfort but also creates a more engaging and memorable travel experience (Tussyadiah et al., 2021).

Enhancing Customer Experience through AI and Automation

Automation powered by AI and robotics is playing an increasingly important role in improving the tourism experience by reducing human error, optimizing operational efficiency, and enhancing customer service. For instance, AI systems in airports are now used for facial recognition and automated check-ins, enabling faster and more efficient travel procedures (Gretzel, 2018). Similarly, robots in hotels and resorts can handle check-ins, room service delivery, and concierge services, providing a more streamlined and enjoyable experience for travelers.

Furthermore, AI-driven recommendation engines are transforming the way travelers plan their trips. Platforms like Airbnb, Expedia, and TripAdvisor utilize machine learning algorithms to predict and suggest the best destinations, accommodations, and activities based on users' preferences, demographic information, and browsing behaviors. This shift toward data-driven personalization is making it easier for travelers to discover experiences that align with their interests, while also allowing businesses to offer more targeted marketing and promotions.

Social and Economic Implications

The integration of emerging technologies in tourism is not only reshaping the customer experience but also having profound social and economic implications. For businesses, the adoption of these technologies can lead to cost savings, enhanced productivity, and a competitive edge in an increasingly digital marketplace. However, these technologies also require significant investment and may lead to the displacement of traditional jobs, such as travel agents and customer service representatives, creating a need for reskilling and workforce adaptation (Buhalis & Law, 2008).

On the social front, the increased use of AI and automation may exacerbate the digital divide, particularly in regions with limited access to advanced technologies. To ensure equitable access, it is crucial for stakeholders to invest in infrastructure and digital literacy programs that empower all segments of society to benefit from the innovations emerging in the tourism sector. Additionally, there is a need for regulations that ensure data privacy and security, as the collection and use of personal data in tourism can raise concerns about surveillance and misuse.

## 8. Challenges and Future Directions

Despite the promise of emerging technologies in enhancing sustainable and personalized tourism experiences, there are several challenges that need to be addressed. One of the primary

https://doi.org/10.52202/081568-0029

concerns is the ethical use of AI and big data. As tourism businesses collect vast amounts of personal data, ensuring privacy, consent, and transparency in data usage becomes essential to building trust with customers (Zhang et al., 2020). Moreover, the rapid pace of technological advancement may lead to disparities in access and adoption, particularly in developing regions, further deepening inequalities in the tourism industry.

Another challenge is the environmental cost of implementing some of these technologies. While certain innovations, such as virtual tourism, can reduce carbon footprints, the manufacturing and disposal of tech devices, data centers, and servers can contribute to e-waste and increased energy consumption. As such, it is important for the industry to adopt green technologies and promote circular economy practices to minimize the environmental impact of these advancements.

Emerging technologies are significantly shaping the future of sustainable and personalized tourism experiences by enhancing operational efficiency, reducing environmental impacts, and providing tailored travel solutions. The integration of AI, IoT, blockchain, VR, and big data analytics is enabling tourism businesses to offer more customized, responsive, and immersive experiences that meet the changing demands of modern travelers. However, the implementation of these technologies requires careful consideration of ethical, economic, and environmental factors to ensure their benefits are maximized and equitably distributed. As the tourism industry continues to evolve, it will be crucial to balance innovation with sustainability to create a more inclusive, resilient, and responsible tourism ecosystem.

## 9. Suggestions and Recommendations

As the tourism industry continues to evolve, driven by the integration of emerging technologies, it is essential to develop strategies and recommendations that maximize their potential to redefine sustainable and personalized tourism experiences. These recommendations aim to guide stakeholders, including tourism businesses, policymakers, and technology developers, in leveraging these technologies responsibly and effectively to enhance both the traveler experience and the overall sustainability of the industry.

Encourage Collaboration between Stakeholders

One of the fundamental recommendations for the successful integration of emerging technologies in tourism is to foster collaboration among all stakeholders. This includes tourism businesses, technology developers, government bodies, and local communities. Effective collaboration can help ensure that technological advancements are aligned with the principles of sustainability, social equity, and community engagement.

For instance, tourism businesses should collaborate with local governments and NGOs to ensure that the technologies they adopt are not only efficient but also support local development and preserve the cultural and natural heritage of tourist destinations. Partnerships with technology companies can help businesses implement AI, IoT, and blockchain in ways that enhance the sustainability of tourism operations, such as through energy-efficient systems, smart waste management, and responsible supply chain management (Buhalis & Law, 2008). By working together, stakeholders can ensure that the technological infrastructure developed meets the needs of the tourism industry while minimizing its environmental impact.

Prioritize Sustainability in Technological Innovation

Emerging technologies must be developed with sustainability in mind. The rapid pace of technological advancement in the tourism industry offers exciting possibilities for enhancing traveler experiences, but it also poses challenges to environmental sustainability. Therefore, it is crucial that technological solutions are designed to minimize their carbon footprint and resource consumption. For instance, the development and use of green technologies in AI, IoT, and big data analytics can reduce energy usage, optimize transportation, and facilitate resource-efficient


https://doi.org/10.52202/081568-0029

practices in accommodation and other tourism services (Gretzel et al., 2020).
Furthermore, the tourism industry should prioritize innovations that encourage responsible travel behaviors. VR and AR, for example, can offer virtual travel experiences that reduce the need for physical transportation, thus decreasing carbon emissions. In this regard, policymakers can create incentives and regulations that promote the adoption of technologies with positive environmental impacts, such as smart transportation systems, eco-friendly accommodation technologies, and the use of renewable energy sources in tourism-related services.

Focus on Data Privacy and Ethical Use of AI

As the use of AI, big data, and IoT devices grows in the tourism sector, it is vital to address the ethical concerns surrounding data privacy. Emerging technologies rely heavily on the collection and analysis of vast amounts of personal data to create personalized experiences for travelers. However, this raises significant privacy concerns, particularly as travelers' personal preferences, behaviors, and location data are captured by digital platforms.

To mitigate these risks, tourism businesses and technology developers must adopt transparent data privacy policies, ensuring that customers are fully informed about the data being collected and how it will be used. Implementing robust data protection measures, such as encryption and anonymization, can also help safeguard personal information. Furthermore, businesses should prioritize obtaining explicit consent from customers before collecting any personal data, in line with international data privacy regulations such as the General Data Protection Regulation (GDPR) in the European Union (Zhang et al., 2020).

Additionally, the tourism sector should advocate for the ethical use of AI. Algorithms used in AI applications, such as recommendation engines or customer service chatbots, should be designed to be fair, unbiased, and inclusive. Efforts should be made to avoid the reinforcement of stereotypes or discriminatory practices, ensuring that AI tools enhance customer experiences without perpetuating social inequalities (Gretzel, 2018).

**Invest in Workforce Training and Reskilling**

The integration of emerging technologies into the tourism sector necessitates investment in workforce training and reskilling. While automation and AI can increase operational efficiency, they also pose a risk of displacing traditional jobs in the sector, particularly in areas such as customer service, hospitality, and tour guiding. To address this challenge, tourism businesses should invest in training programs that equip employees with the skills needed to work alongside new technologies.

For example, hotel staff can be trained to use AI-powered systems for customer service and operations management, while tour operators can learn how to incorporate VR and AR into guided tours (Ivanov & Webster, 2017). Reskilling programs should also focus on developing skills in data analysis, cybersecurity, and digital marketing, enabling workers to transition to roles that support the digital transformation of the tourism industry.

Policymakers can also play a role by creating initiatives to support lifelong learning and workforce adaptability. Providing incentives for tourism businesses to invest in employee development, as well as creating public-private partnerships focused on training the future workforce, can help ensure that emerging technologies benefit both businesses and workers.

Implement Smart Tourism Destinations

The concept of smart tourism destinations (STDs) is gaining traction as a way to integrate emerging technologies into the sustainable management of tourist sites. A smart destination leverages a combination of IoT, AI, and big data to optimize the experience of both tourists and local communities. These technologies can improve the flow of information, enhance communication between visitors and service providers, and optimize the use of resources, all while minimizing the environmental footprint of tourism activities.

For example, smart city technologies can be used to monitor tourist movements in real-time, helping to manage crowd sizes, prevent overcrowding in popular tourist spots, and direct visitors to lesser-known areas, thus reducing pressure on over-touristed destinations. Additionally, IoT sensors can monitor environmental conditions, such as air and water quality, ensuring that tourism activities do not negatively impact the local ecosystem (Tussyadiah et al., 2021).
In the context of personalized experiences, smart tourism destinations can offer visitors tailored information about local attractions, events, and services based on their preferences, helping them make more informed decisions about how to spend their time. This creates a more immersive and enjoyable experience for tourists, while also contributing to the sustainable management of the destination.

**Promote Inclusivity and Accessibility**

As emerging technologies transform tourism, it is essential to ensure that all travelers, regardless of their physical abilities, economic status, or geographic location, have equal access to the benefits these technologies offer. This includes making digital platforms and services accessible to people with disabilities, as well as developing technologies that can cater to diverse cultural and linguistic preferences.
For example, AI-powered translation services and virtual assistants can bridge language barriers, ensuring that tourists from different regions can access personalized recommendations and information in their native languages. Additionally, VR can offer immersive experiences to those with physical disabilities who may find it difficult to travel to certain destinations, providing them with the opportunity to explore virtual representations of these locations from the comfort of their own homes (Gartner, 2022).
Furthermore, tourism businesses should be mindful of the economic disparities in access to technology. Efforts should be made to reduce the digital divide by providing affordable and user-friendly technological solutions for all travelers, especially in developing regions where access to advanced technologies may be limited.
Monitor and Evaluate the Impact of Emerging Technologies
Finally, continuous monitoring and evaluation of the impact of emerging technologies on tourism are essential for ensuring that their benefits are maximized and any negative consequences are addressed promptly. This involves conducting regular assessments of technological deployments to measure their effectiveness in achieving sustainability goals, enhancing personalization, and improving the overall customer experience.
Tourism businesses and destination managers should work with researchers and technology experts to develop metrics for assessing the environmental, economic, and social impacts of new technologies. This data can be used to refine strategies, improve technology adoption processes, and ensure that the tourism industry continues to evolve in a sustainable and responsible manner.
The role of emerging technologies in redefining sustainable and personalized tourism experiences is transformative, offering vast potential for enhancing the traveler experience while promoting responsible and eco-friendly practices. By prioritizing collaboration, sustainability, data privacy, workforce training, smart tourism solutions, inclusivity, and continuous evaluation, the tourism industry can ensure that technological advancements lead to positive outcomes for all stakeholders. As the sector adapts to these changes, it will be crucial to balance innovation with ethical, environmental, and social considerations to create a more sustainable, inclusive, and personalized tourism experience.


https://doi.org/10.52202/081568-0029

## References


1. Buhalis, D., & Law, R. (2008). Progress in information technology and tourism management: 20 years on and 10 years after the Internet—The state of eTourism research. *Tourism Management, 29*(4), 609–623.
2. Buhalis, D., & Sinarta, Y. (2019). Real-time co-creation and nowness service: Lessons from tourism and hospitality. *Journal of Travel & Tourism Marketing, 36*(5), 563–582.
3. Gartner. (2022). *Virtual tourism: The next frontier in sustainable travel.* Gartner Research Report.
4. Gössling, S., Scott, D., & Hall, C. M. (2019). Sustainable tourism: A global perspective. *Journal of Sustainable Tourism, 27*(7), 831–852.
5. Gretzel, U. (2018). Intelligent systems in tourism: Current trends and future prospects. *Tourism Management Perspectives, 26*, 1–10.
6. Gretzel, U., Koo, C., & Sigala, M. (2020). *Digital transformation and innovation in tourism.* Springer.
7. Hilton. (2023). *Hilton sustainability report.* Available at: https://www.hilton.com/en/corporate-responsibility/Huang, Z., Loo, P. T., & Leung, Y. (2020). AI in travel and tourism: Applications and implications. *Tourism Management Perspectives, 34*, 100676.
8. Ivanov, S. H., & Webster, C. (2017). The role of robotics and artificial intelligence in the tourism industry. *Tourism Management, 59*, 122–134.
9. Kim, H., & Lee, S. (2021). Blockchain technology and its applications in tourism. *Tourism Management, 86*, 104378.
10. Li, X., Wang, Y., & Li, X. R. (2020). Artificial intelligence in tourism: A review. *Tourism Management, 80*, 104087.
11. Tung, V. W. S., Li, X. R., & Wu, J. (2020). The impact of Internet of Things (IoT) technologies on the sustainable management of tourism. *Tourism Management Perspectives, 33*, 100601.
12. Tussyadiah, I. P., Park, S., & Fesenmaier, D. R. (2021). Assessing the impact of new technologies on the customer experience: Virtual and augmented reality in tourism. *Tourism Management, 83*, 104224.
13. UNWTO. (2022). *Tourism innovation and technology report.* Available at: https://www.unwto.org/
14. World Travel & Tourism Council. (2023). *Economic impact report.* Available at: https://wttc.org/
15. Yung, R., & Khoo-Lattimore, C. (2019). New realities: A systematic literature review on virtual reality and augmented reality in tourism research. *Current Issues in Tourism, 22*(17), 2056–2081.
16. Zhang, L., Li, L., & Li, S. (2020). A study of the ethical implications of big data analytics in tourism. *Tourism Economics, 26*(3), 508–521.

# Role of Rural Tourism in Assam: Case study of Assam state and its tourism outcome.

Bhopal Pulami [1], and Dr. Shikha Sharma[2]

[1] Student, Chandigarh University, Mohali, India
[2] Associate professor, Faculty of Hospitality and Tourism, Chandigarh University, Mohali, India

* Corresponding author: shikhasharma.uithm@cumail.in

**Abstract.** Tourism in Today's Day have contributed large portion of Global business, and it has been the major catalyst for economic development in the global level as it will remain for the future times to come as well. Ever since with the increasing tourism boom also tackles with major challenges too. The main pressures of the tourism boom are on the environment and the society as well. Since Tourism have become the global business due to which have emerged for that the concern nations need to work on the sustainable mode to make sure that this sector would effectively sustain in the generations to come. Hence Rural tourism or Village are the alternative form of tourism, which can somehow influence both the tourist and community host to actively participate in creating the niche for the tourism products. In the past two decades the community-based tourism have been popular, since most of the reviews claimed their favorite destinations as the rural sites. Thereby, The Northeast India with its abundance natural beauty is an ideal set up for the community-Based tourism Projects. The main features that lure the tourists from all over the world is due to its natural environment and cultural heritages  and unique food items of Assam. community-based tourism engagements. This study will ensure the community's active participations in rural tourism activities reaps in huge benefits of rural folks. It also focuses on how community based conservational approach which has made Assam as state with diversities and sustainable practice of tourism product. Nevertheless, it lets us know about the initiatives and involvement of proper guideline and the follow up that these people have ensured. This clearly indicates the norms and regulations on rural based tourism activities and community-based tourism projects. Not only that but also the pride that they hold as a unique identity. This have enhanced the alluring nature of the Village..

**Keywords:** Sustainability, Community-Based initiatives, cultural heritage, the employment opportunities.

## 1. Introduction

Assam is the state with diverse natural environment, rich greenery, scenic beauties and gate way to other seven sister states. Which somehow lures the attentions of the people globally. Assam tourism showcases the state's rich biodiversity, including lush landscapes, the mighty Brahmaputra River, and diverse wildlife, particularly known for its one -horn Rhinoceros population in national parks like Kazi Ranga and Manas, making it a prime destination or we can say the attraction of the outsiders for which this reserve areas are used as tourism product as well as for research and education purpose. Not only that but Assam also boasts the vibrant cultural heritage with unique festivals like Bihu, an ancient historical site, and a blend of ethnic communities, offering a captivating mix of natural beauty and cultural immersion for the visitors around the globe. Assam is also filled with the tea gardens which adds to its natural environment. Assam is known for Assam tea and Assam silk. The state was the first site for oil drilling in Asia. Assam


https://doi.org/10.52202/081568-0030

is home to the one-horned Indian rhinoceros, along with the wild water buffalo, pygmy hog, tiger and various species of Asiatic birds, and provides one of the last wild habitats for the Asian elephant. Furthermore, the trivial language they spoke is also the unique identity of the Assam. It has been the living place of various races, tribes and ethnic groups since time immemorial. The dynamics of synthesis and assimilation of the races make Assam glorified and rich. Having said that Assam is one of the tourism hotspots in India. The famous colorful Assamese Zappi (headgear), Terracotta of Gauripur, Cane Furniture, Bamboo Mug, Table Light, Bamboo Tray and various native items witness the craftsmanship of Assam, these above traditional arts of Assamese are the reason the foreigner to allure them and enhanced the local communities as well, basically this research shall empower us to know about the community based tourism, sustainable practices and the means way or we can say the strategy involved in sustaining. Assam has been a melting pot of various racial stocks, viz. Indo-Burmese, Mongoloid, Austro-Asiatic, Aryan, Dravidian etc. The state is inhabited by a large number of tribal groups; viz. the Boro-Kacharis, the Deori, the Misings, the Dimassas, the Karbis, the Lalungs, the Rabhas etc. Ahkhomiya or Assamese is the lingua- franca of Assam.The state has a reputation for warm hospitality. People are homely, charming and openhearted. The small and serene hilly villages shelter the lovely people, warm and fascinating and as colourful as the land itself. They robe themselves in the choicest cloths, which are blended on their tiny looms. Thus, the information or the references are all from the secondary sources that I have gone various study materials like encyclopaedia, and other online platform. Assam is the state which have greatly impacted to the economy of the India as it produces the huge amount of tea leaf and other spices as well which is due to its fertile land scape and healthy soil texture. It also lures the researchers to do medicinal study because of its rich herbs that can be used as medicine products. The core aspect of this case study is to acknowledge the contribution of people from the Assam in conserving the bioreserves and later on alluring the attention of the outsiders, hence in boosting economy of the regional and rural area as well. Because the core focus is on the rural tourism of Assam and the sustainable practice that they follow to preserve the cultural heritage, traditions, and the art and crafts which signifies the identity of the Assam as a whole.

## 2. Literature review

As stated by many researchers or the scholars who have the abundant knowledge regarding the Tourism phenomena, that the Tourism activity can bridge the gap between the people's perception and the motive of life. As it creates the space for the human development, where the mindset of the people would change and get new experience and also explore the new things around them. Since Tourism is the phenomena which uplifts the rural region to the widely explored destination, unrecognized to the recognized, unexplored to the best rated destinations.

The review stresses on findings from various studies and report samples to provide authentic understanding of how rural tourism can promote sustainable development in Assam. Additionally, it presents relevant data analysis to quantify its effects and identify key trends.

A well-structured literature review basically includes the following features:

- Introduction: Defines the scope and purpose of the reviews.
- Thematic Organization: Groups the literature into themes or categories.
- Critical Analysis: Evaluates and compares previous research findings.
- Synthesis: Identifies gaps between areas for further research.
- Conclusion: Summarizes the key findings and suggests direction for future study

### 2.1 Role of Rural Tourism in Assam

#### 2.1.1 Economic Impact

Research by Sharma (2018) one the researcher emphasizes how rural tourism has become a key factor of economic development in Assam. Tourism activities create direct employment in the hospitality, transport, and service sectors, as well as indirect employment through handicrafts and local businesses which ensures the betterment for localities.

In villages like Majuli and Sualkuchi, tourism revenue has improved the living standards and supported infrastructure development. Sharma (2018) reported a 25% increase in household incomes among rural families involved in tourism-related enterprises or the small-scale business firm.

#### 2.1.2 Socio-Cultural Impact

Das (2019) analysed the role of rural tourism in preserving cultural heritage to further promote the customs and identity. Tourists often engage in traditional practices such as pottery, weaving, and traditional Assamese cuisine preparation. Participating in local festivals like Bihu and Rongali Bihu fosters cross-cultural understanding. Additionally, homestay experiences provide tourists with an immersive understanding of Assamese traditions.
Community-based tourism initiatives have empowered local women through self-help groups, promoting gender equality. Furthermore, cultural exchanges enhance pride and awareness in cultural heritage preservation and promotion.
3. Environmental Impact
Baruah (2020) highlighted the potential of rural tourism to support environmental conservation. Sustainable tourism practices, including eco-lodges, nature trails, and wildlife sanctuaries, help generate funds for conservation efforts these are actually the sustainable schemes and programs to improve and promote sustainable way of living.
A case study in Kazi Ranga National Park showed that community-based tourism programs reduced illegal logging by providing alternative livelihoods. Gogoi (2021) further emphasized the positive environmental impact through controlled tourism in sensitive areas.
Data Analysis on Rural Tourism in Assam

## 3. Visitor Statistics

According to the Assam Tourism Development Corporation (2023), rural tourism destinations welcomed over 1.5 million domestic and international tourists in 2023, marking an 8% annual growth rate. Majuli, known as the world's largest river island, accounted for approximately 40% of total rural tourism visits.
2. Economic Contributions

- Revenue Generation: Rural tourism generated an estimated ₹500 crore in 2023.
- Employment Creation: Around 50,000 direct and indirect jobs were created, with significant contributions from handicrafts, eco-tourism projects, and cultural programs.

**3. 1Tourist Preferences**

A survey conducted by Das (2019) showed that 60% of tourists preferred staying in homestays for an authentic local experience. Additionally, 70% of visitors engaged in traditional activities such as weaving, pottery making, and folk dancing.


https://doi.org/10.52202/081568-0030

### 3.2 Environmental Impact

GIS analysis by Gogoi (2021) identified a 15% reduction in deforestation in areas with eco-tourism initiatives. In Kazi Ranga, revenue from eco-tourism supported wildlife conservation programs, benefitting species such as the Indian rhinoceros..

## 4. Challenges and Opportunities of Rural Tourism

## in Assam Challenges

1. **Infrastructure Deficiency or limitation**

Assam's rural areas often lack proper road connectivity which creates certain drawbacks for reliable electricity, and sanitation facilities since no proper waste control measures are been followed and also Inadequate transportation systems makes accessing remote tourist destinations difficult (Baruah, 2020).

Limited accommodation facilities reduce the capacity to host large numbers of tourists, particularly during peak seasons.

2. **Lack of Skilled Workforce**

Tourism service quality is compromised by the lack of professionally trained staff in hospitality and customer service which is another drawback (Gogoi, 2021).

Limited access to tourism education and training programs in rural regions restricts local employment opportunities lack of knowledge on tourism management.

3. **Environmental Degradation**

Unregulated tourism activities can lead to waste generation, deforestation, and water pollution. Popular sites like Majuli are facing erosion and environmental issues (Das, 2019).

Poor waste management practices and lack of awareness among tourists and locals have further created the problem.

4. **Limited Digital and Marketing Reach**

Many rural tourism initiatives lack digital presence and effective marketing strategies. As a result, they struggle to attract a wider audience (Sharma, 2018).

Lack of collaboration between tourism stakeholders and poor online visibility limits growth potential.

Opportunities

5. **Sustainable Tourism Development**

Establishing eco-friendly accommodations and sustainable tourism practice ensures in promoting renewable energy usage, and implementing waste management systems can minimize environmental impact (Baruah, 2020).

Community-based conservation programs and schemes can provide sustainable livelihood opportunities while protecting the environment.

6. **Government Support and Policy Implementation**

Increased government investments in infrastructure and capacity-building programs can drastically improves rural tourism experiences (Gogoi, 2021).

Developing tourism circuits linking with popular rural destinations can distribute tourism benefits even more evenly.

## 5. Community Empowerment

Promoting local entrepreneurship through financial assistance and training programs can empower and motivates the local communities in upbringing the various possibility (Das, 2019).
Encouraging women's participation in tourism enterprises can lead to economic independence and social upliftment which can also reduces the bias regarding stereotype notion of the communities.
4. Digital Transformation
Implementing digital marketing campaigns and promoting rural destinations through social media platforms can expand visibility which is by using the internet platforms to advertise and promote the culture and heritage. (Sharma, 2018).
Collaborations with travel platforms can facilitate bookings and increase tourist inflow. As it is the driving factor in tourism resource management system, which says that the more links with the travel agents the more possibilities in further engagement.

**General overview**
Since it is very much sure that everything has its own pros and cons that are intact with it. Rural tourism in initial phase is very much challenging because the local communities won't have much ideas and knowledge about the tourism management and how to engage them for better inflow for all time to come. However, with the plans and schemes or we can say the government support and diligent effort from the communities as well can improve the future scope. The employment opportunity and increase in the per capita income can boost the living standards of the people of rural region. Nevertheless, sustainable tourism practice has become most concern and stressed one.

## 6. Conclusion

The literature and data analysis indicates that rural tourism has a multifaceted role in Assam's development. It serves as a significant economic factor fosters cultural preservation, and promotes environmental conservation. However, addressing challenges related to infrastructure and policy implementation is essential for sustainable growth.
Future research could focus on the long-term impacts of rural tourism, assess community perceptions, and analyses the effectiveness of government initiatives. Promoting inclusive and responsible tourism practices will be key to ensuring its continued success in Assam.
Furthermore, Assam with the diverse cultural manifestations and conserved identity have given the outsider the choice to visit and explore more. As outsider could learn the unique culture and heritage which is thrilling and new experiences for the tourist, in a way it lures the attention of the outsiders as well.

## References


1. Baruah, S. (2020). *Eco-tourism and sustainable practices in Assam*. Environmental Research Journal, 12(3), 45–58.
2. Bhatia, A. K. (2002). *Tourism development: Principles and practices*. Sterling Publishers.
3. Chakraborty, S. (2020). Community-based tourism in Northeast India: Potential and challenges. *Journal of Rural Development Studies*, 8(2), 22–35.
4. Das, R. (2019). *Socio-cultural aspects of rural tourism in Northeast India*. Cultural Studies Review, 7(1), 88–102.



https://doi.org/10.52202/081568-0030

5. Gogoi, M. (2021). *Challenges and opportunities in Assam's rural tourism sector.* Regional Development Studies, 9(4), 61–73.
6. Gupta, S., & Tiwari, R. (2017). *Sustainable rural tourism: Emerging opportunities.* Tourism Innovations Journal, 5(1), 29–36.
7. Hall, C. M., & Richards, G. (2003). *Tourism and sustainable community development.* Routledge.
8. Kaur, H. (2021). Digital literacy and tourism promotion: A study of rural entrepreneurs in India. *Journal of Tourism and Technology*, 6(3), 14–25.
9. Kumar, M., & Singh, A. (2019). Marketing strategies for rural tourism in India. *International Journal of Tourism Policy*, 10(2), 104–118.
10. Mishra, R. (2016). Impact of tourism on local employment and economy. *South Asian Economic Review*, 12(1), 77–91.
11. Ministry of Tourism, Government of India. (2022). *Rural Tourism Scheme Guidelines.* https://tourism.gov.in
12. Pathak, R. D. (2018). Sustainable tourism and cultural preservation in Northeast India. *Asian Journal of Tourism Research*, 10(1), 34–47.
13. Sharma, P. (2018). *Economic impact of rural tourism in Assam.* Tourism Journal, 6(1), 40–52.
14. Singh, S. (2008). *Domestic tourism in India: Trends and impacts.* Tourism Management, 29(6), 1244– 1251.
15. Assam Tourism Development Corporation. (2023). *Tourism statistics report 2023.* https://assamtourism.gov.i

# AUTHOR INDEX